Methods
of Cultivating Parasites
in Vitro

Methods
of Cultivating Parasites
in Vitro

Edited by

ANGELA E. R. TAYLOR

Department of Medical Protozoology,
London School of Hygiene and Tropical Medicine,
Keppel Street, London WC1, England

and

JOHN R. BAKER

MRC Biochemical Parasitology Unit, Molteno Institute
University of Cambridge, Cambridge, England

1978

ACADEMIC PRESS

LONDON   NEW YORK   SAN FRANCISCO

A Subsidiary of Harcourt Brace Jovanovich, Publishers

ACADEMIC PRESS INC. (LONDON) LTD.
24/28 Oval Road,
London NW1

United States Edition published by
ACADEMIC PRESS INC.
111 Fifth Avenue
New York, New York 10003

Science
QL
757
M43

Library of Congress Catalog Number 77–85100
ISBN: 0–12–685550–1

PRINTED IN GREAT BRITAIN BY
WILLIAM CLOWES & SONS LIMITED
LONDON, BECCLES AND COLCHESTER

Contributors

BAKER, J. R., *M.R.C. Biochemical Parasitology Unit, Molteno Institute, University of Cambridge, Downing Street, Cambridge CB2 3EE, England.*

COLEMAN, G. S., *A.R.C. Institute of Animal Physiology, Babraham, Cambridge, CB2 4AT, England.*

EVANS, D. A., *Department of Medical Protozoology, London School of Hygiene and Tropical Medicine, Keppel Street, London WC1E 7HT, England.*

FRIED, B., *Biology Department, Lafayette College, Easton, Pennsylvania 18042, U.S.A.*

HANSEN, E., *561, Santa Barbara Road, Berkeley, California 94707, U.S.A.*

HANSEN, J. W., *561, Santa Barbara Road, Berkeley, California 94707, U.S.A.*

LACKIE, A. M., *Zoology Department, The University, Glasgow G12 8QQ, Scotland.*

LONG, P., *Houghton Poultry Research Station, Houghton, Huntingdon, Cambridgeshire PE17 2DA, England.*

MEEROVITCH, E., *Institute of Parasitology, Macdonald Campus of McGill University, Macdonald College PO, Quebec HOA 1CO, Canada.*

RYLEY, J. F., *I.C.I. Ltd. Pharmaceuticals Division, Alderley Park, Macclesfield, Cheshire SK10 4TG, England.*

TAYLOR, A. E. R., *Department of Medical Protozoology, London School of Hygiene and Tropical Medicine, Keppel Street, London WC1E 7HT, England.*

TRIGG, P. I., *National Institute for Medical Research, The Ridgeway, London NW7 1AA, England.*

VOGE, M., *Department of Microbiology and Immunology, School of Medicine, University of California, Los Angeles, California 90024, U.S.A.*

WILSON, R. G., *I.C.I. Ltd. Pharmaceuticals Division, Alderley Park, Macclesfield, Cheshire SK10 4TG, England.*

Preface

Ten years have passed since our earlier publication on *in vitro* cultivation of parasites ("The Cultivation of Parasites *in vitro*", A. E. R. Taylor and J. R. Baker, Blackwell, Oxford). During that decade there have been some considerable advances: perhaps most notable are the achievement of continuous cultivation of *Plasmodium* (Chapter 5), of haematozoic trypo-mastigotes of *Trypanosoma brucei* (Chapter 4) and of the entire life cycles (not yet consecutively) of *Hymenolepis microstoma* (Chapter 9) and *Cooperia punctata* (Chapter 10). Many other valuable, even if less striking, advances have been made; many procedures and media have been simplified, thus aiding metabolic, biochemical and immunological studies, and the development of insect and snail cell lines *in vitro* by Hansen and others should ease growth of the vector stages of parasites. However, many of the older methods remain in use and probably always will do, e.g. Boeck and Drbohlav's media for *Entamoeba* and the classical NNN for trypanosomatids.

During this decade, too, increasing age has brought to us increasing wisdom (or lethargy), such that we have no longer tried to cover the cultivation field single (or even double) handed but have called on a range of experts to review their special subjects. Inspection of the contents of this book will, we feel sure, justify this decision. Our contributors have done an excellent job; they have done it promptly, willingly and with amazingly good natured tolerance of our editorial quirks and interferences, and we are most grateful to them all. Although we may still often reach for the revolver mentioned on the flyleaf of our previous book, this was not necessary to persuade our contributors to complete their labours; neither, in spite of much provocation, have they turned its barrel towards us. Many of them were our personal friends before they undertook this task and now we are pleased to feel that they all are.

Our earlier aim to provide a laboratory handbook still remains. It is however often difficult to glean full technical details from published accounts, and we make an impassioned plea to all future researchers to be diligent in publishing their methods. We also asked our contributors to review progress in their fields during the last decade or so, in the hope that the book may be of interest to readers, as well as doers only. We have broadly retained the earlier format, though the chapter on general techniques is now the first. There follow sections on Protozoa and Helminths respectively. In response to popular demand, we have included a chapter on the growth of parasitic Protozoa in cell and tissue cultures, and nematode parasites of plants have been included in Chapter 10. Finally there is a list of names and addresses of suppliers mentioned in the text.

In spite of the undoubted advances referred to earlier, it is alas still possible to end this preface as we ended the previous one: ". . . there still remains much to be done before the majority of parasitic species can be grown throughout their life-cycle *in vitro*"; cultivators, at least, need never be out of a job.

January, 1978

ANGELA E. R. TAYLOR
JOHN R. BAKER

Contents

HELMINTHS

Dedication

This book is respectfully dedicated to the memory of FREDERICK G. NOVY and WARD J. MACNEAL, the first workers to cultivate (rather than merely maintain) parasites *in vitro*, on the 75th anniversary of their earliest publications on the subject (*Contributions to Medical Research, dedicated to Victor Clarence Vaughan . . . University of Michigan*, June 1903, pp. 549–577 and *Journal of the American Medical Association*, 21 November 1903, volume 41, pp. 1266–1268).

If you're anxious for to shine...as a man of culture rare
You must get up all the germs...and plant them everywhere.

W. S. Gilbert, *Patience* (1881)

GENERAL METHODS

Chapter 1

Techniques and Media Commonly used for *In Vitro* Culture

JOHN R. BAKER AND ANGELA E. R. TAYLOR

MRC Biochemical Parasitology Unit, Molteno Institute, University of Cambridge, Cambridge, England

and

Department of Medical Protozoology, London School of Hygiene and Tropical Medicine, Keppel Street, London, England.

Many of the general procedures mentioned in this chapter are described in detail in standard works such as those by Hanks *et al.* (1955), Parker (1961), Paul (1970) and Cruickshank *et al.* (1975).

I. CLEANING AND PREPARATION OF APPARATUS

A. GLASSWARE

Glass apparatus used for cultivation of parasites and cell lines should preferably be either borosilicate (e.g. "Pyrex" brand) or soda glass of the type used for British medical prescription bottles ("medical flats"). Suitable plastic ware for cultivation is available sterile from most of the major suppliers (e.g. Flow, Gibco Bio-Cult*) but as it is difficult to clean and resterilize, it can be used only once. Consequently many workers prefer glass since, although more expensive initially, it is usually cheaper in the long term.

1. *Cleaning Glassware*

Cotton plugs and similar objects are removed and the used apparatus is placed in water immediately after use to prevent the drying of proteins etc. onto its surface. If pathogens have been cultivated, the initial immersion should be in an easily removed disinfectant such as sodium hypochlorite solution (e.g. "Chloros", Durham Chemical Distributors) which is afterwards thoroughly washed off with running water. Any encrusted deposits should be scrubbed off at this stage. If the glassware has been coated with grease or paraffin wax, this should be removed with petroleum after (if necessary) scraping with a knife and the article then boiled in sodium metasilicate solution (Section I, A, 1, b); any floating scum of residual paraffin is poured off before removing the article, which is subsequently washed in one of the ways described below. Silicone can be removed from glass by heating in 0·5 N sodium hydroxide, or immersion in cold alcoholic potassium hydroxide; the resultant surface scum should be poured off as described above. Siliconed glassware should either be treated thus or washed separately, lest the silicone be transferred to all the other glassware. Automatic siphon-type pipette washers are helpful in the handling and rinsing of pipettes during cleaning.

After these initial treatments, three main groups of substances may be used to clean glassware: oxidizing acids, alkalis or detergents.

(a) *Oxidizing acids.* Although effective, these are very caustic and should be handled with great care; preferably rubber gloves and apron should be worn. Accidental splashes on clothes or skin should be immediately removed with water and the area washed with sodium bicarbonate before final rinsing with water.

* Names and addresses of suppliers are given in the Appendix

The most common is chromic acid, prepared by dissolving 63 g of potassium or sodium dichromate in 35 ml of hot distilled water contained in a 5 litre beaker, cooling it, and adding concentrated sulphuric acid to bring the volume up to 1 litre. The fresh solution is yellow-brown; when reduced it becomes green and should then be discarded. Twelve to twenty-four hours immersion in chromic acid followed by very thorough rinsing in three changes of tap water (or running water) and three changes of glass distilled water should be adequate.

(b) *Alkalis.* The most favoured is sodium metasilicate since it contains no ion other than those already present in glass. Soft soap, sodium triphosphate or sodium carbonate are also used. Sodium metasilicate stock solution is prepared by dissolving 360 g sodium metasilicate and 40 g "Calgon" (Albright and Wilson) in 4·5 litres (1 British gallon) and filtering the solution. For use, 1 volume is diluted in 100 volumes of tap water. The previously rinsed and scrubbed (if necessary) glassware is immersed in boiling dilute metasilicate solution for 20 min, cooled, soaked for several hours in 0·01 N hydrochloric acid and then rinsed as described in paragraph (a).

(c) *Detergents.* These may be toxic to cells and parasites. However, some are specially formulated for use with culture apparatus and these should be used if possible, according to the makers' instructions. Suitable brands include "Alconox" (Alconox), "7X" (Linbro Chemicals), "RBS-25" (Chemical Concentrates) and "Decon-75" (Medical Pharmaceutical Developments); "RBS-50" (Chemical Concentrates) combines bactericidal and cleansing properties and may be useful when pathogens are involved.

B. CULTURE VESSELS, CLOSURES AND FLEXIBLE TUBING

Parasites may be cultured in a variety of vessels, the more common ones being test-tubes, bottles (e.g. Universal containers) and flasks (e.g. conical, Carrel etc.). These may be closed either with cotton plugs, silicone rubber bungs or (preferably) screw caps lined with white or silicone rubber discs [obtainable from Esco (Rubber)]; the latter, though more expensive, can be sterilized by dry heat whereas white rubber can only be boiled or autoclaved (Section II). Flexible tubing is also usually made of silicone rubber.

C. INSTRUMENTS

New metal instruments should be wiped with cotton soaked in carbon tetrachloride to remove any protective grease. They can be washed in detergent (Section I, A, 1, c) and sterilized either by boiling in water or autoclaving. A quick way to sterilize forceps, scissors, scalpels etc. is to dip the blades into 70% or 90% ethanol and then burn this off in a Bunsen burner flame.

D. SYRINGES AND NEEDLES

Syringes, if glass, can be cleaned as described in Section I, A. Metal needles are treated as described in Section I, C. Many workers prefer to use disposable syringes and needles, obtained sterile from manufacturers such as Becton, Dickinson or Gillette; however, the lubricant used with some brands may be toxic to cells (Nichosia, 1965).

II. STERILIZATION

A. PREPARATION

Apertures in glass apparatus should be plugged with non-absorbent cotton or covered with aluminium foil or kraft paper (tied on). Screw caps should be loosened before sterilizing and tightened as soon as possible afterwards. Graduated pipettes, Pasteur pipettes and Petri dishes may be sterilized in special metal containers. The mouthpieces of graduated pipettes are cotton-plugged and the tips rolled in aluminium foil. Pasteur pipettes are similarly plugged at the broad end; some workers prefer to heat-seal the glass at the narrow end and to scratch it with a glass cutter about 1 cm from the sealed end before sterilization. The tip can then be broken off with sterile forceps just before use. If it is required to maintain sterility of the outer surface of the apparatus, the object can be wrapped in kraft paper or aluminium foil, or sterilized in a bag made from nylon film (autoclave film, Portex) which is permeable to steam but not to bacteria; it is supplied as tubing in a variety of widths and sealed with autoclave tape to form bags. Autoclave tape is an adhesive tape resistant to autoclaving and printed with a heat-sensitive ink which darkens at autoclave temperatures; it is useful to fix a short length to any apparatus which is to be autoclaved (3M autoclave tape number 1222, made by Minnesota Mining and Manufacturing Co.). A similar indicator (3M number 1226) is available for dry heat sterilization; paper or nylon should not be used for wrapping or sealing apparatus to be sterilized in this way, but aluminium foil is satisfactory, although easily punctured.

The various methods of sterilizing apparatus and media are outlined below; more detail is given by Sleigh (1965) and Paul (1970).

B. DRY HEAT

Liquid media cannot be sterilized thus; nor can plastic apparatus, ordinary (i.e. not silicone) rubber or metal apparatus constructed with the aid of solder (e.g. some hypodermic needles). Generally, exposure for 1 h at 160°C is adequate; if the apparatus, plugs, wrappings etc. are adequately thermostable, 20 min at 180°C can be used but this should be avoided if cotton plugs have been used, as these may become charred.

C. MOIST HEAT

This is usually administered under pressure in an autoclave, so that temperatures above 100°C can be achieved. It is suitable for sterilizing most inorganic solutions, ordinary rubber, soldered metal and most other materials (though not all plastics; if in any doubt get advice from the manufacturer). Usually 15 or 20 min exposure to 121°C (pressure of 15 lb in^{-2} or 1·05 kg cm^{-2}) is used (see Sleigh, 1965). Liquids may be sterilized in vessels plugged with non-absorbent cotton (preferably covered also with aluminium foil) or closed with rubber bungs or screw caps; bungs or caps should be loose during autoclaving, and tightened as soon as possible afterwards. Bicarbonate ions decompose into carbon dioxide and water during autoclaving; to prevent loss of the CO_2 and consequent raising of the pH, such solutions can be autoclaved in small sturdy glass containers (e.g. Universal containers) which have been completely filled and had their caps screwed down tightly. Alternatively, the CO_2 can be replaced aseptically afterwards by passing it into the solution through a sterile cotton-plugged tube (e.g. a Pasteur pipette).

D. IRRADIATION

Ultra-violet light may be used to sterilize working surfaces and the exterior of apparatus (see Section III) but its low penetrative power makes it inadequate for sterilizing solutions or apparatus. Exposure to ultra-violet radiation may injure the worker's eyes or skin and must be avoided.

Ionizing radiation (β, γ or X-rays) is used to sterilize plastic apparatus but this is usually done by the manufacturer.

E. FILTRATION

Heat-labile solutions (e.g. those containing serum) can be sterilized only in this way. There are five main types of filter available:

(a) Kieselguehr (= diatomaceous earth; e.g. Berkefeld W)

(b) Porcelain (e.g. Selas 03)
(c) Sintered (= fritted) glass
(d) Asbestos (e.g. Seitz)
(e) Membrane (e.g. Millipore, Sartorius, Nuclepore)

(a) and (b) are useful general purpose filters for sterilizing large volumes; (c) is difficult to clean; (d) is often used but may release toxic substances and should therefore be very thoroughly washed before use by passing a large volume of sterile water through it and then discarding the initial filtrate (House, 1964; Nydegger and Manwell, 1962); (e) is probably the best type, especially for small volumes, though expensive. Pore sizes of $0·22$ μm or less are needed to remove the smaller bacteria. Positive pressure (not suction) should be used if possible when filtering solutions containing serum or carbon dioxide, to avoid frothing of the former or removal from solution of the latter.

F. ANTIBIOTICS

These are often used in media as a safeguard against faulty technique, or to "clean up" contaminated cultures (see Paul, 1970 and the individual chapters of this book). Metazoan parasites and their eggs, or protozoan cysts, may be washed in balanced salt solutions (Section IV, B, 1) containing antibiotics to sterilize their surface before using them to initiate a culture.

Contaminating fungi are particularly difficult to remove from cultures, but Cohen *et al.* (1977) recommend the addition of 10 μg ml^{-1} 5-fluorocytosine (Roche, Ro2-9915) for this purpose. The compound was not toxic to *Leishmania enriettii*, *Trypanosoma cruzi* or Chinese hamster cells in tissue culture; its possible effect on other parasites should be checked before general use.

III. ASEPTIC TECHNIQUE FOR HANDLING CULTURES

The technique of handling cultures to avoid the accidental introduction of contaminating bacteria or fungi is described by Paul (1970) but is best acquired by practice with advice from an experienced worker. It is helpful to work in a sterile room, or with the hands and forearms inserted into a glass-fronted cabinet (e.g. Lumsden *et al.*, 1973, Fig. 18), which has been previously exposed to ultra-violet irradiation (UV) for at least 10 min (remember though that exposure of the operator's eyes and skin to UV is dangerous). Sterile cabinets of this kind are commercially available, some of which function by producing a continual lamina flow of sterile (filtered) air over the working surface (Microflow). For work with pathogens, more complex cabinets which also filter the outgoing air are required (e.g. "Pathoguard" cabinets, Baird and Tatlock). (See also Section V, B.)

Small volumes for microscopical examination only can be removed with a bacteriological nichrome or platinum wire loop, previously sterilized by heating to redness in a Bunsen burner flame and allowed to cool. Subinoculation to fresh medium is often done with a sterile Pasteur pipette (see Section II, A) using aseptic techniques; the cap and neck of the vessel should be briefly passed through a flame before opening, and the neck of the vessel (if glass) heated for rather longer in the flame immediately after removing, and before replacing, the cap; this destroys any bacterial or other spores which may have collected on the outer surface of the vessel. Opened vessels should be held near a flame, and at an angle as near to the horizontal as possible, to minimize the chance of spores falling into the medium.

Cultures containing organisms infective to man (e.g. *Entamoeba histolytica, Balantidium coli, Giardia lamblia, Trichomonas vaginals, Trypanosoma cruzi, T. rangeli, T. brucei rhodesiense, T. b. gambiense, Plasmodium* spp., *Toxoplasma gondii, Schistosoma mansoni, S. japonicum, S. intercalatum, Ascaris, Taenia solium, Echinococcus, Hymenolepis nana*, hookworms, etc.) should be handled only when wearing rubber gloves, should be labelled as a bio-hazard, and should be autoclaved or otherwise sterilized before disposal and washing up. Apparatus used for sampling, examining and subinoculating such cultures must also be disinfected immediately after use—perhaps by immersion in a vessel containing sodium hypochlorite solution (e.g. "Chloros", Durham Chemicals Distributors). Particularly stringent precautions must be taken when handling cultures of *T. cruzi* (see Lumsden *et al.*, 1973, pp. 130–132).

IV. PREPARATION OF MEDIA

The preparation of commonly used media is described below but specialized media are detailed in the relevant chapters. Also, the preparation of some ingredients common to many media is described below; these fall into two main categories—organic and inorganic substances. Many of these are commercially available from, for example, Flow, Gibco Bio-Cult, Oxoid, Wellcome and many other suppliers; but it is often cheaper and sometimes more convenient to prepare them for oneself, especially if the substances are very thermolabile (e.g. embryo extracts).

A. ORGANIC SUBSTANCES

1. *Tissue extracts*

Preparative techniques vary according to different authors, but a basic procedure is that used by Weinstein and Jones (1956) in preparing 50% chick embryo extract (50% CEE, CEE_{50}) for use in cultivating *Nippostrongylus*; Weinstein (1953) used virtually the same technique to extract the raw livers of 3–4 month old rats.

   (i) The egg shell (containing a 9–12 day-old chick embryo) is wiped carefully with 70% ethanol.

   (ii) The blunt end of the shell is cracked by gentle taps with a sterile instrument such as a scalpel handle.

  (iii) Using sterile forceps, the shell above the air sac is removed to expose the membrane.

  (iv) With another pair of sterile forceps the membrane is removed carefully to expose the embryo.

   (v) A sterile pair of curved forceps is slipped under the neck of the embryo, which is gently lifted out of the shell and dropped into a covered sterile beaker embedded in dry ice for rapid freezing.

  (vi) When the required number of embryos has been collected chilled Earle's balanced salt solution (Section IV, B, 1, f) is added (1:1 w/v) and the embryos are homogenized for 1 min in a Waring blender with aseptic precautions.

 (vii) The homogenate is centrifuged (1,000 g, 20 min, 0°C) and the supernatant is dispensed into small sterile bottles. It should be tested for sterility by inoculation into suitable bacteriological media (see Section IV, A, 2). Tissue extracts can be stored at −20°C for up to 6 weeks.

Variations include the performing of the whole procedure at room temperature, washing the embryos thrice in balanced saline to remove blood and yolk, homogenization by expressing the embryos from the barrel of a 20 ml hypodermic syringe, and settling or centrifugation (200 g, 10–20 min) to clarify the extract before use. More drastic methods of homogenization (in a Waring blender or Potter Elvehjem glass tissue homogenizer) completely destroy the individual cells and may liberate toxic substances.

2. *Blood Products*

Blood is usually collected in the laboratory from chickens or rabbits. That of larger mammals can be obtained from suppliers mentioned at the start of Section IV.

(a) *Collection of whole blood*. This must be collected aseptically, since it cannot be sterilized. Clotting can be delayed by adding an anticoagulant (e.g. heparin, minimum effective concentration 0.2 iu ml^{-1}; concentrations of 5–500 times this amount are often used) or prevented by defibrination (the blood is gently swirled for 10–15 min in a sterile vessel containing several small glass beads).

(i) *Heart puncture*. Pentobarbitone sodium ("Nembutal", Abbott Laboratories; "Sagatal", May and Baker) at 25 mg kg^{-1} body weight (intravenous or intramuscular) can be used to anaesthetize rabbits or chickens. Rabbits (not chickens) can also be anaesthetized with ether or—better, since it avoids the risk of explosion—trichloroethylene ("Trilene", Imperial Chemical Industries, or "Hypnorm", Janssen Pharmaceuticals). The skin over the heart should be plucked or shaved and sterilized with 70% ethanol. The technique used with chickens is described by Paul (1970) and illustrated in Fig. 1; the procedure with rabbits is similar.

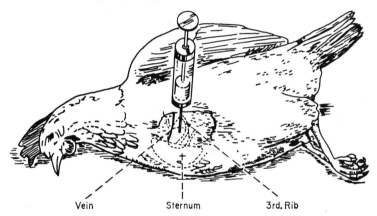

Vein   Sternum   3rd. Rib

FIG. 1. Obtaining blood from a chicken by heart puncture (from Paul, J., "Cell and Tissue Culture", edn 4, 1970).

(ii) *Venepuncture*. The brachial vein or carotid artery of chickens (see Paul, 1970), or the marginal ear vein of rabbits, are usually used. An old fashioned, but still useful method of adding rabbit blood to a series of small culture vessels (e.g. NNN or 4N medium; see Chapter 3), or collecting small volumes of it, was described by Wenyon (1926). The unanaesthetized rabbit is immobilized by wrapping in a towel or other cloth, and held on a table or restrained in a specially made box (Fig. 2). The skin over the marginal ear vein is shaved and sterilized with 70% ethanol; a thin layer of molten paraffin (melting point 45°C) is applied with a glass rod; an incision is made in the vein with a sterile small scalpel blade or Hagedoorn needle and the blood allowed to drip from the paraffin surface into the collecting vessel held just below. Sterile precautions should be observed in opening and closing the vessels; surprisingly, blood collected thus is seldom contaminated once

FIG. 2. Obtaining blood from marginal ear vein of rabbit. Inset shows end of box with hole for rabbit's neck and removable upper section. A false inner back to the box, either spring-loaded or capable of being slotted into different positions, can be used to accommodate rabbits of different sizes (modified from Wenyon, C. M., "Protozoology", 1926).

the worker has gained experience. (A colleague of the authors refers to this as the "drippy-lobe" method.)

The sterility of collected blood can be checked by inoculating a small volume into two tubes of nutrient broth and anaerobic bacterial growth medium [e.g. Oxoid "Nutrient broth number 2" and "Thioglycollate medium (Brewer)", respectively]; ideally one tube of each medium should be incubated at 25–28°C and the other at 37°C for at least 24 h. Cloudiness or precipitation in the tubes indicates bacterial contamination. (The sterility of plasma, serum and prepared culture media should be checked similarly.) Whole blood is best used as fresh as possible, but it can be stored at 4°C for 2 days (if heparinated) or 7 days (if defibrinated).

(b) *Plasma*. This is obtained by centrifugation of blood to remove the cells. Clotting must be prevented either by the addition of an anticoagulant (see Section IV, A, 2, a) or by collecting and processing in siliconed tubes at 0–4°C. If subsequent clotting is required, the second procedure is the best; alternatively, a minimal amount of anticoagulant may be used. Plasma can be sterilized by filtration and stored at −20°C or lyophilized (Strumia, 1954).

(c) *Serum*. Serum is obtained by allowing blood to clot and collecting the supernatant fluid. Clotting occurs faster at 37°C. Once formed, the clot may be gently separated from the side of the vessel with a glass rod. The yield of serum is increased by centrifugation. Serum can be sterilized by filtration and stored similarly to plasma; its complement can be inactivated, if required, by heating at 56°C for 30 min.

B. INORGANIC SUBSTANCES

The preparation of complex media is described where relevant in the text. However certain saline and balanced salt solutions (BSS) are commonly used for washing and suspending parasites before cultivation, as ingredients of many media and in the harvesting of organisms from culture; these often contain buffers to maintain a physiological pH (7·2–7·6).

1. *Salt Solutions*

The purest grade of chemicals obtainable should be used; water should be double or triple distilled from glass. Calcium or magnesium chloride, if required, should be dissolved separately from the other constituents to avoid the risk of precipitation. Since bicarbonate is difficult to autoclave (see Section II, C), it is often omitted from stock solutions and added when required as an aqueous solution of sodium bicarbonate which has been sterilized by membrane filtration under positive pressure. If, as is often the case, phenol red ($0·01$–$0·05$ mg ml^{-1}) is incorporated in the BSS, the bicarbonate solution is added aseptically until the colour indicates that the appropriate pH (usually $7·2$–$7·6$) has been reached; alternatively, the indicator can be incorporated in the bicarbonate solution. The latter is often prepared as a 3% w/v solution of $NaHCO_3$, though other concentrations can be used: sterile solutions are obtainable commercially (e.g. $1·4\%$ w/v from Oxoid, $7·5\%$ w/v from Gibco Bio-Cult or Flow Laboratories). The ingredients and preparative details for some commonly used solutions are given below.

(a) *Physiological saline.* An aqueous solution of $0·95\%$ (w/w) NaCl (ca $0·9$ g litre^{-1}) has an osmotic pressure roughly equivalent to that of mammalian blood serum (Diem and Lentner, 1970, p. 271) and can therefore be used to suspend blood cells and some parasitic protozoa. It is, however, unbuffered and for general use the solution described below is preferable.

(b) *Phosphate buffered saline (PBS).* The following salts are dissolved in distilled water and sterilized by filtration. The final pH is about $7·3$.

	g litre^{-1}
NaCl	8·00
K_2HPO_4	1·21
KH_2PO_4	0·34

Sodium azide, which prevents microbial contamination, may be added ($0·08\%$ w/v) to PBS used to dilute serum or for other purposes but must *not* be added to saline which will be used in the cultivation of cell lines or parasites.

(c) *Ringer's frog solution* (slightly modified from Ringer, 1883, 1895; see Parker, 1961, p. 55. Ringer's original recipe is given in parentheses). This is hypotonic with respect to mammalian sera but approximately isotonic with amphibian sera.

	g litre^{-1}	
NaCl	6·5	(6·75)
KCl	0·14	(0·09)
$CaCl_2$	0·12	(0·115)
$NaHCO_3$	0·20	(0·215)

Dissolved in distilled water and sterilized by filtration.

(d) *Ringer–Locke solution* (Locke, 1901)

	g litre^{-1}
NaCl	9·0
KCl	0·42
$CaCl_2$	0·24
$NaHCO_3$	0·2
Glucose	1·0

Dissolved in distilled water and sterilized by filtration. If used as a constituent of Dobell's media (Dobell and Laidlaw, 1926), sodium bicarbonate and glucose should be omitted (see Taylor and Baker, 1968).

(e) *Tyrode's solution* (1910)

	g litre^{-1}
NaCl	8·0
KCl	0·2
$CaCl_2$	0·2
$MgCl_2.6H_2O$	0·1
$NaH_2PO_4.H_2O$	0·05
$NaHCO_3$	1·0
Glucose	1·0

Dissolved in distilled water and sterilized by filtration. Tyrode's solution is designed for use saturated with oxygen.

(f) *Earle's solution* (1943)

	g litre^{-1}
NaCl	6·8
KCl	0·4
$CaCl_2$	0·2
$MgSO_4.7H_2O$	0·2
$NaH_2PO_4.H_2O$	0·14
$NaHCO_3$	2·2
Glucose	1·0

All except $NaHCO_3$ are dissolved in distilled water (solution 1); the $NaHCO_3$ is made up as 10% (w/v) solution in distilled water with phenol red to give a final concentration of 10 mg litre^{-1} (solution 2). Both solutions are sterilized by filtration. Solution 2 ($NaHCO_3$) is adjusted to pH 7·0 with sterile CO_2 (the phenol red acting as indicator) and 22 ml then added to solution 1. This solution is designed to equilibrate with 5% carbon dioxide.

(g) *Hanks's solution* (Hanks and Wallace, 1949)

	g litre^{-1}
NaCl	8·0
KCl	0·4
$CaCl_2$	0·14
$MgSO_4.7H_2O$	0·2
$Na_2HPO_4.12H_2O$	0·15
KH_2PO_4	0·06
$NaHCO_3$	0·35
Glucose	1·0

Dissolved in distilled water and sterilized by filtration.

2. *Buffer Systems*

The BSS described above (Section IV, B, 1) all depend on carbon dioxide/bicarbonate or phosphate buffers to control their pH, and until recently these were virtually the only buffers used in culture media. However, both have disadvantages: the former are difficult to handle (due to the tendency to lose carbon dioxide from solution) and may require equilibration with a particular concentration of CO_2 in the gas phase (usually 5%), and the latter (phosphates) tend to form insoluble complexes and also may be utilized by cells or parasites in culture. Other buffering systems have therefore been tried. The best for cell cultures seem to be the zwitterionic amino acids developed by Good *et al.* (1966), particularly *N*-2-hydroxyethylpiperazine-*N'*-2 ethanesulphonic acid (HEPES) and *N*-tris(hydroxymethyl)methylaminoethanesulphonic acid (TES). These are metabolically inert and therefore relatively non-toxic. The former is widely used at a concentration of 20 mM (4·7662 g litre^{-1}). To meet the metabolic needs of the cells, 10 mM sodium bicarbonate (0·85 g litre^{-1}) is usually included in the medium. Eagle (1971) suggested combinations of various zwitterionic buffers for precise pH control between 6·5 and 8·3; his paper should be consulted for details. Spendlove *et al.* (1971) found that another zwitterionic amino acid, *N*-tris(hydroxymethyl)methyl glycine (Tricine) was a good non-toxic buffer at 5–10 mM, plus 5 mM bicarbonate. At higher concentrations (23 mM alone, or 15 mM with 5 mM bicarbonate) it had the useful additional function of inhibiting the growth of *Mycoplasma* contaminants. These buffers are available commercially from many suppliers: Flow Laboratories and Gibco Bio-Cult supply HEPES, while the entire range is obtainable from Hopkin and Williams (who have also produced a usefully informative leaflet, number H.150, on the subject).

V. OTHER TECHNIQUES—BAERMANN APPARATUS

This device is used to separate nematode larvae from soil samples. It comprises a funnel connected to rubber tubing which can be closed with a clip. A metal sieve (1 mm mesh) is placed in the funnel and covered with coarse cotton fabric. The soil sample is placed on this and, with the clip closed, the funnel is filled with 0.85% (w/v) sodium chloride solution to a height of about 2·5 cm above the sieve. After several hours any nematode larvae will have fallen through the cloth and sieve into the rubber tube, and can be collected by opening the clip and allowing the liquid in the tube to run into a collecting vessel.

REFERENCES

Cohen, B. A. J., Looij, B. J. and Wittner, M. (1977). 5-Fluorocytosine, a valuable fungicidal agent in specific culture media. *Annales de la Société Belge de Médecine Tropicale* 57, 55–56.

Cruickshank, R., Duguid, J. P., Marmion, B. P. and Swain, R. H. A. (1975). "Medical Microbiology Vol. 2: The Practice of Medical Microbiology." Churchill Livingstone, Edinburgh, London and New York.

Diem, K. and Lentner, C. (eds) (1970). "Documenta Geigy". Geigy, Basle.

Dobell, C. and Laidlaw, C. C. (1926). On the cultivation of *Entamoeba histolytica* and some other entozoic amoebae. *Parasitology* 18, 283–318.

Eagle, H. (1971). Buffer combinations for mammalian cell culture. *Science, New York* 174, 500–503.

Earle, W. R. (1943). Propagation of malignancy *in vitro*. IV. The mouse fibroblast cultures and changes seen in living cells. *Journal of the National Cancer Institute* 4, 165–212.

Good, N. E., Winget, G. D., Winter, W., Connolly, T. N., Izawa, S. and Singh, R. M. M. (1966). Hydrogen ion buffers for biological research. *Biochemistry* 5, 467–477.

Hanks, J. H. and Wallace, R. E. (1949). Relation of oxygen and temperature in the preservation of tissues by refrigeration. *Proceedings of the Society for Experimental Biology and Medicine* 71, 196–200.

Hanks, J. H. and Staff of the Cooperstown Course (1955). "An Introduction to Cell and Tissue Culture." Burgess Publishing Company, Minneapolis.

House, W. (1964). Toxicity of cell culture medium due to filtration through asbestos filter pads. *Nature, London* 201, 1242.

Locke, F. S. (1901). Die Wirking der Metalle der Blutplasmas und verschiedener Zucker auf das isolierte Säugerthierhertz. *Zentralblatt für Physiologie* 14, 670.

Lumsden, W. H. R., Herbert, W. J. and McNeillage, G. J. C. (1973). "Techniques with Trypanosomes." Churchill Livingstone, Edinburgh and London.

Nichosia, M. A. (1965). Water soluble extractions of disposable syringes. Nature and significance. *Journal of Pharmacological Science* **54**, 1379–1381.

Nydegger, L. and Manwell, R. D. (1962). Cultivation requirements of the avian malaria parasite *Plasmodium hexamerium*. *Journal of Parasitology* **48**, 142–147.

Parker, R. C. (1961). "Methods of Tissue Culture" Edn 3. Pitman Medical Publishing Company, London.

Paul, J. (1970). "Cell and Tissue Culture" Edn 4. Livingstone, Edinburgh and London.

Ringer, S. (1883). A further contribution regarding the influence of the different constituents of the blood on the contraction of the heart. *Journal of Physiology* **4**, 29–42.

Ringer, S. (1895). Further observations regarding the antagonism between calcium salts and sodium, potassium and ammonium salts. *Journal of Physiology* **18**, 425–429.

Sleigh, J. D. (1965). Sterilization. *In:* "Medical Microbiology" (R. Cruickshank, ed.), pp. 679–721. Livingstone, Edinburgh and London.

Spendlove, R. S., Crosbie, R. B., Hayes, S. F. and Keeler, R. F. (1971). TRICINE-buffered tissue culture media for control of *Mycoplasma* contaminants. *Proceedings of the Society for Experimental Biology and Medicine* **137**, 258–263.

Strumia, M. M. (1954). The preservation of blood plasma and blood products by freezing and drying. *In:* "Biological Applications of Freezing and Drying" (R. J. C. Harris, ed.), pp. 129–149. Academic Press, New York and London.

Taylor, A. E. R. and Baker, J. R. (1968). "Cultivation of Parasites *in vitro*." Blackwell Scientific Publications, Oxford and Edinburgh.

Tyrode, M. V. (1910). The mode of action of some purgative salts. *Archives Internationales de Pharmacodynamie* (*et de Therapie*) **20**, 205–223.

Weinstein, P. P. (1953). The cultivation of free-living stages of hookworms in the absence of living bacteria. *American Journal of Hygiene* **58**, 352–376.

Weinstein, P. P. and Jones, M. F. (1956). The *in vitro* cultivation of *Nippostrongylus muris* to the adult stage. *Journal of Parasitology* **42**, 215–236.

Wenyon, C. M. (1926). "Protozoology". Ballière, Tindall and Cox, London. (Reprinted 1966: Ballière, Tindall and Cassell, London).

PROTOZOA

Chapter 2

Entamoeba, Giardia, and Trichomonas

E. MEEROVITCH

Institute of Parasitology, Macdonald College of McGill University, Macdonald College Post Office, Province of Quebec, Canada

I. CULTIVATION OF *ENTAMOEBA* SPECIES

A. *E. HISTOLYTICA*

Developments in the techniques of cultivation of *Entamoeba histolytica*, since the publication of reviews on this subject by Neal (1967) and Taylor and Baker (1968), have been towards perfection of more closely defined monoxenic and axenic media. The need for such media resulted from the necessity of obtaining large yields of pure amoebic cells, unaccompanied by other microorganisms, for studies of their biochemistry and pathogenicity, preparing antigens for serodiagnosis and for drug testing. Certainly, an admixture of concomitant culture associates or their components might affect the specificity of the amoebic antigen, and thus render questionable the results of serological testing. However, Schneider and Gordon (1968) found that the undefined components of medium TPS-1 (see Section I, A, 2, 2°) in which the amoebae are grown axenically do not interfere with the specificity of serological reactions. Nevertheless, the search for a defined medium goes on, because elimination of such additives as serum, peptones and tissue digests, and their substitution by known chemical components, would standardize the procedures performed in different laboratories and would help to elucidate more precisely the nutritional requirements of the amoebae.

1. *Monoxenic Cultivation*

The development of the best medium for axenic cultivation of *E. histolytica* (medium TPS-1) available at this time stemmed from the original diphasic axenic medium of Diamond (1961), which consisted of an agar-nutrient broth base with serum, overlaid with the broth supplemented with chick embryo extract and a vitamin mixture. Originally these axenic cultures were derived from amoebae grown monoxenically in association with a species of flagellate of the genus *Crithidia*. Although the yields of axenic amoebae from this medium were quite good (3 to 4-fold increase in 72 h), it was rather complicated to prepare, and, moreover, it was impossible to recover the amoebae in pure suspension because the liquid overlay contained a small percentage of agar.

The next step in the development of an axenic medium for *E. histolytica* was also taken by Diamond (1968a). This resulted in the formulation of a simpler medium, TTY-SB, in which the amoebae had to be grown monoxenically in association with either a *Crithidia* sp. or *Trypanosoma cruzi*. Westphal and Michel (1971) found that none of the several species of parasitic and free-living flagellates that they assayed as food organisms for *E. histolytica* could replace the *Crithidia* sp., probably because none could divide fast enough to maintain adequate numbers to feed the amoebae.

1° TTY-SB medium
(Diamond, 1968a)

MEDIUM

Tryptose (Difco)	1·00 g
Trypticase (BBL)	1·00 g
Yeast extract (BBL)	1·00 g
Glucose......................	0·50 g
L-Cysteine monohydrochloride	0·10 g
Ascorbic acid	0·02 g
Sodium chloride	0·50 g
Potassium phosphate, monobasic ..	0·08 g
Potassium phosphate, dibasic, anhydrous	0·08 g
Distilled water to make	95·00 ml
pH adjusted to 7·2 with N-NaOH	

TECHNIQUE

Although Diamond recommended that the medium should be adjusted to pH 7·2, it was found in the writer's laboratory that better growth was obtained at pH 6·8. After sterilization by autoclaving, the 95 ml of medium are supplemented with 5 ml of sterile horse serum and 0·25 ml of sterile defibrinated rabbit blood in which the erythrocytes have been lysed by freezing and thawing. The complete medium is used to maintain stock cultures of *Crithidia* sp. and for growing the amoeba-*Crithidia* (A-C) cultures. In practice, about 0·2 ml of a 72-h *Crithidia* culture, containing about 3 million flagellates, is added to 15 ml of TTY-SB medium in a 16 × 125 mm screw-capped tube before inoculating with amoebae.

To maintain *T. cruzi* for subsequent use as a food for *E. histolytica* in this medium, Diamond (1968a) used the following modification (TTY-6).

2° TTY-6 medium
(Diamond, 1968a)

MEDIUM

Tryptose (Difco)	0·50 g
Trypticase (BBL)	0·60 g
Yeast extract (BBL)	0·50 g
Sodium chloride	0·50 g
Potassium phosphate, monobasic ..	0·08 g
Potassium phosphate, dibasic, anhydrous	
	0·08 g
Distilled water to make	80·00 ml
pH adjusted to 7·2 with N-NaOH.	

TECHNIQUE

After autoclaving (10 min, 121°C), the 80 ml of medium are supplemented with 19 ml of rabbit serum and 1 ml of frozen and thawed defibrinated rabbit blood. *T. cruzi* is grown in 125 ml flasks containing 12 ml of the complete medium TTY-6-SB. To grow *E. histolytica* with *T. cruzi* (A-T cultures), 2–2·5 million flagellates from a 7d stock culture are added to 15 ml of TTY-SB medium in a 16 × 125 mm tube, before inoculating with amoebae. Amoebic cultures with either *Crithidia* or *T. cruzi* are incubated at 35–37°C in a slanted or almost horizontal position to allow the amoebae to spread more readily along the length of the tube. Transfers are made every 48 and 72 h. In order to prepare the amoebic inoculum, donor cultures are chilled in iced water for 5 min, inverted a few times, and centrifuged (5 min, 850 g). The supernatant fluid is decanted and the sedimented amoebae are inoculated into TTY–SB medium previously seeded with either *Crithidia* or *T. cruzi*.

2. *Axenic Cultivation*

The first axenic medium developed by Diamond (1961) was diphasic, the liquid overlay containing a small amount of agar. As such, it was not suitable for the recovery of amoebae in pure suspension. The next step, the development of medium TTY–SB, was at the same time an improvement and a step backward, since, although it was completely liquid, it was suitable only for monoxenic cultivation of amoebae. The concomitant flagellates were, however, used by amoebae as food, so that after they had been almost completely consumed the amoebae could be recovered in almost pure suspension.

The next step, also taken by Diamond (1968b), was the development of a diphasic medium, TTP–S, which supported axenic growth of amoebae derived from either A–C or A–T cultures. This medium, although diphasic and with agar in the overlay, was, nevertheless, less complex than the original diphasic medium of 1961, since the use of chick embryo extract was no longer required. This was made possible by using Panmede, an ox liver digest, first used in an axenic medium for *E. invadens* by McConnachie (1962). The next, and final improvement to date, was the development, again by Diamond (1968b) and Diamond and Bartgis (1971), of medium TPS–1.

Wittner (1968) proposed a diphasic semi–defined medium (SDM) for axenic cultivation of *E. histolytica*. The slant is made up of 5 ml of 2% (w/v) purified agar in distilled water, to which 1% (v/v) of sterile inactivated horse serum is added when the agar has cooled down sufficiently after autoclaving. The liquid overlay consists of a mixture of 6 components, each of which is sterilized individually by Millipore filtration. These components are: (1) a mixture of 18 amino acids; (2) vitamin mixture 107 (see Section I, A, 2, 3° below); (3) phosphate buffered saline; (4) a mixture of 5 mineral salt solutions; (5) a mixture of glucose, cholesterol and horse serum and (6) a mixture of 5 nucleotides. For details the reader is referred to the original publication (which does not describe the preparation of cholesterol solution; presumably only the portion soluble in serum enters the final medium). Amoebic inocula for this medium were derived from monoxenic A–T cultures. However amoebic growth was not any better in this medium than in Diamond's axenic diphasic medium. This medium is the nearest approach, so far, to a defined medium for *E. histolytica*; it remains to isolate and characterize the serum component(s) necessary for axenic growth of the amoebae.

E. histolytica growing in the monoxenic (TTY-SB) medium in association with flagellates can be axenized when transferred into TPS-1 medium, by repeating such transfers until a successful culture is established. However, according to several published reports not all strains of *E. histolytica* are amenable to axenization. Thus, the establishment of an axenic culture requires the availability of a monoxenic culture of a suitable strain of *E. histolytica.*

In the writer's experience it was possible to achieve axenization of a newly-isolated strain of *E. histolytica* as follows.

(i) A faecal sample containing cysts was cultured in Dobell and Laidlaw's (1926) diphasic medium HSre + S supplemented with about 200 iu penicillin and 200 µg streptomycin ml^{-1}. The established culture was maintained in this medium with twice-weekly transfers for a number of weeks.

(ii) The amoebae were then transferred into MS-F medium with *Bacteroides symbiosus* (Reeves *et al.*, 1959; Taylor and Baker, 1968) which contained, in addition to the requisite amount of 5000 iu of penicillin per tube, 50 µg kanamycin ml^{-1}.

(iii) After a period of culturing in this medium and regular checking of the bacterial flora, the penicillin supplement was increased to 10,000 iu per tube and kanamycin was omitted. Only *Bacteroides symbiosus* remained as the associate, being added at each transfer.

(iv) Next, the amoebae were transferred into TTY-SB medium with *Crithidia* sp., but the addition of penicillin (about 500 iu ml^{-1}) was continued until no more bacteria could be detected.

(v) The monoxenic amoebae were finally transferred into TPS-1 medium, and an axenic culture established.

Dutta and Singh (1975) were able to axenize *E. histolytica* into TPS-1 medium from mixed bacterial culture, bypassing the monoxenic association with trypanosomatids, as follows.

(i) Amoebic cultures in Boeck and Drbohlav medium with mixed bacterial flora were treated for 10–15 passages with gentian violet and acriflavine (0·01% v/v of each). This treatment eradicated some of the bacterial species; a sensitive strain of *Escherichia coli* was added at each subculture.

(ii) Amoebae from the dye-treated cultures were inoculated into TPS-1 medium with 0·2% L-cysteine, but without ascorbic acid. Bovine serum was used instead of horse serum. The medium was supplemented with penicillin, streptomycin, oxytetracycline, chloramphenicol and kanamycin.

(iii) The surviving bacterial species, *Pseudomonas pyocyanea*, was eliminated with gentamycin, penicillin and streptomycin in high doses.

(iv) The surviving amoebae were established in axenic culture in TPS-1 medium.

At this point it may be appropriate to emphasize that, strictly speaking, an axenic culture should be free not only of concomitant bacteria, yeasts or fungi, but also of *Mycoplasma*. It has been found that *E. histolytica* may be infected with viruses which appear to be its normal associates. However, the possible presence of such viruses should not preclude the designation of amoebic cultures as axenic.

Other methods of axenizing amoebae have been used. One is by isolating trophozoites or cysts from amoeba-bacteria cultures by means of micromanipulation, establishing cultures in association with known selected microbial associates, re-isolating the amoebae by micromanipulation, and finally establishing monoxenic A-C or A-T cultures, and eventually axenic cultures in TPS-1 medium (Diamond, 1968a). Bos (1975) established monoxenic and axenic cultures of *E. histolytica* in TTY-SB and TPS-1 media from amoebae growing with bacteria in Jones' medium; the bacteria were eliminated by treating the amoebae with a saline solution containing 20 mg ml^{-1} neomycin sulphate.

Another method used in the writer's laboratory was to inoculate amoebae and residual crithidiae from A-C cultures into a mixture of TTY-SB and TPS-1 media in different proportions, first using more of the former, then more of the latter, and lastly the TPS-1 medium alone. *Crithidia* were added to cultures in the initial transfers, but as the amount of the TPS-1 component was increased the number of crithidiae added was reduced. This procedure resulted finally in an axenic culture of *E. histolytica* in TPS-1 medium.

Raether *et al.* (1973) inoculated *E. histolytica* from A-C cultures in TTY-SB medium into TPS-1 medium without the vitamin supplement, but adding *Crithidia*. Later the vitamin supplement was added and *Crithidia* were omitted. This also resulted in an axenic culture.

1° TTP-S medium
(Diamond, 1968b)

MEDIUM

Tryptose (Difco) .... 1·00 g
Trypticase (BBL) .... 1·00 g
Panmede, liver digest (Paynes and Byrne)
                                1·00 g

Glucose . . . .   0·50 g
L-Cysteine monohydrochloride . . .   0·10 g
Ascorbic acid . . . .   0·02 g
Sodium chloride . . . .   0·50 g
Potassium phosphate, monobasic . .   0·08 g
Potassium phosphate, dibasic, anhydrous
                                    0·08 g
Distilled water, to make . . . .   80·00 ml
pH adjusted to 7·0 with N-NaOH.

TECHNIQUE

To make slants, 0·95 g Ionagar no. 2 (Oxoid) are dissolved in 80 ml of medium, and 4·8 ml portions of the hot mixture are autoclaved in tubes. When the mixture is cooled to 50°C, 1·2 ml of horse serum are added to each tube and a slant is made. The overlay consists of 25 ml of the TTP broth with 0·02 g Ionagar no. 2, diluted with 75 ml of distilled water. Of this, 5·7 ml are added to each slant, and each tube is supplemented with 0·3 ml of vitamin mixture 107 (see Section I, A, 2, 3° below), before inoculating with amoebae.

2° TPS-1 medium
(Diamond, 1968b; Diamond and Bartgis, 1971)

MEDIUM

Trypticase . . . .   1·00 g
Panmede (Paynes and Byrne) . . . .   2·00 g
Glucose . . . .   0·50 g
L-Cysteine monohydrochloride . . .   0·10 g
Ascorbic acid . . . .   0·02 g
Sodium chloride . . . .   0·50 g
Potassium phosphate, monobasic . .   0·06 g
Potassium phosphate, dibasic, anhydrous
                                    0·10 g
Distilled water, to make . . . .   87·50 ml

Adjust pH to 7·0 with N-NaOH. Filter through Whatman no. 1 Paper, autoclave 10 min, 15 lb. When cooled, add under aseptic conditions:

Inactivated serum (bovine or horse)
                                    10·0 ml
Vitamin mixture 107 (see below) .   2·5 ml

TECHNIQUE

Amoebic cultures can be set up in screw-capped tubes of any size, Ehrlenmayer flasks, tissue culture flasks, etc., depending on the purpose of culturing. Sometimes the media TTY-SB and TPS-1 are supplemented with a small amount of a dilute solution of Resazurin, a redox indicator. If the colour change at the top of the fluid in a tube, showing oxidation, is seen to extend lower than about 1·0 cm from the surface, the particular tube is not used to inoculate amoebae. In practice this monitoring of the redox potential in the tube is usually not necessary.

TP Broth Base Powder, as well as the vitamin mixture 107 (frozen solution) can be purchased from North American Biologicals Inc. However, the vitamin mixture 107, originally devised by Evans et al. (1956) can be made up in the laboratory from basic ingredients. The procedure, while somewhat lengthy, is not difficult.

3° Vitamin mixture 107
(Evans et al., 1956)

The mixture consists of 5 stock solutions which are combined in a certain proportion, as described below.

Solution 1. Water-soluble B-vitamins.

(i) Niacin . . . . 12·5 mg
p-Aminobenzoic acid . . . . 25·0 mg
Dissolve in boiling distilled water and bring volume up to . . . . 25 ml
(ii) Niacinamide . . . . 12·5 mg
Pyridoxine HCl . . . . 12·5 mg
Pyridoxal HCl . . . . 12·5 mg
Thiamine HCl . . . . 5·0 mg
Ca pantothenate . . . . 5·0 mg
Iso-inositol . . . . 25·0 mg
Choline chloride . . . . 250·0 mg
Dissolve in distilled water and bring volume up to . . . . 25 ml
(iii) Riboflavin . . . . 10·0 mg
Add to 10 ml of 0·075 N-HCl, then add 5 ml of 0·2 N-NaOH, and apply moderate heat to dissolve; bring volume up to 20 ml with distilled water.

Combine 10 ml of solution (iii) with 25 ml of solution (i) and 25 ml of solution (ii), and bring volume up to 100 ml with distilled water.

Solution 2.
Dissolve 10 mg of D-biotin in approximately 50 ml of distilled water, acidified with 1 ml of N-HCl. Bring volume up to 100 ml with distilled water.

Solution 3.
Dissolve 10 mg of folic acid in 100 ml distilled water at about 80°C.

Solution 4. Lipid-soluble vitamins.

Dissolve 10 mg vitamin D (calciferol) in 2 ml absolute ethyl alcohol in a 100 ml volumetric flask (calciferol may not dissolve completely unless Tween 80 is added). Dissolve 10 mg vitamin A (alcoholic) in the calciferol tincture. Dissolve 10 mg vitamin K (menadione) in 1 ml absolute ethyl alcohol in a separate flask or tube. Add 0·1 ml of this solution to the vitamins D and A solution in the first flask. Add 10 ml of 5% (v/v) aqueous solution of Tween 80 to the pooled alcoholic tinctures. Bring volume up to 100 ml with distilled water.

Solution 5.

Dissolve 10 mg of disodium alpha-tocopherol phosphate (vitamin E) in a small volume of distilled water and bring volume up to 100 ml with distilled water.

The stock solutions 1, 2, 3, 4 and 5 can be kept frozen at $-20°C$ for a very long period of time. In order to make the vitamin mixture 107 the stock solutions are thawed and combined in the following proportions by volume: solution 1, 2; solution 2, 1; solution 3, 1; solution 4, 10; solution 5, 1.

The unused portions of the five stock solutions can be re-frozen until needed again. The vitamin mixture is sterilized by Seitz or Millipore filtration. The filtered vitamin mixture is ready to be added to the TPS-1 medium, or can be frozen at $-20°C$ until required. The complete TPS-1 medium, i.e. the basic TP broth with serum and vitamins (and antibiotics, if necessary) can also be stored frozen at $-20°C$ in 100 ml serum bottles or other similar containers.

3. *Isolation and Diagnostic Cultivation*

The culture methods described in Sections I, A, 1 and 2 presuppose the availability of thriving cultures of amoebae in polyxenic or monoxenic media. However it is often necessary to diagnose amoebiasis by recovery of the parasites in culture, if it is difficult or impossible to do so by direct microscopical examination of faeces. The so-called "classical" methods of isolation of amoebae in a variety of media, diphasic or monophasic, with different bacteria are still in use, but it is also possible to make primary isolates in MS-F medium supplemented with increased amounts of penicillin, streptomycin, kanamycin or other wide-spectrum antibiotics. Myjak (1971) proposed a solid medium for the primary isolation of *E. histolytica* from faeces (see Section I, A, 3, 1° below).

Amoebic liver abscess is usually diagnosed clinically or serologically. However it may be necessary to make primary isolates of amoebae from such abscesses. The abscess aspirate can be inoculated into any of the media used for the cultivation of *E. histolytica*, and if amoebae are present in the inoculum it is reasonable to expect that they would establish a thriving culture, provided that it does not become overgrown with microbial contaminants. Wang *et al.* (1973) inoculated material from liver abscess aspirates into Diamond's TPS-1 medium seeded with either *Trypanosoma cruzi* or *T. conorhini* and obtained monoxenic amoebic cultures. These workers were subsequently able to axenize the cultures in TPS-1 medium if incubation was carried out at 34°C, rather than 37°C (Wang *et al.*, 1974).

1° Primary isolation medium
(Myjak, 1971)

MEDIUM

"Meat extract"	50 ml
Distilled water	50 ml
Bacto-peptone	1·0 g
Purified agar	1·0 g
Glucose	0·5 g
Resazurin, 0·05% (w/v) solution	0·25 ml
Horse serum	10 ml
Penicillin	50,000 iu
Streptomycin	50 mg

TECHNIQUE

The medium, after autoclaving, is poured into Petri dishes and incubated (30–60 min, 37°C) before use. The plate, inoculated with a small faecal sample, is inverted over a filter paper envelope containing a mixture of 0·5 g pyrogallol, 0·25 g potassium carbonate and 1·0 g diatomite, and the whole is sealed with paraffin onto a glass plate. Results could be read under the microscope after incubation for 48 and 72 h; 99% of faecal samples which contained amoebic trophozoites or cysts produced positive cultures.

B. OTHER SPECIES OF *ENTAMOEBA*

The so-called "histolytica-like" amoebae of the Laredo type, that are able to grow at room temperature as well as at 37°C, can be maintained in monoxenic and axenic media used for *E. histolytica*. Cultures of these amoebae usually do not need to be transferred every 3 or 4 d, if kept at 37°C, but weekly transfers are advisable. If incubated at about 25°C transfers can be made every 3–4 weeks.

Entamoeba invadens, commensal in herbivorous reptiles (turtles and some lizards) but pathogenic in carnivorous ones (snakes and some lizards), grows extremely well axenically in TPS-1 medium at about 25°C. The axenization of new isolates of this species can be done through primary isolation in MS-F medium (Taylor and Baker, 1968) with antibiotics and *Bacteroides symbiosus*, but the intermediate monoxenization with haemoflagellates in TTY-SB medium (Section I, A, 1, 1°) can be omitted. On the other hand, *E. terrapinae*, a commensal of turtles, while growing well in MS-F medium, does not, in the writer's experience, thrive axenically in TPS-1 medium; in repeated trials axenic cultures died out after about 10 transfers.

Entamoeba moshkovskii, although not a parasite but mainly an inhabitant of sewage, has also been axenized by Diamond and Bartgis (1970). Monoxenic cultures established in MS-F medium were transferred for 20 passages in TTY-SB medium with *T. cruzi*, and *Bacteroides symbiosus* was eliminated with the aid of an antibiotic mixture. These cultures were then transferred into the TPS-1-GM (gastric mucin) medium in which they grew axenically (see Section I, B, 1° below).

It will be noted that in most cases of axenization of *Entamoeba* species, a period of axenic cultivation in the presence of a small amount of agar in the medium was necessary. The agar seems to substitute in some manner for concomitant organisms, possibly supplying the amoebae with particulate food, until they become adapted to a completely pinocytotic mode of feeding in a particle-free medium.

1° TPS-1-GM medium for *E. moshkovskii*
(Diamond and Bartgis, 1970)

MEDIUM

Trypticase (BBL)	1·00 g
Panmede (Paynes and Byrne)	2·00 g
Glucose	0·50 g
L-Cysteine monohydrochloride . . .	0·10 g
Ascorbic acid	0·50 g
Potassium phosphate, monobasic . .	0·06 g
Potassium phosphate, dibasic, anhydrous	
	0·10 g
Distilled water to make	96·5 ml

Adjust pH to 7·0 with N-NaOH and filter through paper; add:

Gastric mucin powder (Wilson Laboratories, no. 440)	0·10 g
Ionagar no. 2 (Oxoid).	0·01 g

Autoclave, cool to room temperature, and add aseptically:

Horse serum	1·0 ml

Vitamin mixture 107 (Section I, A, 2, 3°)
2·5 ml

Dispense in screw-capped tubes.

After a period of cultivation in this medium, *E. moshkovskii* could be transferred into, and maintained in, TPS-1 medium (Section I, A, 2, 2°).

II. CULTIVATION OF *GIARDIA LAMBLIA*

Recent interest in the pathogenic role of *Giardia lamblia* and success in the achievement of axenic cultivation of *E. histolytica* have given an impetus towards the development of an axenic culture system for this flagellate. The history of cultivation of *G. lamblia in vitro* is not a long one. There have been a number of attempts by several workers in the past to isolate and maintain this flagellate *in vitro*, but without success. Because of the natural habitat of *Giardia* in the upper small intestine it was believed by some that bile or bile salts might play an important part in its metabolism, and various culture media were devised, incorporating such components. None was successful.

The first successful cultivation of *Giardia* was achieved by Karapetyan (1962). His culture method, described by Taylor and Baker (1968), required the presence of *Saccharomyces cerevisiae*, so that the cultures were at best monoxenic. This was an improvement on his original culture method which, in addition to the yeast, required the presence of chick fibroblasts (Karapetyan, 1960).

Meyer (1970) introduced a new medium, M3, for axenic cultivation of *Giardia* from the rabbit, chinchilla and cat and, more recently (Meyer, 1976), media HSP-1 and HSP-2 for its axenic cultivation (see Section II, 2° below). Fortress and Meyer (1976) succeeded in isolating and establishing an axenic culture of *Giardia* from the guinea pig by introducing 0·2 ml of guinea pig small intestinal scrapings into 7 ml of HSP-1 medium in a test tube. *S. cerevisiae* was not added, but yeasts present in the inoculum continued to grow with the flagellates. The medium was changed daily for 24 d, when the organisms were placed in HSP-1 medium with 0·1% agar at the bottom of a 30 cm tube which was left in an inclined position for 4 d at 37°C. At this time, yeast-free *Giardia* trophozoites could be recovered from the top of the tube; they were inoculated in 3 ml amounts into 4 ml of HSP-1 medium in test tubes and incubated at 37°C with daily medium changes for 30 d. Eventually, when the axenic flagellate culture became stabilized, the changes were made twice weekly.

Sharapov and Soloviev (1976) made a study of the growth dynamics of *Lamblia intestinalis* (= *Giardia lamblia*) in Karapetyan's medium in perfusion chambers, in association with *S. cerevisiae* and a concomitant duodenal yeast, *Candida* sp. They found that the rate of growth of the flagellates was a function of their number in the initial inoculum and frequency of changing the medium. The generation time ranged from approximately 11 to 27 h, and maximum growth was obtained with daily changes of medium. They found that for routine maintenance a 48 h schedule of medium changing was suitable. In initial isolations of *Giardia* from intestinal contents, when the number of organisms may be small, even less frequent changing of medium is recommended, until the culture becomes sufficiently rich.

1° M3 medium
(Meyer, 1970)

MEDIUM

There are three components, A, B and a reducing solution.

Component A:

```
Hanks's balanced salt solution
    (Chapter 1, IV, B, 1, f) ...... 140 ml
Yeast extract (Difco) ........... 2·0 g
L-Cysteine hydrochloride ........ 0·2 g
Agar ........................ 1·1 g
Autoclave and cool to 50°C.
```

Component B is made by combining the following sterile ingredients:

```
Medium NCTC-109 .......... 10·0 ml
Seitz-filtered inactivated human serum
                          50·0 ml
Mixture of 100,000 iu of penicillin and 0·2 g
of streptomycin in a volume of 1–2 ml.
```

Components A and B are mixed and pH is adjusted to 6·6 with 0·75% w/v sodium bicarbonate.

Reducing solution is made as follows:

```
Glutathione (reduced) .......... 0·1 g
L-Cysteine hydrochloride ........ 0·1 g
Hanks's balanced salt solution ... 10·0 ml
Adjust pH to 9·0 with N-NaOH, filter
sterilize (Millipore) and store in the re-
frigerator for not more than 4 d.
```

The complete medium M3 is made by combining Components A and B and the reducing solution in the proportions of 67%, 29% and 4% by volume, respectively. The resulting pH is 6·8 and the approximate oxidation-reduction potential is −100 mV.

TECHNIQUE

Giardia from the rabbit and chinchilla, grown in Karapetyan's medium, served as inocula for medium M3. 0·2 ml of *Giardia*-yeast culture was added to 2 ml of M3 medium in a 16 × 125 mm Leighton tube, closed with a rubber stopper and incubated at 37°C. Daily, 1·5 ml of the culture fluid was removed and replaced by 1·5 ml of fresh M3.

Cultures of *Giardia* from the cat were established by inoculating 0·1 ml of *Giardia* suspension from the intestinal contents into 2 ml of M3 and 0·2 ml of yeast suspension in a Leighton tube. Daily withdrawals and replacements of medium were done as above.

Giardia trophozoites were separated from yeasts with the use of U-tubes.* The protozoans and yeasts were inoculated through a rubber stopper into one arm of the tube. As the yeasts settled down and remained at the bottom of the U, and the *Giardia* trophozoites moved across the base and into the other arm, daily withdrawals of medium from the first arm and corresponding additions into the other one were made for about 4 weeks, until axenic *Giardia* trophozoites could be harvested. Axenic trophozoites could not be cultured immediately in the complete absence of yeasts. However, they grew well in the same U-tubes in M3 medium when one of the arms contained a dialysis bag with yeasts in M3 medium. Eventually the protozoans were transferred into conventional culture tubes with M3 medium, and axenic cultures were established.

2° HSP-1 and HSP-2 media
(Meyer, 1976)

MEDIUM (HSP-1)

Phytone peptone (BBL) . . . . 1·0 g
Glucose . . . . 0·05 g
L-Cysteine hydrochloride . . . . 0·1 g
Hanks's balanced salt solution
    (Chapter 1: IV, B, 2, b) . . . . . 85·0 ml

Autoclave 10 min, 15 lb. Before use, add 15 ml sterile inactivated human serum, 50,000 iu penicillin and 0·05 g streptomycin. pH (not adjusted) should be 6·8–7·0.

* 5-Fluorocytosine might be useful here, provided it does not kill *Giardia*; see Chapter 1, Section II, F (Eds).

Medium HSP-2 consists of 100 ml of HSP-1 supplemented with 7·5 ml of tissue culture medium NCTC-135 (Gibco) and 2·5 ml of the glutathione-cysteine reducing solution (see above).

TECHNIQUE

For initial isolation, G. lamblia trophozoites were obtained from the duodenal aspirate from a woman with a history of diarrhoeal disease. The aspirate was inoculated into HSP-1 medium in the proportion of 2 ml aspirate to 5 ml of medium in test tubes. The cultures were incubated on a slant at 37°C. The medium was removed daily with a pipette, taking care not to dislodge the *Giardia* trophozoites attached to the tube wall, and replaced with fresh warm medium. Because some intestinal fungi persisted, the cultures were not axenic. Eventual axenization of cultured G. lamblia trophozoites was made with the aid of U-tubes, as described above, and Medium HSP-2.

III. CULTIVATION OF *TRICHOMONAS VAGINALIS*

Not much progress has been made since 1968 in the axenic cultivation of *Trichomonas*; TYM medium (Diamond, 1957) is still the medium of choice for species of *Trichomonas*. Some relatively minor modifications of this medium have been made in several laboratories, e.g. trypticase can be substituted by tryptose, and the originally recommended sheep serum substituted by either horse or bovine serum. For *T. vaginalis* the pH should be adjusted to 5·8–6·0, for *T. gallinae*, to 7·2 and for *T. foetus*, to 7–7·5.

Rayner (1968) made a comparison of 3 media used for isolation and cultivation of *T. vaginalis*: CPLM medium (Johnson and Trussell, 1943), F-W medium (Feinberg and Whittington, 1957) and S-M medium (Squires and McFadzean, 1962). Rayner found that the CPLM medium established growth of *T. vaginalis* with a very much smaller inoculum than F-W or S-M media, with both stock and freshly isolated strains. This was most probably because CPLM medium is semi-solid and contains cysteine hydrochloride as a reducing agent, while the other two media contain glucose but no other reducing agent, and no agar. F-W medium could be improved greatly if cysteine were added to it and if the cultures were incubated in an atmosphere of 95% N_2 and 5% CO_2.

Lowe (1972) proposed a modification of his earlier medium (Lowe, 1965) for the isolation of *T. vaginalis*. The basic ingredients were the same, but some of the supplements had been changed (see Section III, 1° below).

Barrow and Ellis (1970) recommended a cysteine-peptone-liver infusion-

agar semi-solid medium, without penicillin or streptomycin, but supplemented with $100 \,\mu g \,ml^{-1}$ of chloramphenicol and $50 \,\mu g \,ml^{-1}$ of
nystatin, dispensed in short tubes for direct examination for detecting the
presence of *T. vaginalis* on swabs. These workers and several others
recommended the use of chloramphenicol for primary isolation of *T.
vaginalis*, but Dr. Honigberg (personal communication) cautions against
this and suggests the use of penicillin and streptomycin instead.

Andrews and Thomas (1974) found that *T. vaginalis* would grow in
colonial form on a solid *Mycoplasma*-selective medium containing "conventional constituents" and antibiotics, thallium acetate (1/10,000), urea
(0·1%) and manganese sulphate, $MnSO_45H_2O$ (0·03%). Cultures were
incubated for 7 d at 37°C under 5% CO_2. *T. vaginalis* colonies, up to
0·6 mm in diameter, appeared confluently with *Mycoplasma* colonies.

Most recently, Hollander (1976) proposed a modified agar technique for
plating *T. vaginalis*, which yields clones with distinct characteristics (see
Section III, 2° below).

These solid and liquid media are modifications of the original TYM
medium of Diamond (1957), in that cysteine is omitted and replaced by
additional ascorbic acid as a reducing agent, and that phosphate, rather than
sodium, buffers are used, since an excess of sodium appears to inhibit
growth of *T. vaginalis*.

1° Isolation medium for *T. vaginalis*
(Lowe, 1972)

MEDIUM

Panmede (Paynes and Byrne)	12·5 g
Sodium chloride	2·5 g
Maltose	0·5 g
Water to make	500 ml

Dissolve by heating, adjust pH to 6·2, add 1·25 g agar, steam at 100°C for
30 min while mixing. Cool to 54°C and add:

Sterile inactivated horse serum ...	50 ml
Fildes extract (Oxoid)	0·5 ml
4% (w/v) sterile aqueous solution of ascorbic acid..............	10 ml
1% (w/v) sterile aqueous solution of chloramphenicol...........	5 ml
1% (w/v) sterile aqueous solution of gentamycin sulphide	5 ml
0·01% (w/v) sterile aqueous solution of amphotericin	5 ml

TECHNIQUE

Mix and transfer to 7 ml ($\frac{1}{4}$ oz) "bijoux" bottles, filling to the shoulder. Vaginal swabs are inserted to the bottom and snapped off level with the bottle top. Incubate at 37°C. These primary cultures can be shipped to a diagnostic laboratory; transit time can be as long as 72 h. Inoculated specimens should not be refrigerated.

2° Medium for cloning *T. vaginalis*
(Hollander, 1976)

MEDIUM

Trypticase	20·0 g
Yeast extract	10·0 g
Maltose	5·0 g
Ascorbic acid	1·0 g
KCl	1·0 g
$KHCO_3$	1·0 g
KH_2PO_4	1·0 g
K_2HPO_4	0·5 g
$FeSO_4$	0·1 g
Ionagar (Oxoid)	3·6 g

TECHNIQUE

Dissolve in 900 ml of water, over heat. Dispense 36 ml amounts into 50 ml tubes and autoclave (15 lb, 15 min). When cooled to 40°C, add 4 ml horse serum and 0·1 ml of fluid medium containing the inoculum to the warm melted medium, and pour into a 100 × 15 mm Petri dish, with blotting paper in the cover to prevent moisture condensation. The plates are incubated (not inverted) at 37°C for 5 d in a Brewer jar containing 9·0 g zinc powder to which 50 ml concentrated sulphuric acid is added through a long-stemmed funnel. The jar is sealed and evacuated to about 80 mm of mercury. Anaerobiosis is monitored by observing the decoloration of a tube of dilute methylene blue solution within 12 h. This technique employs soft agar (0·36%) which requires careful handling and incubation of the plates right side up.

The corresponding liquid medium from which the *Trichomonas* inoculum is derived contains 0·05% (w/v) agar instead of 0·36% (Kulda *et al.*, 1970). This medium is recommended for diagnostic cultures and routine maintenance of *T. vaginalis*.

REFERENCES

Andrews, B. and Thomas, M. (1974). Colonies of *Trichomonas vaginalis* on *Mycoplasma* medium. *Lancet* **2** (Aug. 3, 1974), 300.

Barrow, G. I. and Ellis, C. (1970). Direct microscopical examinations of tube cultures for the detection of trichomonads. *Journal of Clinical Pathology* **23**, 91.

Bos, H. J. (1975). Monoxenic and axenic cultivation of carrier and patient strains of *Entamoeba histolytica*. *Zeitschrift für Parasitenkunde* **47**, 119–129.

Diamond, L. S. (1957). The establishment of various trichomonads of animals and man in axenic cultures. *Journal of Parasitology* **43**, 488–490.

Diamond, L. S. (1961). Axenic cultivation of *Entamoeba histolytica*. *Science, New York* **134**, 336–337.

Diamond, L. S. (1968a). Improved method for the monoxenic cultivation of *Entamoeba histolytica* Schaudinn, 1903 and *E. histolytica*-like amebae with trypanosomatids. *Journal of Parasitology* **54**, 715–719.

Diamond, L. S. (1968b). Techniques of axenic cultivation of *Entamoeba histolytica* Schaudinn, 1903 and *E. histolytica*-like amebae. *Journal of Parasitology* **54**, 1047–1056.

Diamond, L. S. and Bartgis, I. L. (1970). *Entamoeba moshkovskii*: axenic cultivation. *Experimental Parasitology* **28**, 171–175.

Diamond, L. S. and Bartgis, I. L. (1971). Axenic cultures for *in vitro* testing of drugs against *Entamoeba histolytica*. *Archivos de Investigación Médica, Mexico* **2** (Suppl. 1), 339–348.

Dobell, C. and Laidlaw, P. P. (1926). On the cultivation of *Entamoeba histolytica* and some other entozoic amoebae. *Parasitology* **18**, 283–318.

Dutta, G. P. and Singh, K. (1975). Axenic cultivation of *Entamoeba histolytica* from amoeba-bacteria cultures without the use of trypanosomatid culture associates. *Programme and Abstracts, International Conference on Amebiasis, Mexico, D.F., 27–29 October, 1975*, p. 139.

Evans, V. J., Bryant, J. C., Fioramonti, M. C., McQuilkin, W. T., Sanford, K. K. and Earle, W. R. (1956). Studies of nutrient media for tissue cells *in vitro*. I. A protein-free chemically-defined medium for cultivation of strain L cells. *Cancer Research* **16**, 77–86.

Feinberg, J. G. and Whittington, M. J. (1957). A culture medium for *Trichomonas vaginalis* Donné and species of *Candida*. *Journal of Clinical Pathology* **10**, 327–329.

Fortress, E. and Meyer, E. A. (1976). Isolation and axenic cultivation of *Giardia* trophozoites from the guinea pig. *Journal of Parasitology* **62**, 689.

Hollander, D. H. (1976). Colonial morphology of *Trichomonas vaginalis* in agar. *Journal of Parasitology* **62**, 826–828.

Johnson, G. and Trussell, R. E. (1943). Experimental basis for the chemotherapy of *Trichomonas vaginalis* infections. I. *Proceedings of the Society for Experimental Biology and Medicine* **54**, 245–249.

Karapetyan, A. E. (1960). Methods of *Lamblia* cultivation. *Tsitologiia* **2**, 379–384.

Karapetyan, A. E. (1962). *In vitro* cultivation of *Giardia duodenalis*. *Journal of Parasitology* **48**, 337–340.

Kulda, J., Honigberg, B. M., Frost, J. K. and Hollander, D. H. (1970). Pathogenicity of *Trichomonas vaginalis*. A clinical and biologic study. *American Journal of Obstetrics and Gynecology* **108**, 908–918.

Lowe, G. H. (1965). A comparison of current laboratory methods and a new semi-solid culture medium for the detection of *Trichomonas vaginalis*. *Journal of Clinical Pathology* **18**, 432–434.

Lowe, G. H. (1972). A comparison of culture media for the isolation of *Trichomonas vaginalis*. *Medical Laboratory Technology* **29**, 389–391.

McConnachie, E. W. (1962). A medium for the axenic culture of *Entamoeba invadens*. *Nature, London* **194**, 603–604.

Meyer, E. A. (1970). Isolation and axenic cultivation of *Giardia* trophozoites from the rabbit, chinchilla, and cat. *Experimental Parasitology* **27**, 179–183.

Meyer, E. A. (1976). *Giardia lamblia*: isolation and axenic cultivation. *Experimental Parasitology* **39**, 101–105.

Myjak, P. (1971). Diagnostics of *Entàmoeba histolytica* by means of cultures on solid media. *Bulletin of the Institute of Marine Medicine in Gdansk* **22**, 165–171.

Neal, R. A. (1967). The *in vitro* cultivation of *Entamoeba*. *Symposia of the British Society for Parasitology* **5**, 9–26.

Raether, W., Uphoff, M. and Seidenath, H. (1973). Adaption von Amoeben-Crithidien-Kulturen (*Entamoeba histolytica*) an axenisches Wachstum in TP-S-1-Medium nach Diamond (1968). *Zeitschrift für Parasitenkunde* **42**, 279–291.

Rayner, C. F. A. (1968). Comparison of culture media for the growth of *Trichomonas vaginalis*. *British Journal of Venereal Diseases* **44**, 63–66.

Reeves, R. E., Meleney, H. E. and Frye, W. W. (1959). The cultivation of *Entamoeba histolytica* with penicillin-inhibited *Bacteroides symbiosus* cells. I. A pyridoxine requirement. *American Journal of Hygiene* **69**, 25–31.

Schneider, C. R. and Gordon, R. N. (1968). The effect of medium components on the specificity of axenic *Entamoeba histolytica* antigen. *Journal of Parasitology* **54**, 711–714.

Sharapov, M. B. and Soloviev, M. M. (1976). Study of *Lamblia intestinalis* growth in culture for improvement of the cultivation method. *Meditsinskaia Parazitologiia i Parazitarnye bolezni (Moscow)* **45**, 655–660.

Squires, S. and McFadzean, J. A. (1962). Strain sensitivity of *Trichomonas vaginalis* to Metronidazole. *British Journal of Venereal Diseases* **38**, 218–219.

Taylor, A. E. R. and Baker, J. R. (1968). "The Cultivation of Parasites *in vitro*." Blackwell Scientific Publications, Oxford and Edinburgh.

Wang, L. T., Jen, G. and Cross, J. H. (1973). Establishment of *Entamoeba histolytica* from liver abscess in monoxenic cultures with hemoflagellates. *American Journal of Tropical Medicine and Hygiene* **22**, 30–32.

Wang, L. T., Jen, G. and Cross, J. H. (1974). Axenic cultivation of four strains of *Entamoeba histolytica* from liver abscesses. *South-East Asian Journal of Tropical Medicine and Public Health* **5**, 365–367.

Westphal, A. and Michel, R. (1971). Versuche zur Adaptation von *Entamoeba histolytica* an verschiedene Protozoenarten im TTY-Medium nach Diamond. *Zeitschrift für Tropenmedizin und Parasitologie* **22**, 149–156.

Wittner, M. (1968). Growth characteristics of axenic strains of *Entamoeba histolytica* Schaudinn, 1903. *Journal of Protozoology* **15**, 403–406.

Chapter 3

Rumen Entodiniomorphid Protozoa

G. S. COLEMAN

*Biochemistry Department, Agricultural Research Council, Institute of Animal Physiology,
Babraham, Cambridge, England*

I. INTRODUCTION

There are two main types of ciliate protozoa present in the rumen, those of
the order Trichostomatida (commonly known as holotrich protozoa)
represented only by *Isotricha intestinalis*, *I. prostoma* and *Dasytricha
ruminantium* and those of the order Entodiniomorphida of which there are
many genera and species. Of the holotrich protozoa only *D. ruminantium* has
been cultured for longer than 60 d *in vitro* (Clarke and Hungate, 1966); as it
was necessary to carry out two daily manipulations separated by an interval
of 2–4 h, the method was very time consuming and tedious. In contrast, all

the entodiniomorphid protozoa tested have been cultured *in vitro* with varying degrees of success, although it is still necessary to feed the cultures every day. The methods available will be considered below. The rationale of these is based on reproducing conditions in the rumen, which provides a warm, constant temperature and anaerobic environment deficient in readily metabolizable compounds but rich in many particles such as bacteria, starch grains, chloroplasts and cellulose fibres. The media used therefore are all oxygen-free, contain viable bacteria but none of the usual bacteriological medium constituents such as peptone, are incubated at $39 \pm 2°C$ and have the minimum amount, necessary to support protozoal growth, of particulate food such as starch grains added to them each day. As these food particles are readily engulfed by the protozoa but are comparatively resistant to bacterial attack, they promote protozoal growth and depress bacterial growth. No rumen ciliate protozoon has been grown axenically.

II. IDENTIFICATION AND CLASSIFICATION

As these protozoa cannot be freeze-dried and there are no type collections of living organisms, anyone wishing to culture them must start with crude rumen contents obtained from the host animal by stomach tube or rumen cannula. The investigator must therefore be able to identify the protozoon that he wishes to isolate and this is not always easy, especially with protozoa containing many food particles, which make it impossible to see any internal detail.

The literature on the classification of these protozoa is difficult to follow as some have had five or six names since they were first recognized. The beginner is advised to consult Kofoid and MacLennan (1930, 1932, 1933) and Dogiel (1927). These papers contain many line drawings although the last is written in German. The presence, size and distribution of caudal spines is very variable in clone cultures of rumen ciliate protozoa and is a poor taxonomic character. Both Hungate (1966) and the present author believe that there are many fewer species than have been described but that each species has several variant forms.

III. MATERIALS AND GENERAL METHODS

A. BASAL MINERAL SALTS SOLUTIONS

1. *Hungate-type*

This salts medium is based on that of Hungate (1942) and contains (g litre^{-1}): NaCl, 5·0; $CH_3.COONa$, 1·5; K_2HPO_4, 1·0; KH_2PO_4, 0·3.

2. Coleman-type

This salts solution is based on that of Coleman (1958) and is usually made up at double the final concentration. The concentrated solution contains (g litre^{-1}): K_2HPO_4, 12·7; KH_2PO_4, 10·0; NaCl, 1·3; $CaCl_2$ (dried), 0·09; $MgSO_4.7H_2O$, 0·18. When this solution is prepared, a precipitate will form unless care is taken not to add the calcium salt (dissolved in 2 litres H_2O) until the phosphates have been dissolved in most of the water. This concentrated solution can be made up in bulk as it will keep indefinitely.

3. Caudatum-type

This consists of 50 ml concentrated Coleman-type salts solution, 50 ml H_2O, 0·5 ml 15% (w/v) $CH_3.COONa$ and 1·0 ml 2% (w/v) L-cysteine hydrochloride (neutralized immediately before use). If required, the medium without cysteine can be autoclaved (115°C, 20 min) without a precipitate forming. Before use the complete medium is gassed vigorously with 95% (v/v) N_2 + 5% (v/v) CO_2 (or N_2 in established cultures—see Section III, E) for 2 min before sealing with a rubber bung.

4. Simplex-type

This consists of 40 ml concentrated Coleman-type salt solution, 60 ml H_2O, 15 ml 5% (w/v) $NaHCO_3$ and 1·1 ml 2% (w/v) L-cysteine hydrochloride (neutralized). The whole is bubbled vigorously with CO_2 for 2 min before sealing with a rubber bung.

B. RUMEN FLUID FOR MEDIA

All rumen fluid for media is taken from sheep via a rumen cannula 1–5 h after feeding and strained through one layer of muslin to remove the larger food particles. This strained fluid is centrifuged (500 g, 5 min) to remove the protozoa and the supernatant used. If the rumen fluid is to be used at once this supernatant fluid must be filtered through a Ford "Sterimat CC/01" to remove any residual protozoa. If it is not required for 24 h, storage at 4°C will kill the protozoa. Prepared fresh rumen fluid can be stored for up to 2 weeks at 4°C. Autoclaved rumen fluid is prepared by autoclaving the prepared fresh fluid (115°C, 20 min) under N_2.

C. DRIED GRASS

Dried grass is prepared from lawn mowings that have been air dried in thin layers on the bench for 3 days. They are then ground in a Lee Attrition Mill (Lee Engineering) to produce a coarse powder containing particles up to 5 mm long ("coarse dried grass") or in a ball mill for 16 h to produce a fine powder ("powdered dried grass"). These preparations can be stored indefinitely in the dark but, if sterilized in an oven (160°C, 1 h), no longer support protozoal growth. Dried grass should be added to cultures from the end of a spatula: with practice the correct amount can be judged fairly accurately. In the author's laboratory where grass has to be added, as quickly as possible, to over 200 tubes each day, experience has shown that even when adding powdered dried grass alone, the amount added to each 50 ml tube in any one day can vary from 1–20 mg without affecting the protozoal population density although over several days the amount added should approximate to the optimum.

D. STARCH

Rice starch (British Drug Houses) or wholemeal flour (W. Prewett) is added as a 1.5% (w/v) suspension in water. The wholemeal flour suspension (0.1 to 0.5 ml) must be added to the medium rapidly (6×0.15 ml samples in 3 s) through a 1 ml calibrated pipette with a large hole at the tip as the solid matter tends to sink quickly to the bottom of the pipette if pipetting is carried out in the conventional way. In cultures of *Entodinium caudatum*, the use of rice starch sterilized by heating dry grains (160°C, 1 h) resulted in death of the protozoa. However, the protozoa grew well in the presence of rice starch that had been heated dry (120°C, 24 h) before use.

E. GASES

"Medical quality" CO_2 and 95% $N_2 + 5\%$ CO_2 (British Oxygen) can be used without prior treatment. Removal of traces of oxygen by red-hot copper or chromous sulphate solution does not improve the growth of *En. caudatum*. As 95% $N_2 + 5\%$ CO_2 is expensive compared with O_2-free N_2 or CO_2, the caudatum-type salts medium required for the dilution of *established* cultures can be bubbled instead with O_2-free N_2. After feeding cultures, CO_2 can be used to gas over the top of the medium, without harming the protozoa. This modification of previously published methods (e.g. Coleman, 1960) is satisfactory provided that it is not used during the initial isolation of the protozoa or with cultures in which the protozoa are growing poorly.

F. CULTURE VESSELS

The cultures are normally maintained in 50, 100 or 250 ml centrifuge tubes almost completely filled with medium. In these, the protozoa are present as a loose pellet on the bottom of the tube and are therefore unaffected by any leakage of air into the vessel due to accidental displacement of the rubber bung. For the safe continuation of a culture it is essential to divide it between at least 6 tubes as all the protozoa in some tubes occasionally die for no obvious reason.

G. FIXATION AND COUNTING OF PROTOZOA

The best fixative for entodiniomorphid protozoa is iodine which also stains any starch-like material in the organisms and makes them easier to count: protozoa fixed with formalin frequently disintegrate after 24 h. Provided that sufficient protozoa are present, the best method is to add 0·5 ml protozoal culture to 4·5 ml 0·02M-I_2. This material will keep indefinitely stoppered in the dark. If a larger proportion of protozoal material is added or if crude rumen contents are used, it is necessary to increase the iodine concentration to 0·1M; the final product must be well coloured with iodine.

To sample a protozoal culture, it should be well mixed and then, using a 1 ml graduated pipette with a large hole at the tip, the material should be sucked up and blown into the iodine as quickly as possible. To count all the protozoa, except for *Polyplastron multivesiculatum* and *Metadinium medium*, 0·1 ml of the mixture is expelled onto a microscope slide from a 0·2 ml graduated pipette calibrated to the shoulder and cut off at that calibration. After putting on a cover slip, all the protozoa are counted under the microscope using a × 4 or × 10 objective by tracking back and forth across the area of the liquid. With the large protozoa (*P. multivesiculatum* and *M. medium*) when less than 100 organisms ml^{-1} are present in culture, all the protozoa in the original fixed material are counted by placing the tube (preferably 100 × 12 mm) horizontally on the stage of an inverted microscope (Olympus Optical, model CK) and moving the tube across the field of vision over a × 4 objective. Only those protozoa showing no signs of disintegration should be counted.

IV. SPECIAL TECHNIQUES

A. INITIAL ISOLATION OF A PROTOZOAL SPECIES

1. *Picking protozoa from a mixed population*

No selective medium is available, except for *En. caudatum*. It is therefore necessary, in order to isolate most entodiniomorphid protozoa, to pick out individual protozoa from rumen contents or a mixed culture *in vitro*. A suitably diluted drop of the crude material is placed on a slide under an inverted microscope and individual protozoa sucked up through a micropipette attached to a LeFonbrune suction and force syringe (Beaudouin). The position of the micropipette is adjusted by attaching it to a Brinkman CP5 micromanipulator (Camlab).

The 2 main difficulties, apart from the manual dexterity involved, are (a) the necessity for speed as the protozoa are strict anaerobes and are exposed to air during the procedure and (b) the risk of removing small transparent *Entodinium* spp. along with the larger protozoon that it is hoped to isolate. Three drops of caudatum-type salts medium containing 10% prepared fresh rumen fluid are placed on a slide next to the crude material. The required protozoa are then removed from the crude material without worrying overmuch if other protozoa are picked up as well. All the material is then placed in 1 of the drops of medium. This process is continued through the drops of medium until individuals of just one protozoal species can be picked up easily in the pipette. It should take less than 10 min from the time the crude material is placed on the slide until the protozoa are placed in a tube of growth medium. If it takes longer the whole process should be begun again as the protozoa are unlikely to survive.

The chance of successfully culturing a protozoal species from an individual picked out in this way is improved if 5–10 protozoa can be inoculated into the medium. This is a safe procedure if the wanted protozoon is morphologically distinct from the others present, e.g. where the protozoon is the only *Ophryoscolex* sp. present, but it may result in a mixed culture when trying to grow one *Diplodinium* sp. morphologically similar to others present.

2. *Treatment of newly inoculated cultures*

The newly picked protozoa are inoculated into 3 ml of the appropriate salts medium (see Table 1 under individual species) containing 10% prepared fresh rumen fluid in a 125 × 12 mm tube. For protozoa which use starch, one drop of a dilute suspension of wholemeal flour (obtained by allowing a 1·5% w/v suspension of wholemeal flour to settle for 3–5 min and removing the liquid from near the surface) is added together with a few mg of coarse dried grass. It is essential that most of the grass sinks to the bottom of the tube or the protozoa will not survive. For cellulolytic protozoa (whose preferred substrate is grass alone) add only a few mg of powdered dried grass. The tubes are then gassed with the appropriate gas, sealed with a rubber bung and incubated at 39°C.

The tubes are opened daily, starch and/or grass added as before and the tubes resealed. After a week the protozoa that survived the isolation process should have increased in numbers sufficiently to be easily visible under the microscope. If an inverted microscope is used the protozoa can be observed without opening the tubes. If no protozoa are visible the process should be continued for a further week and then, if there are still none present, the tubes discarded. In the author's experience the success rate varies from 0–100%, depending on the protozoon, but is usually about 50%. A week after inoculation the cultures should be diluted with an equal volume of the same medium and the daily feeding continued for a further week.

After this time the procedure adopted will depend on how well the protozoa are growing. If they are growing well (i.e. over 50 active protozoa can be seen at first glance under the microscope), the contents of the tube can be poured into 30 ml of the same medium in a 50 ml centrifuge tube and biweekly dilution of the cultures carried out as described below. If the protozoa are growing less well (i.e. 10–50 protozoa visible), pour the contents of the tube into an equal volume of medium in a 15 ml centrifuge tube and a week later dilute this again with an equal volume of fresh medium. If only a few protozoa are visible, suck off the top 80% of the culture medium and replace it with an equal volume of fresh medium. Then treat as above when sufficient protozoa are present. It must be emphasized that during all these procedures daily feeding must be continued and that the amount added, especially of starch, should be increased as the number of protozoa increases.

Unfortunately at any stage in the first month in the life of an isolate, the protozoa may suddenly all die. If several apparently successful isolates have been made, it is inadvisable to discard any until the chosen one is established in at least 6 tubes.

Until the protozoa are growing well and reproducibly it is advisable to continue to add 10% prepared fresh rumen fluid to the media, but after this (1–3 months depending on the species) it can be omitted with those species where it is known not to be stimulatory (Table 1). Similarly with those species where autoclaved rumen fluid (ARF) is stimulatory to growth, the change from fresh rumen fluid can be made at this time.

3. *Contamination*

As it is usual when isolating individual protozoa to inoculate a series of tubes, care must be taken when feeding the cultures each day not to transfer protozoa from one tube to another. If this precaution is not taken it is easy for a vigorously growing contaminant protozoon such as *En. simplex* to be transferred from one tube, into which it was accidently inoculated originally, to all the others in a batch. The necessary precautions include always using the same bung for a tube, flaming the pipette used to gas the tubes before insertion into each and never allowing the pipette used to add the starch suspension to touch the side of a tube.

4. *Temperature*

The optimum temperature for all species is 39°C and they will not grow above 42°C or below 35°C. A few individuals in a culture of *En. caudatum* will still be alive after 12 h at room temperature but this treatment kills all the other species. Growth of *En. caudatum* at 41°C produces distorted protozoa that do not separate normally during binary fission and grow in chains.

B. MAINTENANCE OF ESTABLISHED CULTURES

Each day every tube of culture is opened and food is added. If the protozoon is growing on wholemeal flour or rice starch, the amount given in Table 1 is added (as a 1·5% w/v suspension in water) together with a few mg coarse dried grass per tube. If the substrate is powdered dried grass, 0·4 mg ml^{-1} culture is added. The tubes are then gassed with CO_2, sealed with a rubber bung and returned to the incubator (39°C) as soon as possible. Twice a week the number of tubes containing each protozoal species is doubled by division of the culture. Except for *En. caudatum* (q.v.), the culture is mixed, half poured into another tube and each tube made up to the original volume with fresh medium. Food is then added as usual and the tubes gassed and sealed with a rubber bung.

    When protozoa are required for metabolic experiments the additional tubes produced at the biweekly division can be used for this purpose.

C. MAINTENANCE OF INDIVIDUAL PROTOZOAL SPECIES

The basic methods for the isolation and subsequent growth and maintenance of these protozoa are given above and in Table 1. The following notes are intended to supplement these and emphasize points of technique relevant to one genus or species. References to published work with the organism are given at the end of the section.

1. *Diplodinium monacanthum*

This protozoon has been isolated only once (on wholemeal flour and coarse

Protozoon	Isolation method	Preferred source of starch[a]	Growth on grass alone in absence of starch[b]	Stimulation of growth of established cultures by rumen fluid[c]	Salts medium preferred	Method of increasing culture	Requirement for other protozoa	Maximum population density (protozoa ml^{-1})	Maximum life of culture[e]
1. Diplodinium monacanthum	Single cell	WF 0·05	+	—	Simplex	Dilution	—	6,000	13 months
2. Diplodinium pentacanthum	Single cell	(WF 0·05)	++	—	Caudatum or simplex	Dilution	—	800	>1 year
3. Diploplastron affine	Single cell	WF 0·05	++	—	Caudatum	Dilution	—	200	2¼ years
4. Enoploplastron triloricatum	Single cell	(WF 0·05)	++	—	Simplex	Dilution	—	150	2 years
5. Entodinium bursa	Single cell	WF 0·03	—	FRF	Caudatum	Replace medium and dilution	+	150	>1 year
6. Entodinium caudatum	Enrichment	RS 0·15	—	ARF	Caudatum + CAP[d]		—	32,000	>17 years
7. Entodinium longinucleatum	Single cell	WF 0·1	—	FRF, ARF	Caudatum	Dilution	—	1,900	1¼ years
8. Entodinium simplex	Differential centrifugation	WF 0·2	—	ARF	Simplex	Dilution	—	27,000	8½ years
9. Epidinium ecaudatum caudatum (bovine)	Single cell	WF 0·1	?	—	Caudatum or simplex	Dilution	—	500	1½ years
10. Epidinium ecaudatum caudatum (ovine)	Single cell	WF 0·1	+	—	Caudatum or simplex	Dilution	—	500	2 years
11. Epidinium ecaudatum tricaudatum	Single cell	WF 0·1	?	—	Caudatum or simplex	Dilution	—	1,000	1½ years
12. Eremoplastron bovis	Single cell	(WF 0·03)	++	—	Simplex	Dilution	—	1,000	1¼ years
13. Eudiplodinium maggii	Single cell		++	—	Caudatum or simplex	Dilution	—	700	2½ years
14. Metadinium medium	Single cell	(WF 0·03)	++	—	Caudatum	Dilution	—	2	6 months
15. Ophryoscolex caudatus	Single cell	WF 0·25	—	FRF	Caudatum	Dilution	—	50	2 years
16. Ostracodinium obtusum bilobum	Single cell	(WF 0·03)	++	—	Simplex	Dilution	—	300	1¾ years
17. Polyplastron multivesiculatum (+)	Single cell	WF 0·05	—	—	Caudatum	Dilution	+	50	3 years
18. Polyplastron multivesiculatum (−)	Single cell	WF 0·03	+	—	Hungate	Dilution	—	5	1 year

[a] WF = wholemeal flour; RS = rice starch; the figures are the optimum amount of the material (mg ml^{-1}) to be added each day; parentheses indicate that although the protozoon will grow on starch for a limited time on starch, this is not the preferred substrate.

[b] + = protozoon will grow on grass alone; ++ = grass is the preferred substrate for growth; ? = not determined.

[c] FRF = prepared fresh rumen fluid; ARF = autoclaved rumen fluid.

[d] Chloramphenicol 48 µg ml^{-1}.

[e] > means that at the time of writing a culture was still alive and growing well.

dried grass) and proved difficult to culture. However much care was taken, some tubes in a batch always contained very few organisms. Once established it grew better on powdered dried grass than on wholemeal flour and coarse dried grass (6000 compared with 2000 protozoa ml^{-1}) (Coleman *et al.*, 1976).

2. *Diplodinium pentacanthum*

This protozoon can be isolated equally well on powdered dried grass or wholemeal flour and coarse dried grass but once established it grows better on the former (800 ciliates ml^{-1}) than the latter (400 ml^{-1}) (Coleman *et al.*, 1976).

3. *Diploplastron affine*

This protozoon can be isolated equally well on powdered dried grass or wholemeal flour and coarse dried grass and once established grows equally well on both substrates (Coleman *et al.*, 1976).

4. *Enoploplastron triloricatum*

This protozoon must be isolated on simplex-type salts medium containing powdered dried grass or wholemeal flour and coarse dried grass. However, once established it grows equally well on caudatum- or simplex-type salts medium and better in the absence of prepared fresh rumen fluid than in its presence. It grows better and lives longer in the presence of powdered dried grass than on wholemeal flour (Coleman *et al.*, 1976).

5. *Entodinium bursa*

En. bursa engulfs *En. caudatum* and will not grow or survive *in vitro* except in the presence of living protozoa from a culture of *En. caudatum* grown as described below. It is isolated by transferring protozoa from crude rumen contents into caudatum-salts medium containing 10% prepared fresh rumen fluid and fed initially and daily thereafter with wholemeal flour, coarse dried grass and a drop of *En. caudatum* culture. As the *En. bursa* grows, and provided that most of the *En. caudatum* have disappeared, it is essential to add fresh and increasing amounts of *En. caudatum* every day even though a few are still apparently present in the culture because, after 1–3 weeks exposure to *En. bursa*, those *En. caudatum* that survive engulfment grow caudal spines. These caudate protozoa are engulfed much less readily than the normal culture form and although *En. bursa* survives in their presence the population density is much lower. *En. bursa* will not grow on simplex-type salts medium or in the absence of rumen fluid.

To obtain large numbers of *En. bursa*, 8 ml of a stock culture is inoculated into 100 ml of a culture of *En. caudatum* obtained 3 d previously from the division of a culture and containing approximately 20,000 protozoa ml^{-1}. The inoculated culture is fed daily with 0·075 mg rice starch and 0·015 mg wholemeal flour ml^{-1}. After 4 (occasionally 3 or 5) d all the *En. caudatum* have disappeared and about 900 *En. bursa* organisms ml^{-1} are present. Occasionally, under these conditions, the size of *En. caudatum* increases to such an extent that *En. bursa* cannot engulf these organisms. If this tends to happen the culture should be fed only 0·05 mg wholemeal flour ml^{-1}. (Coleman *et al.*, 1977.)

6. *Entodinium caudatum*

En. caudatum is isolated by a 0·3% v/v inoculum of rumen contents into caudatum-type salts medium containing 10% prepared fresh rumen fluid and 48 μg ml^{-1} chloramphenicol (in a 50 ml tube); rice starch (0·03 mg ml^{-1}) and a few mg dried grass are added daily to each tube. Twice a week 80% of the supernatant fluid is removed and replaced by fresh medium. This process is continued until sufficient protozoa are present (20,000 ml^{-1}) for the culture to be divided into two, by first replacing the supernatant fluid and then, after mixing, pouring half into another tube and making both up to the original volume with fresh medium.

During the first week after inoculation, both the number of protozoal species present and the total number of protozoa declines until few can be seen. The number of *En. caudatum* then increases steadily and after 2 weeks the amount of rice starch added each day can be increased to 0·25 mg ml^{-1}. After several months in culture, maintenance can be facilitated by replacement of the fresh rumen fluid in the medium by autoclaved rumen fluid provided that the daily addition of rice starch is decreased to 0·15 mg ml^{-1}. However, the rather complicated manipulations involved in the division of a culture cannot be altered. The presence of chloramphenicol is essential during the initial isolation in order to kill the other protozoa present but it cannot be omitted from established cultures without the *En. caudatum* dying. The protozoa in cultures that have been maintained for over 10 years metabolize starch more slowly than freshly isolated organisms and need be fed only every 2 days. The characteristic caudal spines of *En. caudatum* become shorter and disappear after a few months in culture and never reappear unless the protozoa are exposed to *En. bursa* (Coleman, 1958; 1960).

7. *Entodinium longinucleatum*

This protozoon can be isolated equally well on wholemeal flour or rice starch in the presence of coarse dried grass but established cultures grow more reliably on the former than the latter. The prepared fresh rumen fluid added for the isolation can be replaced by autoclaved rumen fluid without change in the population density; omission of any rumen fluid results in the death of the culture (Owen and Coleman, 1976).

8. *Entodinium simplex*

A number of small *Entodinium* spp. (under 45 μm long) without caudal spines, e.g. *En. exiguum*, *En. nanellum*, *En. simplex* and *En. parvum*, were described by Dogiel (1927); the only one apparently present in Clun Forest sheep at Babraham is *En. simplex*. In this circumstance it can be separated from all the larger protozoa by differential centrifugation. Some rumen contents are removed 6 h after feeding the animal and strained through one layer of muslin to remove the larger food particles. The fluid is then allowed to stand in a 200 × 25 mm tube at 39°C for 2 h to allow the larger protozoa to fall to the bottom. The fluid from the top of the tube is then centrifuged at 600 g for about 3 min or until, on microscopic examination, *En. simplex* is the only protozoon present in a drop of the supernatant fluid. One drop of this fluid is then inoculated into the medium. If at any stage a protozoon with a caudal spine is observed, that culture must be discarded. As all rumen ciliates in culture tend to lose their caudal spines, it is possible to be certain that the culture is not that of another species that has lost its spine, only if care is taken to ensure that no caudate protozoa have ever been present.

The fresh rumen fluid in the isolation medium can be replaced by autoclaved rumen fluid without change in the population density; omission of rumen fluid decreases the density to 10–15% (Jarvis and Hungate, 1968; Coleman, 1969).

9. *Epidinium ecaudatum caudatum (bovine)*

This protozoon, found only in cattle, has a single, straight and heavy caudal spine which tends to get smaller and shorter with time in culture (Coleman *et al.*, 1972).

10. *Epidinium ecaudatum caudatum (ovine)*

This subspecies, found only in sheep with a "B-type" population (Eadie, 1962, 1967), has a single, short, curved caudal spine which disappears after 4–5 months in culture. The form of the spine in this and the bovine form must be determined in living material as they become distorted on fixation. This protozoon is one of the easiest to isolate by picking individual organisms, as it is the least sensitive to exposure to air. It grows equally well on caudatum- or simplex-type salts medium in the presence or absence of prepared fresh rumen fluid when fed with wholemeal flour and coarse dried grass. It will also grow on powdered dried grass but only at a population density of 20 ml^{-1} compared with 500 ml^{-1} on wholemeal flour.

The life of *Ep. e. caudatum in vitro* is limited to about 2 years irrespective of the medium on which it is grown. The reason for this is unknown but after $1\frac{1}{2}$ years in culture there is a "loss of vigour" shown by a general and progressive fall in population density and erratic growth in different tubes of an apparently identical set (Coleman *et al.*, 1972).

11. *Epidinium ecaudatum tricaudatum*

This protozoon has three caudal spines which gradually disappear in culture but at an intermediate stage both caudate and non-caudate organisms are present.

12. *Eremoplastron bovis*

This protozoon can be isolated and grown on powdered dried grass in simplex-type salts medium. Its growth on wholemeal flour has not been tested.

13. *Eudiplodinium maggii*

Different strains of *Eu. maggii* behave differently with respect to their preferences for caudatum- or simplex-type salts media. Most can be isolated in simplex-type salts medium but once established some die unless transferred to caudatum-type salts medium. However, some will never grow on simplex-type salts medium and must be isolated and grown subsequently on caudatum-type salts medium. This protozoon grows best at all times on powdered dried grass although it can be isolated initially on wholemeal flour and coarse dried grass. However, if this feeding regime is continued some strains will die after a few months unless they are fed on powdered dried grass instead. The growth of most established strains is not stimulated by the addition of 10% prepared fresh rumen fluid to the medium (Coleman *et al.*, 1976).

14. *Metadinium medium*

The author has isolated this protozoon only once, when he found it in the rumen of a cow. It grew best in caudatum-type salts medium in the absence of rumen fluid on powdered dried grass but proved difficult to keep alive and died after 6 months.

15. *Ophryoscolex caudatus*

Established cultures of this protozoon can be grown for 3–4 weeks in the absence of rumen fluid but for reliable maintenance over long periods it is essential to add 10% prepared fresh rumen fluid. This can be replaced by autoclaved rumen fluid but growth is less consistent under these conditions.

The morphology of *Op. caudatus* changes markedly after 2–4 weeks in culture, the long main caudal spine becoming much shorter and twisted so that it is at right-angles rather than parallel to the long axis of the protozoon (for photograph of culture form see Coleman, 1971). This morphological change is also associated with some physiological change because some strains that had grown well after initial isolation suddenly died at this time. If they survive this change they grow well for $1\frac{1}{2}$ years.

16. *Ostracodinium obtusum bilobum*

All isolates grow well with the daily addition of powdered dried grass but only some grow equally well on wholemeal flour and coarse dried grass. After a few weeks in culture this protozoon loses its two caudal lobes and comes to resemble *Os. obtusum obtusum*.

17. *Polyplastron multivesiculatum* (with *Epidinium* spp.)

Cultures of this protozoon are initiated by inoculating single organisms into 3 ml of a culture of *Ep. ecaudatum caudatum* (ovine) growing on caudatum-type salts medium containing 10% prepared fresh rumen fluid and fed daily, before and after the inoculation, with wholemeal flour and coarse dried grass. *P. multivesiculatum* has, when grown on caudatum-type salts medium, an obligate requirement for one of a number of other species of rumen protozoa, e.g. *Ep. ecaudatum caudatum* (ovine or bovine), *Ep. e. tricaudatum*, *Eu. maggii* and *Diplodinium monacanthum* but not *Entodinium* spp. or *Diploplastron affine*. *Ep. e. caudatum*, which is the best food protozoon, is engulfed by *P. multivesiculatum* at a rate of 1–10 organisms per protozoon per day. When those *Epidinium* spp. initially present in the culture have disappeared, it is essential to add more (in the form of a culture of *Epidinium* spp. at the rate of 1 ml per 30 ml *P. multivesiculatum* culture) daily. About one *Epidinium* $Polyplastron^{-1}$ d^{-1} gives good growth of the latter protozoon. Prepared fresh rumen fluid can be omitted from established cultures but this protozoon will not grow on simplex-type salts medium.

Some cultures of *P. multivesiculatum* will engulf only protozoa from the strain of *Ep. e. caudatum* on which they were isolated and die when this strain dies. However others can engulf *Epidinium* sp. from a new isolate initiated when the original dies, although in the author's experience no culture of *P. multivesiculatum* has ever survived the death of a second isolate of *Epidinium* sp. The maximum life of a culture (Table 1) therefore depends on the life of the food protozoon (Coleman *et al.*, 1972).

18. *Polyplastron multivesiculatum* (without *Epidinium* spp.)

If *P. multivesiculatum* organisms picked from rumen contents are placed in Hungate-type salts medium (inoculated the previous day with a drop of prepared fresh rumen fluid) and wholemeal flour and coarse dried grass added each day, about 10% of the protozoal cultures become established. As the maximum number of protozoa found under these conditions is never more than 5 ml^{-1}, the amount of wholemeal flour added must be very carefully controlled and never exceed 0.03 mg ml^{-1} each day. If an excess of starch is added the protozoa become very sluggish and are liable to die unexpectedly. The exact amount to add each day can be judged only by experience and it is always better to add too little than too much. Rumen fluid is not stimulatory to growth and the Hungate-type salts medium cannot be replaced by caudatum- or simplex-type salts media without killing the protozoa. In the author's experience it was possible only on one occasion to leave out the wholemeal flour and obtain protozoal growth for a few months on dried grass alone (Coleman *et al.*, 1972).

REFERENCES

Clarke, R. T. J. and Hungate, R. E. (1966). Culture of the rumen holotrich ciliate *Dasytricha ruminantium* Schuberg. *Applied Microbiology* **14**, 340–345.

Coleman, G. S. (1958). Maintenance of oligotrich protozoa from the sheep rumen *in vitro*. *Nature, London* **182**, 1104–1105.

Coleman, G. S. (1960). The cultivation of sheep oligotrich protozoa *in vitro*. *Journal of General Microbiology* **22**, 555–563.

Coleman, G. S. (1969). The cultivation of the rumen ciliate *Entodinium simplex*. *Journal of General Microbiology* **57**, 81–90.

Coleman, G. S. (1971). The cultivation of rumen entodiniomorphid protozoa. *In* "Isolation of Anaerobes". (Shapton, D. A. and Board, R. G., eds), Academic Press, London.

Coleman, G. S., Davies, J. I. and Cash, M. A. (1972). The cultivation of the rumen ciliates *Epidinium ecaudatum caudatum* and *Polyplastron multivesiculatum in vitro*. *Journal of General Microbiology* **73**, 509–521.

Coleman, G. S., Laurie, J. I., Bailey, J. E. and Holdgate, S. A. (1976). The cultivation of cellulolytic protozoa isolated from the rumen. *Journal of General Microbiology* **95**, 144–150.

Coleman, G. S., Laurie, J. I. and Bailey, J. E. (1977). The cultivation of the rumen ciliate *Entodinium bursa* in the presence of *Entodinium caudatum*. *Journal of General Microbiology* **101**, 253–258.

Dogiel, V. A. (1927). Monographie der Familie Ophryoscolecidae. *Archiv für Protistenkunde* **59**, 1–227.

Eadie, J. M. (1962). Inter-relationships between certain rumen ciliate protozoa. *Journal of General Microbiology* **29**, 579–588.

Eadie, J. M. (1967). Studies on the ecology of certain rumen ciliate protozoa. *Journal of General Microbiology* **49**, 175–194.

Hungate, R. E. (1942). The culture of *Eudiplodinium neglectum*, with experiments on the digestion of cellulose. *Biological Bulletin, Marine Biological Laboratory, Woods Hole, Massachussetts* **83**, 303–319.

Hungate, R. E. (1966). "The Rumen and its Microbes." Academic Press, London and New York.

Jarvis, B. D. W. and Hungate, R. E. (1968). Factors influencing agnotobiotic cultures of the rumen ciliate *Entodinium simplex*. *Applied Microbiology* **16**, 1044–1052.

Kofoid, C. A. and MacLennan, R. F. (1930). Ciliates from *Bos indicus* Linn. 1. The genus *Entodinium* Stein. *University of California Publications in Zoology* **33**, 471–544.

Kofoid, C. A. and MacLennan, R. F. (1932). Ciliates from *Bos indicus* Linn. 2. A revision of *Diplodinium* Schuberg. *University of California Publications in Zoology* **37**, 53–152.

Kofoid, C. A. and MacLennan, R. F. (1933). Ciliates from *Bos indicus* Linn. 3. *Epidinium* Crawley, *Epiplastron* gen. nov. and *Ophryoscolex* Stein. *University of California Publications in Zoology* **39**, 1–33.

Owen, R. W. and Coleman, G. S. (1976). The cultivation of the rumen ciliate *Entodinium longinucleatum*. *Journal of Applied Bacteriology* **41**, 341–344.

Chapter 4

Kinetoplastida

D. A. EVANS

Department of Medical Protozoology, London School of Hygiene and Tropical Medicine, Keppel Street, London, England

I. INTRODUCTION

The Order Kinetoplastida contains a vast assemblage of Protozoa, a great many of which are parasitic. The diversity of organisms parasitized by kinetoplastids is likewise vast, including vascular plants as well as vertebrate and invertebrate animals. It is therefore hardly surprising that the nutritional requirements of the parasitic kinetoplastid flagellates can differ greatly one from another, and these differences are mirrored in the variety of culture media employed in their *in vitro* cultivation.

The aim of this chapter is to bring the section covering "Trypanosomatidae" in Taylor and Baker (1968) as far as possible up to date. It makes no pretence of being fully exhaustive, and is bound to omit someone's favourite recipe for growing some trypanosomatid or other; but hopefully it will provide a practical guide to the most successful methods of cultivation. Except for 4N medium (Section III, C) culture media and methods described by Taylor and Baker (1968) are not repeated here. This does not imply that they are no longer applicable; indeed, a great many are still in routine day-to-day use, e.g. NNN medium (Wenyon, 1926), are extremely useful, and are likely to remain so in the foreseeable future.

Note on the use of antibiotics in culture media
In some laboratories it has become almost routine practice to include antibiotics (usually penicillin and streptomycin) in media designed for the cultivation of kinetoplastid flagellates. This routine addition is, I am convinced, a bad thing. Certainly antibiotics have their use in culture media, especially when one is attempting to isolate an organism from a microbiologically dirty site such as a skin lesion or the hindgut of an insect. Once the isolation has been made and the organism is growing in culture, then the sooner the antibiotics can be omitted from the medium the better it is for the organism. Most antibiotics, with the probable exception of penicillin, are likely to have some effect or other on the metabolism of the organism one is trying to cultivate; streptomycin, for instance, one of the most commonly added antibiotics, is a known inhibitor of mitochondrial protein synthesis. Another good reason for omitting antibiotics is that it encourages the use of good aseptic technique when handling the organisms.

II. MONOXENOUS TRYPANOSOMATIDAE

These are the so-called monogenetic or insect trypanosomatids (genera *Herpetomonas*, *Blastocrithidia*, *Leptomonas* and *Rhynchoidomonas*); they inhabit the guts of insects (and a few other invertebrates) and have no other host. They have the reputation of being easy to grow *in vitro*; some are indeed very easy, but others can be quite difficult. As with other kinetoplastid flagellates, different isolations of the same species may show different growth characteristics which can be rather disconcerting. Also some of these organisms may contain endosymbionts, e.g. *Crithidia oncopelti*, *C. deanei* and *Blastocrithidia culicis*, and the endosymbionts can have a great effect on the nutrition of the host flagellate. In the absence of endosymbionts, Guttman (1967) has shown that the nutritional requirements of the "lower Trypanosomatidae" are quite similar; they all require similar inorganic ions, a carbohydrate (usually), 12 amino acids, a purine, seven vitamins and haem.

A. DEFINED MEDIA

There are now several defined media available for insect flagellates with or without endosymbionts. Generally, the relatively simple defined media which support the growth of the flagellates with endosymbionts are useless for cultivating those which naturally have no endosymbiont or have been "cured" of their symbiont. These media usually lack essential nutrients (e.g. haemin) which the flagellate is unable to build up from simple precursors. When an endosymbiont is present, additional synthetic processes become available (presumably as part of the metabolism of the endosymbiont) and the infected flagellate can survive and grow in the more simple medium.

Usually one cultures these organisms in the temperature range 25–28°C, in stationary culture, without stirring or forced aeration; in Ehrlenmeyer flasks or "medical flat" bottles for volumes in excess of 10 ml, and capped or plugged test-tubes or small Universal bottles for the routine maintenance of small volumes. It is possible to grow some of these flagellates at 30–37°C, but this may require modification of the growth medium, especially of the osmolarity (Guttman, 1963; Roitman et al., 1972).

1. Trypanosomatids Without Endosymbionts

The following organisms have been grown (Guttman, 1967) on the defined medium of Guttman (1963):

> Crithidia fasciculata (*Anopheles* strain) ATCC 11745
> C. fasciculata (*Culex pipiens* strain) ATCC 12857
> C. fasciculata var. noelleri ATCC 12858
> Crithidia from Euryophthalmus ATCC 14766
> Crithidia from a syrphid
> C. rileyi
> C. acanthocephali
> C. luciliae ATCC 14765
> Blastocrithidia culicis ATCC 14806
> B. leptocoridis

With little modification, the medium also grew: *Leptomonas collosoma*—add glycine 0·1% (w/v), choline Cl 0·15 mg dl^{-1} and Ca pantothenate 0·1 mg dl^{-1}; *L. mirabilis*—add glycine 0·1% (w/v); *Leptomonas* from *Dysdercus*—add glycine 0·1% (w/v) and choline Cl 0·15 mg dl^{-1}.

O'Connell (1968) devised a defined medium for *Crithidia fasciculata* and an almost identical medium was used by Roitman *et al.* (1972) for the cultivation of *Herpetomonas samuelpessoai* (= *Leptomonas pessoai*). The latter workers omitted alanine, asparagine, glutamic acid, glycine, guanylate, choline and nicotinic acid from the medium and also showed that sorbitol was not the ideal main carbon source for *H. samuelpessoai*; raffinose (10 mg ml^{-1}) instead of sorbitol gave the best growth at 28°C and 37°C, with sucrose (10 mg ml^{-1}) being almost as good, and of course a great deal cheaper!

1° Defined medium for *Crithidia fasciculata*
(O'Connell, 1968)

	mg dl^{-1}		mg dl^{-1}
K$_3$ citrate H$_2$O	100	L-Phenylalanine	20
Citric acid H$_2$O	50	L-Threonine	20
L-Malic acid	20	L-Tryptophan	10
Succinic acid	100	L-Tyrosine ethyl ester	
MgCO$_3$	100	(free base)	20
MgSO$_4$.3H$_2$O	10	L-Valine	20
Trace elements (see below)	20	Ca pantothenate	1
Fe(NH$_4$)$_2$(SO$_4$)$_2$.6H$_2$O	1	Nicotinamide	0·5
CaCO$_3$	2	Nicotinic acid	0·2
DL-Alanine	10	Thiamine HCl	0·5
L-Arginine (free base)	40	Na riboflavin PO$_4$	0·3
L-Arginine-L-glutamate ½H$_2$O	400	Choline H$_2$ citrate	3
L-Asparagine H$_2$O	40	Biotin	0·002
L-Glutamic acid	200	Pyridoxamine 2HCl	0·1
Glycine	20	Folic acid	0·2
L-Histidine (free base)	250	Adenine	2
L-Histidine HCl.H$_2$O	50	Na$_2$guanylate PO$_4$	10
L-Isoleucine	20	β-Na glycerophosphate 5H$_2$O	2000
L-Leucine	20	Sorbitol	1500
L-Lysine HCl	20	Haemin (in 50% aqueous	
L-Methionine	20	triethanolamine)	2

Trace elements (mg dl^{-1}): Fe, 0·5 as Fe(NH$_4$)$_2$(SO$_4$)$_2$.6H$_2$O; Mn, 0·25 as MnSO$_4$.H$_2$O; Cu, 0·2 as CuSO$_4$ (anhydrous); Zn, 0·25 as ZnSO$_4$.7H$_2$O; Mo, 0·1 as (NH$_4$)$_6$Mo$_7$O$_{24}$.4H$_2$O; V, 0·02 as NH$_4$VO$_3$; B, 0·0005 as H$_3$BO$_3$; Co, 0·005 as CoSO$_4$.7H$_2$O.

The pH of the medium is adjusted to 6·5 with HCl or KOH and surprisingly it may be autoclaved (121°C, 15 min).

2° Carbohydrate-free medium for *Crithidia fasciculata*
(Tamburro and Hutner, 1971)

	mg dl^{-1}		mg dl^{-1}
Nitrilotriacetic acid	30	L-Phenylalanine	22
1,2,3,4-Cyclopentanetetracarboxylic acid		L-Threonine	25
	250	Tween 80	20
NH_4HCO_3	30	L-Tryptophan	15
KH_2PO_4	40	L-Tyrosine	8
$MgCO_3$	60	L-Valine	25
NaCl	200	Orotic acid	0·6
$CaCO_3$	2	Uracil	0·3
Trace elements (see below)	12	Thiamine	0·1
Adenine	6	Nicotinamide	0·075
Guanine	2	Ca pantothenate	0·045
L-Arginine HCl	30	Na riboflavin $PO_4.2H_2O$	0·075
L-Cysteine HCl	4	Pyridoxamine 2HCl	0·04
L-Histidine (free base)	30	Thiamine HCl	0·045
L-Isoleucine	50	Folic acid	0·045
L-Leucine	75	Haemin (see below)	1·4
L-Lysine HCl	30	Biotin	0·0035
L-Methionine	10	Biopterin	0·0045

Trace elements: grind together the following (all in grams): $Fe(NH_4)_2(SO_4)_2.6H_2O$: 42; $MnSO_4.H_2O$: 1·8; $CuSO_4$ (anhydrous): 2·5; NH_4VO_3: 2·3; $CoSO_4.7H_2O$: 4·8; H_3BO_3: 0·57; $NiSO_4.6H_2O$: 0·45; $CrK(SO_4)_2.12H_2O$: 0·96. The resulting mix is used at 12 mg dl^{-1}.

Haemin is supplied from a 10 mg ml^{-1} solution in 50% aqueous Quadrol [(ethylenedinitrilo)-tetra-2-isopropanol].

The pH of the complete medium is adjusted to 4·2–4·6 with solid Tris, and it is sterilized by autoclaving (121°C, 15 min).

2. *Trypanosomatids with Endosymbionts*

Crithidia oncopelti requires only one amino acid, methionine, in the relatively simple defined medium originally formulated for its growth (Newton, 1956). Some stocks (= "strains")* of this organism refuse to grow in Newton's defined medium (Roitman, I., personal communication) but will grow on a somewhat richer defined medium such as that of Guttman (1967).

The nutritional requirements of *C. deanei* differ from those of *C.*

---

* *Stock(s)*, defined as "populations derived by serial passage *in vivo* or *in vitro* from a primary isolation, without any implication of homogeneity or characterization" (Lumsden, 1977, p. 29 footnote), is used throughout this chapter (Eds).

oncopelti: one amino acid, tyrosine, is required in addition to methionine (growth is, however, very slow in the absence of tryptophan and phenylalanine); also *C. oncopelti* requires adenine, while *C. deanei* grows well without a purine source (Mundin *et al.*, 1974).

1° Guttman's IIB Medium for *Crithidia oncopelti*
(Guttman, 1967)

	mg dl^{-1}
K_3PO_4	20
$MgSO_4.7H_2O$	80
NaCl	50
Triethanolamine	500
L-Arginine	30
L-Glutamic acid	100
L-Histidine	20
DL-Methionine	20
Sucrose	1000
Salts solution (see below)	0·1 ml
Vitamin solution (see below)	0·1 ml

SALTS SOLUTION

	mg dl^{-1}
$CaCl_2$	300
$CuSO_4.5H_2O$	40
$Fe(NH_4)_2(SO_4)_2.6H_2O$	200
$ZnSO_4.7H_2O$	100
$MnSO_4.H_2O$	100
$CoSO_4.7H_2O$	10
H_3BO_3	10
Ammonium molybdate	5

VITAMIN SOLUTION

	mg dl^{-1}
Thiamine HCl	100
Nicotinic acid	100
Ca pantothenate	100
Pyridoxamine HCl	100
Biotin	1
p-Aminobenzoic acid	1

The pH of the medium is adjusted to 7·2 with KOH or HCl and sterilized by autoclaving (121°C, 15 min).

2° Defined medium for *C. deanei*
(Mundin *et al*., 1974)

	mg dl^{-1}		mg dl^{-1}
Sucrose	2000	Trace elements	2
β-Glycerophosphate.5H$_2$O Na salt	2000	L-Methionine	20
KCl	1000	L-Tyrosine ethyl ester	40
K$_3$ citrate H$_2$O	100	L-Phenylalanine	20
Citric acid H$_2$O	50	L-Tryptophan	20
Malic acid	20	Nicotinamide	0·2
Succinic acid	100	Folic acid	0·2
MgCO$_3$	100	Thiamine HCl	0·6
CaCO$_3$	2	Biotin	0·008
Fe(NH$_4$)$_2$(SO$_4$)$_2$.6H$_2$O	10		

Trace elements to yield the following final concentrations (mg dl^{-1}): Fe, 600 as Fe(NH$_4$)$_2$(SO$_4$)$_2$.6H$_2$O; Mn, 500 as MnSO$_4$.H$_2$O; Cu, 40 as CuSO$_4$ (anhydrous); Zn, 500 as ZnSO$_4$.7H$_2$O; Mo, 200 as (NH$_4$)$_6$Mo$_7$O$_{24}$.4H$_2$O; V, 40 as NH$_4$VO$_3$; B, 10 as H$_3$BO$_3$; Co, 10 as CoSO$_4$.7H$_2$O.

The pH of the complete medium is adjusted to 7·0 with HCl or KOH and it is sterilized by autoclaving (121°C, 15–20 min).

B. COMPLEX MEDIA

The majority of the monoxenous trypanosomatids will grow to a greater or lesser extent in most of the complex media that have been devised for nutritionally more exacting members of the Kinetoplastida (Sections III–IV), but a number of complex media have been specifically formulated for monoxenous trypanosomatids. These are usually cheap and simple to prepare and therefore are the complex media of choice for mass cultivation when chemically defined media are not required.

1° Medium for *C. fasciculata*
(Edwards and Lloyd, 1973)

MEDIUM

	mg dl^{-1}
Proteose peptone (Difco)	2000
Liver digest (Oxoid)	100
Glycerol	1000
Tween 80	500
Triethanolamine	600
Haemin	2·5
Folic acid	0·25

pH 8·0–8·2

Preparation: half the ethanolamine, and all of the peptone, liver digest, glycerol and Tween 80 are dissolved in a sixth of the final volume of distilled water, heated to 100°C, then centrifuged (ca 2000 g, 15 min) to remove suspended solids. The supernatant liquid is decanted and 25 mg litre^{-1} haemin dissolved in the remainder of the triethanolamine (made up as a 50% v/v aqueous solution) is added. Folic acid is dissolved in the minimum volume of 1·0 M-KOH and added to give 2·5 mg litre^{-1}. The medium is adjusted to pH 8·0 to 8·2 with 2·0 M-HCl, made up to the final volume with distilled water and autoclaved (121°C, 20 min).

TECHNIQUE

Culture volumes of up to 200 ml may be contained in litre conical flasks and incubated at 29°C in a rotary orbital shaker (150 rev min^{-1}). Larger volumes (up to 6 litres) have been grown in a laboratory fermentor with stirring (90 rev min^{-1}) and forced aeration (1 litre air per litre of medium min^{-1}), with sterile silicone MS antifoam RD (0·05% v/v; Dow Corning) added to prevent foaming.

2° Medium for *Crithidia fasciculata* and *Leptomonas* spp.
(Goldberg *et al.*, 1974)

	mg dl^{-1}
N-Z-Amine AS (pancreatic digest of casein)	300
Yeast hydrolysate (ICN, Nutritional Biochemicals)	500
Cane sugar	500
MgSO$_4$.7H$_2$O	10
Haemin (Sigma type III—equine)	0·01
pH 7·4	

The haemin is added from 50% (v/v) aqueous Quadrol [2,2′,2‴(ethylene dinitrilo) tetraethanol] after filter sterilization.

The medium minus the haemin is autoclaved (121°C, 15 min) and the filter-sterilized haemin solution added aseptically to the cooled medium.

3° *Crithidia* medium
(American Type Culture Collection)

	mg dl^{-1}
Trypticase (BBL)	600
Yeast extract	100

mg dl^{-1}

Liver extract concentrate (1:20) or
Liver L (Nutritional Biochemicals) . 10
Sucrose 1500
Haemin in triethanolamine
(see below) . . . . 2·5
pH 7·8

HAEMIN SOLUTION

Triethanolamine . . . . 0·25 ml
Haemin . . . . 2·5 mg
Distilled water . . . . 0·25 ml

An almost identical medium is given by Bacchi *et al.* (1968).

4° Medium for *Herpetomonas samuelpessoai*
(Roitman *et al.*, 1972)

mg dl^{-1}

Sucrose 2000
Trypticase (BBL) 500
Folic acid . . . . 0·2
Haemin 2
Yeast extract . . . . 500.
pH 6·5

Sterilize by autoclaving (121°C, 15 min).

III. *LEISHMANIA*

A. DEFINED MEDIUM

Until recently the sole member of this genus successfully cultivated in a defined medium was *Leishmania tarentolae* (Trager, 1957; see Taylor and Baker, 1968). But now a defined medium (RE1) based on Basal Medium (Eagle) tissue culture medium (BME) has been devised which supports the growth of *L. donovani* and *L. braziliensis* (Steiger and Steiger, 1976).

Krassner and Flory (1971), working with Trager's defined medium C (Trager, 1957), showed the following amino acids to be essential for the continuous cultivation of *L. tarentolae*: L-alanine, L-histidine, L-tryptophan, DL-phenylalanine, DL-serine, L-tyrosine, L-threonine, L-valine, L-leucine and L-lysine. They considered L-proline also to be essential, as culture in its absence was impossible if any amino acid other than glycine was deleted from medium C as well as the proline.

A variety of chemically defined tissue culture media, together with Cross's HX25 medium (a medium supporting growth of *Trypanosoma brucei* spp. and *T. cruzi*), have been used in attempts to cultivate *L. donovani* (Berens *et al.*, 1976), but without success.

Unless very specialized culture systems are employed, *Leishmania* grows *in vitro* as the promastigote, the form occurring in the midgut of the insect vector. The forms of the organism which grow in the vertebrate host can sometimes be cultivated, but they require the presence of live tissue cells (see Chapter 6).

1° Medium RE1
(Steiger and Steiger, 1976)

	mg litre^{-1}			mg litre^{-1}
(i) NaCl	8000	(iii)	NaHCO$_3$	1,000
KCl	400		HEPES	14,250
MgSO$_4$.7H$_2$O	200			
Na$_2$HPO$_4$.2H$_2$O	60	(iv)	Adenosine	20
KH$_2$PO$_4$	60		Guanosine	20
CaCl$_2$	70			
Glucose	2000	(v)	D-Biotin	1
Na-Acetate	600		Choline Cl	1
			Folic acid (see below)	11
(ii) L-Arginine HCl	200		*i*-Inositol	2
L-Cysteine HCl	50		Niacinamide	1
L-Cystine	50		D-Ca Pantothenate	1
L-Glutamic acid	300		Pyridoxal HCl	1
L-Glutamine	300		Riboflavin	0·1
L-Histidine	100		Thiamine HCl	1
L-Isoleucine	100			
L-Leucine	300	(vi)	Lipoic acid	0·4
L-Lysine HCl	250		Menadione	0·4
L-Methionine	50		Vitamin A	0·4
L-Phenylalanine	100			
L-Proline	300	(vii)	Ascorbic acid	0·2
DL-Serine	200		Vitamin B$_{12}$	0·2
L-Threonine	400		Bovine albumin (defatted; see below)	15
L-Tryptophan	50		Haemin (see below)	10
L-Tyrosine	50		Phenol red	10
L-Valine	100			

The pH is adjusted to 7·3–7·4 with M-NaOH and the medium filter-sterilized.

TECHNIQUE

(i) and (ii) are conveniently prepared as 2 × and 5 × concentrated stock solutions which are stored frozen. (v) can be purchased as a $100 \times$ concentrated stock solution (BME vitamin solution $100 \times$—Gibco), but extra folic acid (10 mg litre^{-1} of medium) must be added; this is best dissolved with the haemin in N-NaOH. The fat soluble vitamins (vi) are first dissolved in 1 ml ethanol. Defatted bovine albumin (Fraction V from bovine plasma, METRIX, Armour Pharmaceutical) is prepared according to Cross and Manning (1973).

B. SEMI-DEFINED MEDIA

Berens *et al.* (1976) formulated a medium for *Leishmania* spp. in which the only undefined component was serum. The medium (HO-MEM) supported good growth (and amastigote to promastigote transformation) of *L. donovani* and *L. tarentolae*; the Costa Rica stock of *T. cruzi* also grew well. However, not all haemoflagellates tested in this medium grew satisfactorily: *T. scelopori* and the Corpus Christi stock of *T. cruzi* grew very poorly.

Hendricks (1975) obtained growth of eleven different haemoflagellates, including *L. brasiliensis* and *L. donovani*, in commercially available tissue culture media, with the addition of 30% (v/v) foetal calf serum. The media used were: tissue culture medium 199, Grace's insect tissue culture medium and Schneider's *Drosophila* medium (revised).

Vessel *et al.* (1974) used Eagle's tissue culture medium enriched with 10% (v/v) heat-inactivated foetal calf serum for the "mass production" of promastigotes of *L. tropica major*, *L. t. minor*, *L. donovani* and *L. enriettii*. This medium was not used for long-term cultivation, but just for one large batch culture of each. So we do not know whether the medium will, on its own, support the growth of *Leishmania* spp. as a large number of promastigotes growing in modified NNN medium were used as the inoculum.

1° Medium HO-MEM
(Berens *et al.*, 1976)

litre^{-1}

SMEM minimal essential medium
(Eagle) for suspension culture
with Spinner's salts (Gibco) ... 10·6 g
MEM amino acids $50 \times$ (Gibco). 10·0 ml
MEM non-essential amino acids
$100 \times$ (Gibco)............. 10·0 ml
Na-pyruvate $100 \times$ (Gibco;
100 mM in British catalogue) .. 11·0 ml

$$\text{litre}^{-1}$$

Na-bicarbonate .... 2·2 g
Glucose .... 1·0–2·0 g
Biotin .... 0·1 mg
p-Aminobenzoic acid .... 1·0 mg

Adjust the pH of the above medium to 7·2–7·4, filter-sterilize and then add 5–10% (v/v) foetal calf serum.

Additional buffering capacity may be given by the inclusion of any of the following buffers:

HEPES (Sigma) 0·025M, pH 7·2–7·4
or
MOPS (Sigma) 0·025M, pH 7·2–7·4
or
HEPES, EPPS, Tricine (Sigma) 0·01M of each, pH 7·2–7·4
or
$NaPO_4$ 0·025M, pH 7·2–7·4

TECHNIQUE

Culture vessels are gassed with 5% (v/v) CO_2 in air, the tops sealed and then incubated at 26°C.

C. COMPLEX MEDIA

Recipes for these monophasic and biphasic "Macbethian" brews are legion (Taylor and Baker, 1968), and biphasic blood agar media are still the most common media used in the diagnosis of leishmaniasis by cultural methods. One such medium in everyday use is 4N nutrient agar-blood medium (Baker, 1966).

In common with most members of the Kinetoplastida, the genus *Leishmania* contains species that are more difficult to cultivate than others (e.g. *L. brasiliensis brasiliensis*).

Also various stocks of the same species can differ quite significantly in ease of cultivation, and one medium which will grow all *Leishmania* with equal facility is still an elusive ideal. The closest to this ideal that I have found is a monophasic medium (EBLB medium) devised for the cultivation of *T. brucei* sspp. (Section V, C, 1, 1°); preliminary studies have shown it to be very suitable for culturing large volumes of several different isolates of *L. tropica* and *L. donovani*. The only *L. brasiliensis* tried grew very well; this was a diagnostic cultivation direct from the primary lesion of a patient infected with *L. b. guyanensis*. Promastigotes developed in this medium several days before they appeared in 4N culture medium inoculated at the same time. Likewise, a positive diagnosis from a patient with cutaneous leishmaniasis contracted in Ethiopia, was obtained in two days with EBLB medium; 4N took one week.

If a particular stock of *Leishmania* proves difficult to cultivate in the standard media, it can be worth trying a biphasic medium with a solid phase of nutrient agar and blood such as 4N, using a tissue culture medium (e.g. 199) enriched with 10% (v/v) foetal calf serum as the liquid phase. This I have found to work on several occasions when all else had failed.

1° 4N nutrient agar-blood medium
(Baker, 1966)

(i) Add 40 g blood agar base no. 2 (Oxoid) to 1 litre of distilled water, mix and dissolve by steaming or autoclaving. Dispense 5 ml aliquots while molten into screw-capped glass bottles (30 ml capacity) and autoclave if necessary.

(ii) When cooled to about 45°C, add aseptically to each bottle 20 drops (*ca* 1 ml) of fresh rabbit blood and allow to set in a slant at the base of the bottle.

(iii) Add aseptically to each bottle 1 ml of sterile modified Locke's solution (Tobie *et al.*, 1950) containing (g litre^{-1}): NaCl, 8·0; KCl, 0·2; CaCl$_2$, 0·2; KH$_2$PO$_4$, 0·3; glucose, 2·5. Antibiotics (penicillin, 200 iu and streptomycin, 2 μg ml^{-1}) can be included if desired.

The classical NNN recipe is similar, but contains plain agar dissolved in 0·9% (w/v) saline with no added overlay (Wenyon, 1926).

IV. *TRYPANOSOMA* OTHER THAN SALIVARIA

As with *Leishmania*, we still mainly rely on undefined media for the cultivation of these organisms, and again the culture forms obtained are similar to those which develop in the insect vector: epimastigotes, trypomastigotes and occasionally promastigotes. So far only three of these trypanosome species have been cultivated in chemically defined media: two, *T. mega* and *T. ranarum*, are parasites of poikilotherms and one, *T. cruzi*, the causative agent of Chagas' disease, is from homoeotherms.

A. DEFINED MEDIA

Guttman (1967) devised a defined medium for the growth of *T. mega*, essentially an enriched version of that for monoxenous trypanosomatids (Guttman, 1963). The first successful cultivation of *T. cruzi* on a defined medium was achieved by Anderson and Krassner (1975). These workers grew the Costa Rica stock of *T. cruzi* in medium HX25, a defined medium devised by Cross and Manning (1973) for the cultivation of *T. brucei* sspp. (Section V, A). *T. cruzi* grew well if a large inoculum was used, and the cultures were transferred to fresh medium within 1 or, at most, 2 days of their peak growth. They also found that the linoleic acid–albumin complex called for in the formulation of HX25 could be omitted, but that growth was adversely affected. The cultures were grown at 26°C in 25 and 125 ml flasks plugged with "parafilm" covered cotton wool.

Mundin *et al.* (1976) published a preliminary account of the successful cultivation of the Y strain of *T. cruzi* in medium HX25 from which coenzymes Q_6 and Q_{10} had been omitted. Azevedo has continued work on the modification of medium HX25 and has kindly allowed me to quote the medium which he is now using (personal communication).

1° Medium HX25—modified
(Azevedo and Roitman, unpublished)

	mg dl^{-1}		mg dl^{-1}
Glucose	200	ATP	5
β-Glycerophosphate	2000	Albumin, bovine, defatted	20
Sodium stearate (see below)	5	Choline chloride	1·2
NaCl	400	Nicotinamide	1
Na$_3$PO$_4$.12H$_2$O	500	Nicotinic acid	0·0125
KCl	40	D-Calcium pantothenate	1
Trisodium citrate 2H$_2$O	60	Pyridoxine HCl	0·0125
Sodium acetate 3H$_2$O	79	Pyridoxal HCl	1
Sodium succinate	27	Thiamine HCl	1
D(+)Glucosamine	22	Inositol	2
EDTA, disodium salt	8	Riboflavin	0·1
Adenosine	2	D-Biotin	1
Adenine HCl	5	Ascorbic acid	0·025
Guanosine	2	p-Aminobenzoic acid	0·025
Guanine HCl	0·15	L-Alanine	28·5
Cytidine	2	L-Arginine	55
Uridine	2	L-Aspartic acid	55
Hypoxanthine	0·15	L-Asparagine	10
Uracil	0·15	L-Cysteine	10
Xanthine	0·15	L-Cystine	14
Thymine	0·15	L-Glutamic acid	104·5

	mg dl^{-1}		mg dl^{-1}
Haemin	1·5	L-Glutamine	10
Tween 40	0·5	Glycine	35
Folic acid	3	L-Histidine HCl	26
D-α-Tocopherol succinate	0·4	L-Isoleucine	40
Vitamin B-12	0·1	Glutathione (reduced)	0·25
DL-α-Lipoic acid (oxidized form)	0·04	L-Hydroxyproline	5
Menadione	0·045	L-Leucine	76
Trans-retenoic acid	0·04	L-Lysine HCl	75
Vitamin A	0·05	L-Methionine	17·5
Tween 80	1	L-Phenylalanine	37·5
Cholesterol	0·1	L-Proline	78
Calciferol	0·05	L-Serine	28·5
Ribose	0·25	L-Threonine	27
Deoxyribose	0·25	L-Tryptophan	14
Ferric nitrate 9H$_2$O	0·05	L-Tyrosine	36
AMP	0·1	L-Valine	48·5

The pH is adjusted to 7·5 and the medium minus the sodium stearate sterilized by filtration. Sodium stearate is then added as a freshly prepared solution in hot distilled water made alkaline with a few drops of 10% (w/v) solution of KOH and sterilized by filtration.

2° Defined medium for *T. mega*
(Guttman, 1967)

This is a version of Guttman's (1963) defined medium for monoxenous trypanosomatids (see Taylor and Baker, 1968) enriched by the following additions:

	mg litre^{-1}		μg litre^{-1}
Glycine	5000	Biotin	0·006
L-Arginine HCl	29	Ca pantothenate	0·066
L-Histidine	31	Folic acid	0·01
L-Isoleucine	53	Nicotinic acid	0·2
L-Leucine	53	Pyridoxine HCl	0·034
L-Lysine	85	Riboflavin	0·1
DL-Methionine	28	Thiamine HCl	0·1
DL-Phenylalanine	44	Vitamin B$_{12}$	0·0001
L-Threonine	20		
DL-Valine	90		

B. PARTIALLY DEFINED MEDIA

Yoshida (1975) devised a medium containing no macromolecules, in which the only undefined component was dialysate of a liver infusion. *T. cruzi* stocks Y, MR and CL all grew well in this medium, and there was some differentiation into trypomastigotes, the proportion of which varied with stock and number of passages. Another liquid medium, with foetal calf serum as the sole non-defined component, in which *T. cruzi* grows well and produces up to 90% trypomastigotes by the time the stationary growth phase is reached, was devised by O'Daly (1975).

Finally, *Trypanosoma (Megatrypanum) theileri* has the distinction of being the first mammalian trypanosome to be cultivated serially in a blood- and cell-free medium at 37°C (Sollod and Soulsby, 1968).

1° Macromolecule-free partially defined medium for *T. cruzi*
(Yoshida, 1975)

mg dl^{-1} of liver dialysate		mg dl^{-1} of liver dialysate	
Na$_2$HPO$_4$	800	Glycine	65
NaCl	400	L-Alanine	22·5
KCl	40	L-Proline	35
Glucose	200	L-Methionine	5
Haemin (see below)	2	L-Threonine	17·5
Cholesterol (see below)	0·2	L-Tryptophan	10
Sodium stearate (see below)	0·25	L-Isoleucine	5
5% (v/v) Tween 80 (see below)	0·1 ml	L-Leucine	7·5
L-Arginine	70	L-Phenylalanine	15
L-Lysine	62·5	L-Tyrosine	5
L-Histidine	100	L-Valine	10
L-Aspartic acid	35	DL-Serine	110
L-Asparagine	35	L-Cysteine	8
L-Glutamic acid	60	L-Cystine	5
L-Glutamine	60		

Liver infusion dialysate: a mixture of 8 g of liver infusion (Oxoid) in 15 ml of water is dialysed twice against 250 ml of water for 12 h each time to prepare about 600 ml; the solid ingredients listed above are added to this dialysate in the proportions given.

The pH is adjusted to 7·2 and the complete medium sterilized by filtration.

Haemin is added from a solution prepared by dissolving 20 mg in 2 drops of M-NaOH and 10 ml distilled water. The lipid mixture is added from a solution prepared as follows: 0·2 ml of M-NaOH, 1 ml of 5% (v/v) Tween 80 and 5 ml of distilled water are added to 2 mg of stearic acid. The mixture is heated to 70°C for 15 min and stirred. 2 mg cholesterol, dissolved in 2 drops of ethanol and distilled water to make 10 ml, are then added. Autoclave the final mixture (15 min, 110°C).

2° Semi-defined medium for *T. cruzi*
(O'Daly, 1975)

Eagle's Minimal Essential Medium (1 ×), with Earle's salts, L-glutamine and non-essential amino acids (all from Gibco) is dissolved in 800 ml of demineralized distilled water with 2·2 g $NaHCO_3$, 10 ml vitamin solution 100× (Gibco), 1 ml of nucleotide mixture (see below) and antibiotics (10^5 iu penicillin and 10^5 μg streptomycin, both of which may be omitted). The above mixture is further supplemented with 10 mM EPPS (N-2-hydroxyethyl piperazine propane sulphonic acid), 10 mM HEPES (N-2-hydroxyethyl piperazine-N'-2-ethane sulphonic acid), 5% (v/v) foetal calf serum and 100 ml of haemin solution (prepared by dissolving 15 mg of haemin in 100 ml of distilled water containing 1·8 g Tricine [N-tris (hydroxymethyl) methyl glycine], the pH of which is adjusted to 11·5 using NaOH, and sterilized by filtration).

NUCLEOTIDE MIXTURE

	mg dl^{-1}
2'-deoxy-5-GTP	2
2'-deoxy-5'-ATP	2
2'-deoxy-5'-ITP	2
ATP	200
D-ribose	50
Deoxyribose	50
Deoxy-CTP	2
Deoxy-UTP	2

3° Serum- and cell-free medium for *T. theileri*
(Sollod and Soulsby, 1968)

This medium was achieved by a progressive simplification of an initial culture medium containing whole blood, as follows (v/v):

Tissue culture medium NCTC	35%
Newborn calf serum	10%
Veal infusion broth	35%
Heparinated blood	20%
pH 7·2–7·4	

The heparinated blood was taken from a cow infected with *T. theileri* and acted as the inoculum for the culture as well as a medium ingredient. The final medium contained:

Haemin . . . .   0·001% (w/v)
Proteose peptone (2·5% solution
   in water) . . . .   19% (v/v)
McCoy 5A medium (modified)
   (Gibco Bio-Cult) . . . .   77% (v/v)
Transfer from the previous
   culture . . . .   4% (v/v)

TECHNIQUE

The entire medium (excluding the transfer) was adjusted to pH 7·2–7·4 and then autoclaved (15 min, 121°C). The progressive simplification of the initial medium was achieved by the transfer of 4% of the initial culture into the final medium, 4% of this culture to fresh final medium and so on until the initial medium was diluted out. The peak population ($5·8 \times 10^6$ organisms ml^{-1}) was reached in 3–4 d. The actively dividing forms were epimastigotes, whilst over 90% of the non-dividing population were trypomastigotes. After cultivation for a number of passages at 37°C the organisms lost their ability to grow at 25°C.

C. COMPLEX MEDIA

T. cruzi can usually be grown in a variety of complex media (Taylor and Baker, 1968). In those such as NNN, growth is largely as epimastigotes, but for some investigations it has become increasingly important to have as high a percentage of trypomastigotes in the culture as possible. Therefore, a number of media which allow growth of epimastigotes and differentiation into trypomastigotes have been devised.

One of the most widely used media for *T. cruzi* cultivation is Yaeger's Liver Infusion Tryptose Medium (LIT); not only do most stocks of *T. cruzi* grow very well in it, but a degree of morphological differentiation into trypomastigotes also occurs. In fact, LIT is one of a family of media designed for growth and differentiation of *T. cruzi*.

Pan (1971) developed two liquid media in which the "Brazil strain" (stock) of *T. cruzi* grew well and differentiated into trypomastigotes and amastigotes. Baker and Price (1973) obtained growth of the "Sonya" stock of *T. cruzi* initially mainly as amastigotes when cultured at 28°C in a liquid medium (L4N), the composition of which was based on that of the diphasic medium 4N (Section III, C, 1°). L4N was originally used for the growth of *T. (Schizotrypanum) dionisii* from bats, but is suitable for a variety of other less fastidious trypanosomatids also.

For a single batch culture of *T. cruzi* the simple autoclavable monophasic medium cited by Mattei *et al.* (1977) is very convenient. It is not, however, suitable for long-term maintenance of *T. cruzi*, as growth diminishes rapidly on subculture, but is cheap, easy to prepare, and for one growth cycle produces large numbers of trypanosomes (mainly epimastigotes), provided a heavy inoculum of *T. cruzi* growing in a rich medium is used.

1° Yaeger's LIT medium
(Cited by Castellani *et al.*, 1967)

BASIC MIXTURE

NaCl	4 g
KCl	0·4 g
Na_2HPO_4	8 g
Glucose	2 g
Calf serum (heat inactivated)	100 ml
10% Haemoglobin solution	
(see below)	20 ml
Distilled water	750 ml

TECHNIQUE

The addition of 100 ml litre^{-1} of a 5% (w/v) solution of ox liver infusion (Oxoid) and 5 g litre^{-1} of tryptose (see below) to the basic mixture completes the medium. The pH is adjusted to 7·3 with 0·1 M-HCl or KOH and the complete medium filter sterilized.

The haemoglobin solution may be made in a variety of ways: "add 90 ml of distilled water to 10 ml packed bovine red cells" (Yoshida, 1975), or wash the centrifuged cells from defibrinated rabbit or sheep blood twice with isotonic saline, pack the cells by centrifuging at 100 *g* for 5 min, lyse the red cells by adding 10 ml of water per ml of packed cells, wait 10 min, centrifuge (1000 *g*, 10 min), decant the supernatant and filter-sterilize (method used at the Instituto Oswaldo Cruz, Rio de Janeiro, Brazil).

A useful modification of the medium is to replace the haemoglobin solution with a solution of 25 mg haemin dissolved in 5 ml of 50% (v/v) triethanolamine in water per litre of LIT (Gutteridge *et al.*, 1969).

Tryptose from either Oxoid or Difco may be used. Growth is often more rapid with the Difco brand (I. Roitman, personal communication).

LIVER INFUSION LACTALBUMIN MEDIUM (LIL)

This medium is made by adding 5 g lactalbumin hydrolysate and 100 ml of 5% (w/v) ox liver infusion to 1 litre of LIT basal mixture.

HEART INFUSION LACTALBUMIN MEDIUM (HIL)

5 g of lactalbumin hydrolysate and 100 ml 5% dog heart infusion are added to 1 litre of LIT basal mixture. The infusion is prepared by removing the heart from a freshly killed dog; the pericardium and large vessels are removed and the organ homogenized in a Waring blender for 5 min in 5 volumes of LIT basal mixture with 5 g litre^{-1} of lactalbumin hydrolysate added (pH 6·7). The homogenate is transferred to a boiling water bath, where it is left for about 10 min (5 min after a deep change in colour), then cooled and filtered through cheese cloth. The infusion may be stored at $-20°C$.

Of these three related media, LIT at pH 7·2 gives the best growth, but not the highest percentage of trypomastigote forms; this is given in HIL medium at pH 6·7, where the Y stock of *T. cruzi* produces up to 70% trypomastigotes after 5 d *in vitro*.

2° Media F-29 and F-32 for the cultivation and morphogenesis of *T. cruzi*
(Pan, 1971)

COMPOSITION OF BASAL MEDIUM—common to F-29 and F-32

	ml
Tissue culture medium 199 (10 × concentrate)	10
0·8% (w/v) glucose solution	25
5% aqueous "Trypticase" solution (BBL)	10
2·8% NaHCO$_3$ solution	1·5
Distilled water	53·5

The glucose and bicarbonate solutions are sterilized by filtration and the trypticase solution by autoclaving (121°C, 15 min).

COMPOSITION OF F-29

	ml
Basal medium	85
Foetal calf serum (heat-inactivated)...	10
Haemin solution (see below)	5
pH 7·5	

COMPOSITION OF F-32

	ml
Basal medium	85
Foetal calf serum (heat-inactivated)....	5
Haemin solution (see below)	5
Chicken plasma	5
pH 7·5	

The haemin solution (0.5 mg ml^{-1}) is prepared in 0.04% (w/v) NaOH and filter-sterilized.

The original recipe calls for the addition of penicillin (100 iu ml^{-1}) and streptomycin (100 μg ml^{-1}); these can be omitted. The media can be stored at $-15°C$ to $-20°C$ for a maximum of three months.

The inocula used to seed these two media were grown in modified NNN medium (Pan, 1968) and contained over 99% epimastigotes. During ten serial passages in medium F-29 at $29.5°C$ the proportions of morphological types did not change significantly, but in the second passage in F-29 at $35.5°C$ most of the organisms became amastigote, and after several more subcultures over 92% were amastigotes. In F-32 medium at $35.5°C$, over 90% of the organisms in the original passage were trypomastigotes with terminal kinetoplasts. After several more passages at either $29.5°C$ or $35.5°C$, over 94% became amastigotes.

It is important, when using these media to produce cultures with high percentages of trypomastigotes or amastigotes, to keep the culture volume small. The author used culture volumes of 5 ml; if the volumes are increased much above this, the proportions of amastigotes and trypomastigotes are not obtained with the same predictability.

3° Medium L4N
(Baker et al., 1972)

(i) Basal solution

	g litre^{-1}
Proteose peptone (Oxoid L46)	15
Liver digest (Oxoid L27)	2.5
Yeast extract (Oxoid L20)	5
NaCl	5

(ii) Rabbit serum—heat inactivated (56°C, 30 min).
(iii) 10% erythrocyte lysate—prepared by lysing 1 ml of packed erythrocytes from defibrinated rabbit blood in 9 ml distilled water.

Complete medium:	basal solution	(i)	10 parts by volume
	serum	(ii)	1 part by volume
	lysate	(iii)	2 parts by volume

N.B.: it is convenient to prepare the basal solution (i) at twice the above concentration and store it at $-20°C$. When required, thaw (i), add (ii) and (iii) as above; clarify the solution by centrifugation (4°C, 1 h, 23,000 g); mix with a volume of distilled water equal to that of basal solution (i) and further clarify the medium by passage through membrane filters, 0·45 μm pore size followed by a sterile 0·22 μm membrane filter. It was later found that adequate growth of many stocks and species was obtained using a modification of this medium (designated L4NHS) containing only half the above concentrations of serum and erythrocyte lysate (Baker *et al.*, 1976).

4° Simple monophasic medium for *T. cruzi*
(Mattei *et al.*, 1977)

	g litre^{-1}
NaCl	4·0
KCl	0·4
Na$_2$HPO$_4$	8·0
Glucose	2·0
Tryptose	10·0
Liver infusion	2·0
Haemin (see below)	0·01
Distilled water	1 litre

pH 7·2

Sterilize by autoclaving (121°C, 15 min).

Presumably the haemin is added as a solution in triethanolamine or aqueous Quadrol (see Section II, B, 2°); the authors do not state how the addition is made.

V. THE SALIVARIA

The salivarian trypanosomes have proved to be among the most difficult members of the order Kinetoplastida to cultivate *in vitro*. Only within the last three years has a chemically defined medium been devised which will support the growth of any of them. Success or failure in culturing this group seems to depend on many factors, but the history of the trypanosome stock before its attempted cultivation must be one of the most important. In general, freshly isolated material or else cryopreserved material from a stock that has had a minimum number of syringe passages through laboratory animals is the most successful starting material for culture. Old laboratory stocks of these trypanosomes which have been maintained by syringe passage are frequently very difficult or impossible to cultivate *in vitro*; such stocks (certainly of *Trypanozoon*) tend to lose their bloodstream pleomorphism and the ability to activate their mitochondria, this latter function being essential for development in the tsetse fly vector or in culture. For similar reasons it is impossible to cultivate the naturally occurring monomorphic members of *Trypanozoon* such as *T. evansi* and *T. equiperdum*, or any of the dyskinetoplastic forms.

The method of preparing trypanosomes for cultivation described in detail below is used routinely in my laboratory for *T. b. brucei*, *T. b. gambiense*, *T. b. rhodesiense* and *T. congolense*; it has a high success rate, and cultures with $> 10^7$ trypanosomes ml^{-1} are usually obtained.

Bloodstream infections of the trypanosomes to be cultivated are raised in the appropriate laboratory rodent: rat or mouse for *T. b. brucei* and *T. b. rhodesiense*, nursling rat for *T. b. gambiense* and mouse for *T. congolense*. The stage of parasitaemia at which the trypanosomes are removed for subsequent cultivation is important, especially with *T. brucei* sspp. A population comprising mostly slender forms takes significantly longer to establish in culture than one which has a high percentage of intermediate and short stumpy forms; consequently when working with *T. brucei* sspp. it is better to harvest the trypanosomes 1 d after a peak of parasitaemia, when the proportion of intermediate and short stumpy trypanosomes is great, rather than during an ascending phase or at the peak of parasitaemia when long slender forms predominate.

The strictest aseptic precautions are employed throughout all procedures. The trypanosome-infected animals are bled by cardiac puncture (right ventricle) into syringes containing heparin (a minimum of 5 iu ml^{-1} of blood) and the blood is immediately diluted with three volumes of sterile phosphate buffered saline glucose at pH 8·0 (PSG; Lanham, 1968). Normally, diluted blood is placed on ice, but in this case this has proved not to be a good idea, as a cold shock at this stage slows down the subsequent transformation from bloodstream to culture forms. The diluted trypanosome-containing blood is next poured on to the top of a column of sterile DEAE cellulose equilibrated to pH 8·0 with sterile PSG (Lanham, 1968). The eluate is collected in sterile centrifuge tubes [plastic disposable 1 oz (ca 30 ml) universal containers with cone-shaped bottoms from Sterilin are convenient] and then centrifuged (ca 1000 *g*, 10 min) at room temperature (cooling at this stage can again slow transformation and initial growth of the trypanosomes). The supernatant liquid is removed and the pellet resuspended in a small volume of sterile PSG. The trypanosome suspension is now used to inoculate monophasic human blood lysate broth medium EBLB (Section V, C, 1°).

I have mentioned before that I like to keep antibiotics as far as possible out of my cultures, but in this primary culture penicillin 100 iu ml^{-1} is added routinely. Streptomycin is best avoided: it is a known inhibitor of mitochondrial protein synthesis (causing abnormal codon/anticodon interaction at the acceptor sites) and adding it to a system where active mitochondrial proliferation is to take place therefore seems unwise. The trypanosome inoculum is adjusted so that each ml of medium receives approximately 2×10^6 trypanosomes. The most satisfactory culture vessels are glass "medical flat" bottles, or for larger volumes of medium, Thomson bottles: 100 ml "medical flats" containing 15–20 ml of medium, 500 ml "medical flats" 100–120 ml and Thomson bottles 500 ml. The culture bottles containing the inoculated medium are incubated lying flat at 27°C. When screw-topped bottles are used the caps must be slightly loose to allow some gaseous interchange between the air phase above the medium and the surrounding atmosphere, but not the ingress of contaminating micro-organisms.

Samples are taken from the primary cultures after 24 h and examined fresh by phase contrast microscopy. Further samples are examined after 48 h and, provided the trypanosomes are alive and the cultures are not contaminated, the primary cultures are subinoculated into fresh medium, approximately 1 part of primary culture into 9 parts of fresh medium. The primary culture is also subinoculated into biphasic medium (an extensively modified Tobie's medium, Section V, C, 2, 1°). This last step is essential as certain trypanosome stocks transform and grow well initially in the monophasic medium, but then go through a short phase where they grow poorly if at all in this medium, while growing well in the biphasic medium.

A. DEFINED MEDIA FOR *TRYPANOZOON*

So far the only defined media for trypanosomes of this group are HX25 (Cross and Manning, 1973), and its modifications HX25M and HX28 (Cross *et al.*, 1975) for *T. brucei* sspp.

1° Medium HX25M
(Cross *et al.*, 1975)

	mg dl^{-1}		mg dl^{-1}
HEPES	1900	L–Asparagine	10
Glucose	362	L–Aspartic acid	40
NaCl	100	L–Cysteine HCl	10
KH_2PO_4	90	L–Cystine	4
$NaHCO_3$	80	L–Glutamic acid	22·3
Trisodium citrate $2H_2O$	60	L–Glutamine	10
Sodium acetate $3H_2O$	54	Glycine	10
Sodium succinate $6H_2O$	27	L–Histidine HCl	16
D(+)Glucosamine HCl	7·3	L–Isoleucine	30
EDTA, disodium salt	8	L–Leucine	46
Adenosine	2	L–Lysine HCl	40

	mg dl^{-1}
Cytidine	2
Uridine	2
Haemin	2
Tween 80	0·5
Folic acid	1
DL-α-Tocopherol	0·4
Vitamin B$_{12}$	0·1
DL-α-Lipoic acid (oxidized form)	0·04
Menadione	0·04
Coenzyme Q$_6$	0·04
Coenzyme Q$_{10}$	0·04
Trans-retenoic acid	0·04
L-Alanine	16
L-Arginine HCl	40

	mg dl^{-1}
L-Methionine	10
L-Phenylalanine	25
L-Proline	19·3
L-Serine	16
L-Threonine	36
L-Tryptophan	9
L-Tyrosine	16
L-Valine	36
Medium 199 (powder—Wellcome Reagents)	930
Vitamin solution 100 × concentrate (see below)	10 ml litre^{-1}
Linoleic acid–albumin complex (see below)	6 ml litre^{-1}

VITAMIN SOLUTION

	mg dl^{-1}
D-Biotin	0·1
D-Calcium pantothenate	0·1
Choline chloride	0·1
Folic acid	0·1
L-Inositol	0·2
Nicotinamide	0·1
Pyridoxal HCl	0·1
Riboflavin	0·01
Thiamine HCl	0·1

LINOLEIC ACID–ALBUMIN COMPLEX

Preparation: wash activated charcoal in 1 M-HCl followed by distilled water, then air dry. Dissolve 10 g albumin in 100 ml distilled water and mix 5 g charcoal into the solution, cool to 0°C, then bring the pH to 3·0 by adding 0·2 M-HCl; stir at 4°C for 1 h; remove the charcoal by centrifugation (30 min, 10 000 g) followed by filtration through a nitrocellulose filter (pore size 0·45 μm); raise the pH to 7·0 using 1 M-NaOH and dialyse against distilled water; adjust the volume to give a protein concentration of 50 mg ml^{-1}. The solution may be stored at −15°C. Thirty ml of the defatted albumin solution are warmed to 37°C and added to 42 mg of linoleic acid contained in a 100 ml beaker and stirred for 1 h at room temperature. The solution is filtered and stored at −15°C.

The pH of the complete medium is adjusted to 7·40 with NaOH and it is sterilized by filtration. Growth will occur even in the absence of the linoleic acid–albumin complex, though the final yield of organisms is less. Coenzyme Q$_6$ can also be omitted from the medium with no apparent effect on growth.

B. SEMI-DEFINED MEDIA FOR *TRYPANOZOON*

Fromentin (1971) devised a semi-defined medium for *T. b. gambiense* based on medium 199, with either one part of the liquid overlay from Tobie's medium (Tobie *et al.*, 1950) added to 20 parts of medium 199, or 1 part of the liquid overlay from Nöller's medium (Nöller, 1917) added to 10 parts of medium 199 or 199 plus washed rat red blood cells. Many more modifications were tried, but in none did *T. b. gambiense* grow as well as did *T. brucei* sspp. in the media developed by Cross and Manning (1973). The latter were also based on medium 199; the simpler version is HX12 from which HX12V was derived by the inclusion of the vitamin mixture of HX25M (Section V, A, 1°). In HX12V, consistent growth of *T. b. brucei* is obtained, with doubling time 24 h, yielding up to 10^7 organisms ml^{-1}. More recently, Brun and Jenni (1977) have reported a semi-defined medium for *T. brucei* sspp. based on Eagle's MEM and medium 199 and designated SDM-77; this supported good growth of the *T. brucei* sspp. stocks tested but *T. congolense* died out after 2–3 passages.

1° Media HX12 and HX12V
(Cross and Manning, 1973)

	mg dl^{-1}		mg dl^{-1}
HEPES	1900	EDTA, disodium salt	8
Casein hydrolysate	500	Adenine	2
KH_2PO_4	136	Adenosine	2
Glucose	100	Guanine	2
$NaHCO_3$	80	L-Methionine	2
L-Proline	57·5	Haemin	1
Trisodium citrate $2H_2O$	58·8	Folic acid	1
Sodium acetate $3H_2O$	54·4	Medium 199 (10× concentrate)	8·8 ml
Sodium succinate $6H_2O$	27	Linoleic acid–albumin complex prepared	
D(+)Glucosamine HCl	22	as for HX25M above	0·6 ml

Addition of 10 ml $litre^{-1}$ of the 100 × concentrated vitamin solution from medium HX25M (Section V, A, 1°) to this medium converts it into HX12V.

2° Medium SDM-77
(Brun and Jenni, 1977)

	$litre^{-1}$		$litre^{-1}$
Minimum Essential Medium (MEM) (Eagle) for suspension culture:		HEPES (=40 mM) (Calbiochem)	9·53 g
		MOPS (=20 mM) (Calbiochem)	4·18 g
F-14 powder (Gibco)	8·00 g	$NaHCO_3$	1·00 g

	litre^{-1}		litre^{-1}
Medium 199 TC 45 powder		L-Methionine	50 mg
(Wellcome Reagents)	2·00 g	L-Proline	650 mg
MEM amino acids (50 ×) (Gibco)		L-Threonine	200 mg
	8·00 ml	Adenosine	6 mg
MEM nonessential amino acids		Guanosine	6 mg
(100 ×) (Gibco)	6·00 ml	Glucosamine HCl	10 mg
Na-pyruvate (100 mM)	6·00 ml	Biotin	0·10 mg
Glucose	2·50 g	p-Aminobenzoic acid	1·00 mg

The components are dissolved in 900 ml glass-distilled water, the pH is adjusted to 7·25 with 5 N-NaOH and the volume is made up to 1 litre. Finally SDM-77 is filter sterilized (Millipore, 0·22 µm) and 10% inactivated (30 min, 56°C) foetal calf serum and 0·25 ml dl^{-1} haemin stock solution (see below) are added before use. The medium is kept frozen in 20 ml batches at −20°C. Osmolarity without serum = 380 milliosmol.

250 mg haemin are dissolved in 50 ml 0·05 N-NaOH and made up to 100 ml with glass-distilled water. The pH is adjusted to 8·0 with N-HCl and the solution is autoclaved.

C. COMPLEX MEDIA FOR *TRYPANOZOON* AND OTHERS

1. *Monophasic Media*

The most commonly used monophasic media are blood lysate broths based on that originally devised by Pittam (1970). One modification of this medium is that described by Cross and Manning (1973), medium MCM. Another, used routinely in the author's laboratory, is a further modification of Pittam's medium described by Brown *et al.* (1973), the preparation of which is described below. Other monophasic complex media for *T. brucei* sspp. have been described by Balber (1971) and Hanas *et al.* (1975).

1° EBLB Medium
(Evans, unpublished)

	litre^{-1}
EDTA, disodium salt	0·4 g
NaOH	1·7 g
KH$_2$PO$_4$	6·8 g
L-Proline	1·5 g
Tryptose (Oxoid L47)	15·0 g
Casein hydrolysate (acid)	
(Oxoid L41)	10·0 g
Liver digest (Oxoid L27)	10·0 g
Blood lysate (see below)	125 ml

The pH is adjusted to 7·4 with NaOH

Blood lysate is prepared from human blood (out-dated transfusion stock)
from which the bulk of the plasma has been removed; it is frozen ($-20°C$)
and then thawed, the pH adjusted to 7·4 with NaOH or HCl, 20 ml of a 4%
(w/v) solution of $CaCl_2$ added per litre of crude blood lysate, and the
mixture is allowed to clot. The mixture is shaken vigorously, transferred to
centrifuge pots and spun (ca 20,000 g, 1 h). The supernatant liquid is
carefully decanted off the pellet and used as the blood lysate in the above
recipe.

The complete medium is clarified by passage through graded nitrocellulose membrane filters (1·2, 0·8, 0·45 and 0·22 μm pore sizes), sterilized by
passage through a sterile 0·22 μm pore size membrane filter, dispensed into
sterile bottles and stored, preferably at $-20°C$ (but the medium will keep
for up to nine months at 4°C).

TECHNIQUE

In addition to *T. brucei* sspp. this medium supports excellent growth of the
following organisms (Evans, unpublished observations): *Trypanosoma
cruzi, T. lewisi, T. musculi, T. legeri, T. rangeli, Leishmania tropica major, L.
tropica minor, L. aethiopica, L. donovani, L. enriettii, L. brasiliensis guyanensis*
and *L. mexicana amazonensis.*

Preliminary work indicates that the medium may be lyophilized then
reconstituted simply by the addition of sterile distilled water to the
lyophilized powder. This may be of great value for organism isolation in the
field, if the shelf-life of the lyophilized medium is sufficiently long.

2. Biphasic Media

Most biphasic media are various forms of nutrient blood agar overlaid with a
liquid phase consisting of a balanced salts solution such as Earle's, Hanks's
etc. (Chapter 1, Section IV, B, 1). Certain stocks of salivarian trypanosomes
are very difficult to cultivate on the standard formulations, e.g. those of
Tobie *et al.* (1950) and Weinman (1960), but most difficulties can be
overcome by using the biphasic medium described below.

1° **Modified Tobie's Medium**
(Evans, unpublished)
SOLID PHASE

g litre^{-1}

Beef extract (Oxoid
   Lab-Lemco L29) 3
Bacteriological peptone
   (Oxoid L37) 5

g litre^{-1}

NaCl 8
Agar 20

This is autoclaved (15 min, 121°C) and cooled to 56°C. To it are added horse red blood cells (from defibrinated blood), washed and resuspended in an equal volume of the liquid phase of the medium. 15 ml of the washed red cell suspension are added to 85 ml of the molten solid phase (at 56°C), and blood agar slants or flats made in the usual way.

LIQUID PHASE

mg dl^{-1}

KCl 40
Na$_2$HPO$_4$.12H$_2$O 6
KH$_2$PO$_4$ 6
CaCl$_2$.2H$_2$O .... 18·5
MgSO$_4$.7H$_2$O 10
MgCl$_2$.6H$_2$O 10
NaCl 800
L-Proline 100
Phenol red .... *qs*

The pH is adjusted to 7·2 with solid Tris, and the complete mixture sterilized by autoclaving (121°C, 15 min). Immediately before use, 5% (v/v) sterile foetal calf serum is added to the mixture; this completes the liquid overlay of the medium.

D. *TRYPANOSOMA VIVAX*

This extremely important pathogen of cattle and other ruminants has still not been successfully grown axenically *in vitro*. Isoun and Isoun (1974) achieved an initial 3-fold increase in numbers 24 h after inoculating *T. vivax*, separated from the blood of infected cattle, into tissue culture medium 199 buffered to pH 7·4 with HEPES and supplemented with foetal calf serum and plasma from animals of the same variety as those from which the parasites had been isolated. Trypanosome numbers decreased after 48 h in the medium, but if passaged every 72 h, some trypanosomes were still motile 10 d after initiation of the culture.

There is obviously much more work to be done on the *in vitro* cultivation of *T. vivax*, but this is at least a start in the right direction.

E. HAEMATOZOIC TRYPOMASTIGOTES OF *T. BRUCEI*

A major breakthrough has occurred recently in the cultivation of African trypanosomes, this being the first successful long-term cultivation *in vitro* of the bloodstream forms of *T. brucei* (Hirumi *et al.*, 1977). All previous attempts to cultivate the bloodstream forms of *T. brucei* sspp. have ended either in the death of the trypanosomes within a few days or with their transformation into forms not infective to mammals.

Hirumi and his co-workers cultivated the bloodstream forms of *T. brucei* over bovine fibroblast-like cells in RPMI 1640 medium containing HEPES buffer and foetal calf serum. The trypomastigotes grew and maintained their bloodstream morphology and infectivity for at least 415 d in this culture system (Hirumi, personal communication).

1° Cultivation of haematozoic *T. brucei*
(Hirumi *et al.*, 1977)

(i) Establishment of fibroblast-like cell lines. Pulmonary fluid was collected from cattle infected with *Theileria parva* (Muguga stock) shortly after death and used to initiate cell cultures at 37°C in 250 mm² Falcon plastic culture flasks (T-25) with RPMI medium plus 25 mM HEPES and 20% (v/v) heat inactivated foetal calf serum. Confluent fibroblast-like cell monolayers were subcultured every 5–7 days. The cell line was designated ILR-BPF-376. Another fibroblast cell-line, ILR-BHF-476, was prepared from peripheral blood of a healthy yearling steer (Boran cross); this line differed morphologically from ILR-BPF-376 and formed multiple cell layers.

(ii) Cultivation of *T. brucei*. Twelve different media, four different concentrations of foetal calf serum and the presence or absence of ILR-BPF-376 cells were tested. Haematozoic *T. brucei* were inoculated into the culture system (2×10^3 trypanosomes ml^{-1}). The system containing RPMI 1640 with HEPES and 20% foetal calf serum gave the best results, with an increase in trypanosome numbers from 2×10^3 to $3 \cdot 5 \times 10^5$ ml^{-1} after 96 h. The initial successful cultures of bloodstream *T. brucei* were used to inoculate fresh ILR-BPF-376 cultures, and also ILR-BHF-476. The trypanosomes maintained by rapid passage (24 h) kept the typical appearance of bloodstream long slender trypomastigotes. In culture systems where the trypanosome numbers were diminishing following a period of maximum growth, mainly short stumpy trypanosomes were seen.

(iii) Subculture. The most successful subcultivations were made by 1:3 "splits" of previous culture fluid containing mainly long slender forms (i.e. the contents of 1 flask were used to inoculate 3 fresh flasks). At the time the paper describing this method was written, 1246 subcultures had been made into T-25 flasks (6 ml medium per flask) or T-75 (18 ml medium per flask).

Under optimum conditions trypanosomes increased in number 16-fold within 24 h (population doubling time 6 h).

F. METACYCLIC TRYPOMASTIGOTES OF *T. BRUCEI*

As this chapter was going to press, a further breakthrough was reported—the cultivation in tsetse fly tissue cultures of forms of *T. brucei brucei* infective to mice, and morphologically resembling metacyclic trypomastigotes (Cunningham and Honigberg, 1977).

Medium for metacyclic *T. brucei*
(Cunningham, 1977; Cunningham & Honigberg, 1977)

COMPOSITION

	mg dl^{-1}		mg dl^{-1}
NaH_2PO_4	53	L-Glutamic acid	25
$MgCl_2.6H_2O$	304	L-Glutamine	164
$MgSO_4.7H_2O$	370	Glycine	12
KCl (anhydrous)	298	L-Histidine	16
$CaCl_2.2H_2O$	15	DL-Isoleucine	9
Glucose	70	L-Leucine	9
Fructose	40	L-Lysine	15
Sucrose	40	DL-Methionine	20
L-Malic acid	67	L-Phenylalanine	20
α-Ketoglutaric acid	37	L-Proline	690
Fumaric acid	5·5	DL-Serine	20
Succinic acid	6	L-Taurine	27
β-Alanine	200	DL-Threonine	10
DL-Alanine	109	L-Tryptophan	10
L-Arginine	44	L-Tyrosine	20
L-Asparagine	24	DL-Valine	21
L-Aspartic acid	11	Vitamin mixture (BME × 100)	0·2 ml
L-Cysteine HCl	8	Phenol red (0·5% w/v)	0·4 ml
L-Cystine	3		

The solids are dissolved in double distilled water as follows: all inorganic salts except $CaCl_2$, in 10 ml; $CaCl_2$ in 5 ml; sugars in 10 ml; organic acids in 5 ml; amino acids in 65 ml. Mix solutions and adjust pH to 7·4 using 2N-NaOH. Sterilize by filtration through Millipore 0·22 μm filter. Store at −20°C. Thaw before use and add foetal bovine serum ("either non-inactivated or heat-inactivated") to 20% v/v.

TECHNIQUE

Puparia of *Glossina morsitans morsitans* containing flies about to emerge were surface sterilized in White's solution ($HgCl_2$ 0·25 g, NaCl 6·5 g, HCl 1·25 ml, ethanol 250 ml, distilled water 750 ml), washed several times in insect balanced salt solution (IBSS: NaCl 7·5, KCl 0·2, $CaCl_2$ 0·2, $NaHCO_3$ 0·2, glucose 2·0, phenol red 0·02, all g litre^{-1}) containing penicillin 100 iu ml^{-1}, streptomycin 100 µg ml^{-1} and Fungizone (Squibb) 10 µg ml^{-1}, and dissected in IBSS. Heads with attached salivary glands were placed on 9 × 35 mm coverglasses in Leighton tubes (5 explants per tube) or in 25 cm^2 Falcon flasks (13–24 explants per flask) containing respectively about 0·2 or 0·5 ml medium.

Procyclic trypomastigotes of *T. b. brucei* were cultivated in the above medium containing explants of alimentary canal of *G. m. morsitans* dissected as described above (Cunningham, 1977): 1·0 ml of medium containing 4–5 × 10^6 haematozoic trypomastigotes was placed in a 25 cm^2 Falcon flask and incubated at 28°C; after 4 days, 1 ml of medium was added; 4 days later, the trypanosomes were subinoculated to further flasks at initial concentrations of about 1·5 × 10^6 ml^{-1} and similar passages were made every 5–6 days. These procyclic forms were used as inocula for the salivary gland tissue cultures, either 0·5 or 1·0 ml of culture supernatant containing 10^6 ml^{-1} being added to tubes or flasks respectively. Cultures were maintained at 28°C; every 2 days the medium was removed and replaced by an equal volume of fresh medium. Parasite numbers increased about 25 fold during 48 h, and about 10% of the salivary glands became invaded by trypanosomes after 12–17 d *in vitro*; most of the parasites were of the procyclic trypomastigote type, but a small number resembled epimastigotes or metacyclic trypomastigotes.

REFERENCES

Anderson, S. J. and Krassner, S. M. (1975). Axenic culture of *Trypanosoma cruzi* in a chemically defined medium. *Journal of Parasitology* 61, 144–145.

Bacchi, C. J., Hutner, S. H., Ciaccio, E. I. and Marcus, S. M. (1968). O_2-polarographic studies on soluble and mitochondrial enzymes of *Crithidia fasciculata*: glycerophosphate enzymes. *Journal of Protozoology* 15, 576–584.

Baker, J. R. (1966). Studies on *Trypanosoma avium*. IV. The development of infective metacyclic trypanosomes *in vitro*. *Parasitology* 56, 15–19.

Baker, J. R. and Price, J. (1973). Growth *in vitro* of *Trypanosoma cruzi* as amastigotes at temperatures below 37°C. *International Journal for Parasitology* 3, 549–551.

Baker, J. R., Green, S. M., Chaloner, L. A. and Gaborak, M. (1972). *Trypanosoma (Schizotrypanum) dionisii* of *Pipistrellus pipistrellus* (Chiroptera): intra- and extracellular development *in vitro*. *Parasitology* 65, 251–263.

Baker, J. R., Liston, A. J. and Selden, L. F. (1976). Trypomastigote dimorphism and satellite deoxyribonucleic acid in a clone of *Trypanosoma (Schizotrypanum) dionisii*. *Journal of General Microbiology* 96, 113–115.

Balber, A. E. (1971). Pleomorphism and the physiology of *Trypanosoma brucei*. PhD Thesis, The Rockefeller University, New York.

Berens, R. L., Brun, R. and Krassner, S. M. (1976). A simple monophasic medium for axenic culture of hemoflagellates. *Journal of Parasitology* 62, 360–365.

Brown, R. C., Evans, D. A. and Vickerman, K. (1973). Changes in oxidative metabolism and ultrastructure accompanying differentiation of the mitochondrion in *Trypanosoma brucei*. *International Journal for Parasitology* 3, 691–704.

Brun, R. and Jenni, L. (1977). A new semi-defined medium for *Trypanosoma brucei* sspp. *Acta Tropica* 34, 21–33.

Castellani, O., Ribeiro, L. V. and Fernandes, J. F. (1967). Differentiation of *Trypanosoma cruzi* in culture. *Journal of Protozoology* 14, 447–451.

Cross, G. A. M. and Manning, J. C. (1973). Cultivation of *Trypanosoma brucei* sspp. in semi-defined and defined media. *Parasitology* 67, 315–331.

Cross, G. A. M., Klein, R. A. and Linstead, D. J. (1975). Utilization of amino acids by *Trypanosoma brucei* in culture: L-threonine as a precursor for acetate. *Parasitology* 71, 311–326.

Cunningham, I. (1977). New culture medium for maintenance of tsetse tissues and growth of trypanosomatids. *Journal of Protozoology* 24, 325–329.

Cunningham, I. and Honigberg, B. M. (1977). Infectivity reacquisition by *Trypanosoma brucei brucei* cultivated with tsetse salivary glands. *Science, New York* 197, 1279–1282.

Dixon, H. and Williamson, J. (1970). The lipid composition of blood and culture forms of *Trypanosoma lewisi* and *Trypanosoma rhodesiense* compared with that of their environment. *Comparative Biochemistry and Physiology* 33B, 111–128.

Edwards, C. and Lloyd, D. (1973). Terminal oxidases and carbon monoxide-reacting haemproteins in the trypanosomatid, *Crithidia fasciculata*. *Journal of General Microbiology* 79, 275–284.

Fromentin, H. (1971). Contribution à l'étude comparée des besoins nutritifs chez diverses espèces du genre *Trypanosoma*. Possibilités et limites de la culture à 27°C en milieu semi-synthétique liquide. *Annales de Parasitologie Humaine et Comparée* 46, 337–445.

Goldberg, B., Lanpros, C., Bacchi, C. J. and Hutner, S. H. (1974). Inhibition by several antiprotozoal drugs of growth and O_2 uptake of cells and particulate preparations of a *Leptomonas*. *Journal of Protozoology* 21, 322–326.

Gutteridge, W. E., Knowler, J. and Coombes, J. D. (1969). Growth of *Trypanosoma cruzi* in human heart tissue cells and effects of aminonucleoside of puromycin, trypacidin and amenopterin. *Journal of Protozoology* 16, 521–525.

Guttman, H. N. (1963). Experimental glimpses at the lower Trypanosomatidae. *Experimental Parasitology* 13, 129–142.

Guttman, H. N. (1967). Patterns of methionine and lysine biosynthesis in the Trypanosomatidae during growth. *Journal of Protozoology* 14, 267–271.

Hanas, J., Linden, G. and Stuart, K. (1975). Mitochondrial and cytoplasmic ribosomes and their activity in blood and culture forms of *Trypanosoma brucei*. *Journal of Cell Biology* 65, 103–111.

Hendricks, L. D. (1975). Liquid media for the rapid and quantitative cultivation of *Leishmania*. *50th Annual Meeting, The American Society of Parasitologists*, p. 68 (abstract).

Hirumi, H., Doyle, J. J. and Hirumi, K. (1977). African trypanosomes: *in vitro* cultivation of animal-infective *Trypanosoma brucei*. *Science, New York* 196, 992–994.

Isoun, T. T. and Isoun, M. J. (1974). *In vitro* cultivation of *Trypanosoma vivax* isolated from cattle. *Nature London* 251, 513–514.

Krassner, S. M. and Flory, B. (1971). Essential amino acids in the culture of *Leishmania tarentolae*. *Journal of Parasitology* 57, 917–920.

Lanham, S. M. (1968). Separation of trypanosomes from the blood of infected rats and mice by anion-exchangers. *Nature, London* 218, 1273–1274.

Lumsden, W. H. R. (1977). Horae subsecivae inter trypanosomata. *Protozoology* 3, 25–32.

Mattei, D. M., Goldberg, S., Morel, C., Azevedo, H. P. and Roitman, I. (1977). Biochemical strain characterisation of *Trypanosoma cruzi* by restriction endonuclease cleavage of kinetoplast DNA. *Federation of European Biochemical Societies Letters* 74, 264–268.

Mundin, M. H., Roitman, I., Hermans, M. A. and Kitajima, E. W. (1974). Simple nutrition of *Crithidia deanei*, a reduviid trypanosomatid with an endosymbiont. *Journal of Protozoology* 21, 518–521.

Mundin, M. H., Azevedo, H. P., Roitman, C., Gama, M. I. C., Manaia, A. C., Previato, J. O. and Roitman, I. (1976). Cultivation of *Trypanosoma cruzi* in defined medium. *Revista do Instituto de Medicina Tropical de São Paulo* 18, 143.

Newton, B. A. (1956). A synthetic growth medium for the trypanosomatid flagellate, *Strigomonas (Herpetomonas) oncopelti*. *Nature, London* 198, 210–211.

Nöller, W. (1917). Blut- und Insektenflagellatenzüchtung auf Platten. *Archiv für Schiffs- und TropenHygiene* 21, 53–94.

O'Connell, K. M. (1968). Development of screening techniques for antiprotozoal agents with special reference to drug toxicity studies and freeze preservation techniques. PhD Thesis, Fordham University Department of Biological Sciences, Bronx, New York.

O'Daly, J. A. (1975). A new liquid medium for *Trypanosoma (Schizotrypanum) cruzi*. *Journal of Protozoology* 22, 265–270.

Pan, C. (1968). Cultivation of the leishmaniform stage of *Trypanosoma cruzi* in cell-free media at different temperatures. *American Journal of Tropical Medicine and Hygiene* 17, 823–832.

Pan, C. (1971). Cultivation and morphogenesis of *Trypanosoma cruzi* in improved liquid media. *Journal of Protozoology* 18, 556–560.

Pittam, M. D. (1970). Medium for the *in vitro* culture of *Trypanosoma rhodesiense* and *Trypanosoma brucei*. Appendix to Dixon and Williamson (1970), q.v.

Roitman, C., Roitman, I. and De Azevedo, H. P. (1972). Growth of an insect trypanosomatid at 37°C in a defined medium. *Journal of Protozoology* 19, 346–349.

Sollod, A. E. and Soulsby, E. J. L. (1968). Cultivation of *Trypanosoma theileri* at 37°C in a partially defined medium. *Journal of Protozoology* 15, 463–466.

Steiger, R. E. and Steiger, E. (1976). A defined medium for cultivating *Leishmania donovani* and *L. braziliensis*. *Journal of Parasitology* 62, 1010–1011.

Tamburro, K. M. and Hutner, S. H. (1971). Carbohydrate-free media for *Crithidia*. *Journal of Protozoology* 18, 667–672.

Taylor, A. E. R. and Baker, J. R. (1968). "Cultivation of Parasites *in vitro*." Blackwell Scientific Publications, Oxford.

Tobie, E. J., Brand, T. von, and Mehlman, B. (1950). Cultural and physiological observations on *Trypanosoma rhodesiense* and *Trypanosoma gambiense*. *Journal of Parasitology* 36, 48–54.

Trager, W. (1957). Nutrition of a hemoflagellate (*Leishmania tarentolae*) having an interchangeable requirement for choline or pyridoxal. *Journal of Protozoology* 4, 269–276.

Vessel, H., Rezai, H. R. and Pakzad, P. (1974). *Leishmania* species: fatty acid composition of promastigotes. *Experimental Parasitology* 36, 455–461.

Weinman, D. (1960). Cultivation of the African sleeping sickness trypanosomes from the blood and cerebrospinal fluid of patients and suspects. *Transactions of the Royal Society of Tropical Medicine and Hygiene* 54, 180–190.

Wenyon, C. M. (1926). "Protozoology." Baillière, Tindall and Cox, London.

Yoshida, N. (1975). A macromolecule-free partially defined medium for *Trypanosoma cruzi*. *Journal of Protozoology* 22, 128–130.

Chapter 5

Plasmodiidae

P. I. Trigg

National Institute for Medical Research, Mill Hill, London, England

I. INTRODUCTION

During the last few years increased interest in the cultivation of all stages of malaria parasites has arisen for several reasons. Many current biological, biochemical, chemotherapeutic and immunological problems in malaria research can be answered adequately only by the use of reliable cultivation techniques. Also, experimental vaccines have been produced from various stages of the parasite that will produce some protection in animals and man. Although no one claims that a practical vaccine for human use is around the corner, one possible source of antigen for such a vaccine could be developed from techniques involving the cultivation of the parasite *in vitro*.

Most work since the previous review by Taylor and Baker (1968) has been on the cultivation of erythrocytic stages, on which most of this chapter will concentrate. Tissue culture has recently yielded results which may be applied to the cultivation of tissue stages of malaria parasites and these are also briefly discussed.

II. CULTIVATION OF ERYTHROCYTIC STAGES

A. ASEXUAL STAGES

1. *Dilution Techniques*

(a) *Plasmodium falciparum*. Until recently it has been impossible to subculture any species of malaria parasite through more than 3–4 cycles *in vitro* (see Bertagna *et al.*, 1972). However, in the last year, two laboratories have independently reported the successful long-term culture of *Plasmodium falciparum in vitro*.

Trager (1976) and Trager and Jensen (1976) first reported the continuous cultivation of a chloroquine-resistant strain of *P. falciparum* from *Aotus* monkeys, using both a continuous flow system (Trager, 1971) and a simple dilution system in Petri dishes. The basic principle of both methods is to have a shallow stationary layer of erythrocytes covered by a shallow layer of medium. In the former (Section II, A, 2, 1°), the medium flows slowly and continually over the cell layer, whilst in the latter (Section II, A, 1, 1°) the medium has to be changed manually. The growth medium was RPMI 1640 supplemented with 25 mM HEPES buffer, 0·2% (w/v) sodium bicarbonate and 10% (v/v) AB group human serum. The cultures were incubated at 38°C. The initial parasite material was diluted by the addition of AB Rh + human erythrocytes to give an initial parasite level of around 0·1–0·2%, i.e. ca 10^6 parasites ml^{-1} and 10^9 rbc ml^{-1}. The medium was changed every day and fresh erythrocytes added every 3rd or 4th day so that the initial parasite level after each subculture was 0·1–0·2%. Multiplication, at least four-fold and often more, was observed during each cycle; the parasites were still multiplying *in vitro* after two months; they retained their normal morphology and were infective to an *Aotus* monkey when inoculated intravenously after 35 days *in vitro*. However, the parasites lost *in vitro* some of the synchronicity characteristic of *P. falciparum in vivo*. Other strains have now been grown continuously *in vitro*, with variation in their growth characteristics. Sometimes multiplication rates were low during the first four weeks but then increased. The reasons for this are unknown, but may involve adaptation to growth *in vitro*. Cultures have so far been maintained for nearly two years (Trager and Jensen, personal communication).

Haynes *et al.* (1976) also reported the continuous cultivation of *P*.

falciparum in vitro. In their experiments, cryopreserved chimpanzee blood infected with *P. falciparum* was thawed (Section II, A, 4), diluted with modified medium 199, and incubated at 38°C in a simple dilution system. Medium was changed every 1–2 d and when the parasites had grown to the schizont stage they were reinoculated into cultures containing fresh human erythrocytes. Parasites continued multiplying for 22 d.

Previous attempts to subculture erythrocytic stages of malarial parasites have been restricted to *P. falciparum* and *P. knowlesi* except for a study of *P. gallinaceum* (Anderson, 1953) which appears not to have been repeated. In these studies, using similar dilution techniques to those described above, the parasites survived for only three or four asexual cycles with a progressive decrease in multiplication at each cycle (Anfinsen *et al.*, 1946; Trigg, 1969a; Trigg and Gutteridge, 1971; Phillips *et al.*, 1972). There appear to be several important differences in the more recent work which may account for its success. How far these points may be applied to the subculture of other species awaits further experimentation.

(i) *Buffering.* One of the major technical problems encountered is that the large amount of lactic acid produced by the parasite lowers the pH of the medium. The maintenance of pH between 7·3 and 7·5 is critical (Geiman *et al.*, 1966). Trager and Jensen (1976) and Haynes *et al.* (1976) have used several modifications to counteract this problem. In both studies the cultures were initiated at very low parasite densities, approximately $\frac{1}{2}-\frac{1}{4}$ that used earlier, which reduces the lactate production in each culture. In addition, zwitterionic buffers (Chapter 1: IV, B, 2) were employed to improve the buffering capacity of the medium. Earler work by Geiman *et al.* (1966) had indicated that the addition of the zwitterionic buffer glycylglycine (pKa 7·9 at 37°C) at 5 mM improved growth of *P. knowlesi in vitro.* However, it is better to use a zwitterionic buffer with a pKa near to and slightly below the pH desired (7·3–7·5). Trager and Jensen (1976) used 25 mM HEPES (pKa 7·31 at 37°C) which appears theoretically to be the best buffer for these systems. However, Siddiqui and Schnell (1973) report marginally better multiplication of *P. falciparum* during one cycle *in vitro* using 40 mM TES (pKa 7·16 at 37°C) rather than 40 mM HEPES. Haynes *et al.* (1976) successfully used TES in their system. Zwitterionic buffers have the further advantage that they ensure pH stability during the time cultures are manipulated outside the CO_2-rich atmosphere used to grow the malarial parasites *in vitro.* Cultures are also buffered with a bicarbonate/CO_2 system which requires adequate exchange between gaseous and liquid phases, thus a thin layer of cells covered by a shallow layer of medium (Trager and Jensen, 1976) improves buffering; Trager (1971) found that the thickness of the cell and medium layer was critical for development of *P. falciparum* in short-term cultures.

(ii) *Gas phase.* The gas phase is important in the growth of asexual erythrocytic stages of malarial parasites. High oxygen concentrations are inhibitory but oxygen appears to be required since complete maturation to the schizont stage of *P. knowlesi* does not occur under anaerobic conditions (Trigg, 1969b). Earlier work employed a gas phase containing 20% O_2 (Taylor and Baker, 1968) but a reduction in oxygen tension to 5% or below appeared to produce better multiplication although the results were highly variable (Trigg, 1969b; Butcher and Cohen, 1971). The successful long-term cultivation of *P. falciparum* in dilution systems was achieved at oxygen tensions of 5–6%. Trager and Jensen (1976) reported that 1% oxygen gave better development in flow vials than 5% oxygen. Maintenance of cultures at low O_2 tension is one method of maintaining a low redox potential in the cultures. Trigg (1969b) showed that a redox potential of ca 140 mV and ca 170 mV was maintained in cultures of *P. knowlesi* in 1% and 5% oxygen respectively, compared to a value rising to ca 230 mV in 20% oxygen. Experiments designed to test the effect of agents which maintain a low redox potential have given highly variable results but it is of interest to note that the modified growth medium 199 of Haynes *et al.* (1976) contained the reducing agents 2-mercaptoethanol and vitamin E (α-tocopherol).

(iii) *Susceptibility of host cells.* Recent work has indicated that the susceptibility of the host erythrocyte is an important factor. Initial experiments on the cultivation of *P. falciparum* from *Aotus* monkeys resulted in limited multiplication (about 2-fold) during a single cycle *in vitro* (Siddiqui *et al.*, 1970). Greater multiplication resulted from mixing parasitized blood from *Aotus* with human erythrocytes. Care must be taken in doing this since Trager (1971) found that human group A, but not human group B or AB erythrocytes, will agglutinate with *Aotus* erythrocytes. Trigg (1975), using a subculture technique, showed that *P. falciparum* originally grown *in vitro* in *Aotus* erythrocytes would invade human (group O) erythrocytes to a greater extent than *Aotus* erythrocytes. However, Haynes *et al.* (1976) showed that *P. falciparum* invaded chimpanzee erythrocytes as well as human erythrocytes (blood group not indicated), but were unable to draw any conclusions as to the relative susceptibility of *Aotus* and human erythrocytes, since insufficient experiments were performed. The successful long-term cultivation of *P. falciparum* was obtained in both instances when the parasites were grown in human erythrocytes; all the human ABO blood groups appear to be suitable. The significance of this is not clear since no one has reported attempts to cultivate *P. falciparum* serially in non-human erythrocytes but it seems possible that under the conditions used human erythrocytes maintain their viability and/or susceptibility to infection *in vitro* better than non-human erythrocytes. Trigg and Shakespeare (1976a,b) have shown that normal rhesus monkey erythrocytes incubated for periods up to 48 h *in vitro* are less susceptible to invasion by *P. knowlesi* than are normal erythrocytes taken directly from the monkey. These observations may not contradict the finding by Trager and Jensen (1976) that successful subculture of *P. falciparum* could be obtained in time-expired blood obtained from blood transfusion centres since even after incubation of normal erythrocytes for 48 h *in vitro* multiplication of parasite numbers exceeded fourfold.

(iv) *Removal of leucocytes.* The successful use of stored time-expired blood for the culture of erythrocytic stages of malarial parasites may be linked with the reduction in number of leucocytes found in such blood. Bass and Johns (1912) stressed the necessity for removing leucocytes from infected blood for the successful cultivation *in vitro* of *Plasmodium*. However, this aspect has been largely ignored by later workers, with the possible exception of Richards and Williams (1973) who removed leucocytes from malaria infected blood by passing it through a column of CF 11 cellulose powder. It is not clear whether this technique results in improved multiplication of the parasite *in vitro* since no comparative studies have been reported, but it is interesting to note that in both successful attempts at the continuous cultivation of *P. falciparum* the leucocytes were removed either by aspiration (Trager and Jensen, 1976) or by passage through a CF 11 cellulose column (Haynes *et al.*, 1976).

(v) *Plasma and serum.* Short-term cultures have shown that plasma and sera from infected and normal monkeys vary considerably in their ability to support growth *in vitro* of *P. knowlesi* (Butcher and Cohen, 1971). The reason for this is unknown but it could be due to the absence of certain growth factors resulting from different dietary or physiological states of the animals, or to the presence of inhibitory factors. Short-term cultures also indicate that homologous sera are not required; Butcher *et al.* (1973) and Trigg (1975) showed that with *P. knowlesi* and *P. falciparum* the host specificity resides in part in the susceptibility of the erythrocyte to invasion by the merozoite and that serum factors do not appear to play a major part in host specificity. Heat-inactivated foetal calf serum has been used successfully in our laboratories for the short-term culture of *P. knowlesi* (Trigg, unpublished observations) and this was the serum source in the initial cycles of the long-term cultivation of *P. falciparum* by Haynes *et al.* (1976). However, this serum was replaced by autologous human serum during later stages of their work; it is not clear whether this was because human serum was required but it has recently been shown that commercial foetal calf serum is not adequate for long-term cultivation (Trager and Jensen, personal communication). Variable success has been obtained using fresh citrated plasma obtained chiefly from blood banks but plasma from outdated blood is apparently inhibitory. Serum should be stored at $-20°C$

and repeated freezing and thawing avoided; 15% serum may be required to initiate long-term cultures.

(b) *Other mammalian malarial parasites.* It is not possible to grow all species of malaria parasites *in vitro* with equal ease; there is no published report of the long-term cultivation of parasites other than *P. falciparum*. Good growth and multiplication in a variety of growth media can be obtained with *P. knowlesi* (Trigg, 1968, 1969b; Butcher and Cohen, 1971; Trigg and Gutteridge, 1971); recently this species has been grown continually by methods similar to those used for *P. falciparum* (Carter and Diggs, personal communication). Rodent malaria parasites have proved extremely difficult to grow *in vitro*. Good growth from ring to late trophozoite and schizont stage has been obtained with *P. vinckei chabaudi* in a modified Harvard medium (Trigg, 1968) and modified Eagle's MEM (Coombs and Gutteridge, 1975). In the former study a small but significant amount of reinvasion, and therefore increase in parasite numbers, was observed using a perfusion technique. In the dilution cultures of Coombs and Gutteridge (1975) no increase in parasite numbers was observed although reinvasion did take place; Coombs and Gutteridge attributed the good growth through a single cycle *in vitro* to the high buffering capacity of the medium produced by the addition of 10 mM glycylglycine, 10 mM HEPES and 40 mM TES in conjunction with 9·7 mM sodium phosphate buffer and the low oxygen tension (1%) of the gaseous phase. *P. berghei* is also relatively difficult to grow *in vitro*. Growth to the schizont stage has been obtained in modified medium 199, supplemented with (g litre^{-1}): glucose, 2; sodium bicarbonate, 1·76; inosine, 0·02; ATP, 0·01; heparin, 0·002 and (iu litre^{-1}) penicillin, 5000. No significant reinvasion or multiplication occurred at 37°C but at 15°C some multiplication was observed (Smalley and Butcher, 1975). Trigg and McColm (1976) considered one of the problems associated with the cultivation of *P. berghei* to be the availability of susceptible cells, since the KSP 11 strain which they used showed a predilection for reticulocytes. These authors obtained consistent slight multiplication of *P. berghei* by diluting the infected cell inoculum with uninfected blood from rats made reticulocytaemic either by the daily injection of 1 ml of 0·25% (v/v) phenylhydrazine or by repeated daily bleeding. Cultures were grown in the modified medium 199 described by Smalley and Butcher (1975) at very low parasite densities (ca 10^6 parasites ml^{-1}) and incubated at 37°C in an atmosphere of 5% CO_2 (v/v) in air. The technique of setting up the cultures is essentially that described by Trager and Jensen (1976) for the long term culture of *P. falciparum*. Subsequent experiments (Trigg *et al.*, unpublished observations) have shown that similar results can be obtained using the candle jar technique and RPMI 1640 plus 10% foetal calf serum or 10% normal rat serum as the growth medium.

1° Continuous cultivation of *P. falciparum*
(Trager and Jensen, 1976)

MEDIUM

Dehydrated RPMI 1640 (Grand Island Biological), supplemented with 25 mM HEPES buffer and 10% (v/v) type AB human serum after hydration in double distilled water and sterilization by filtration.

TECHNIQUES

Preparation of uninfected blood. Blood can be stored at 4°C with the addition of either $\frac{1}{4}$ volume of ACD (disodium hydrogen citrate, 2·0 g; dextrose, 3·0 g; distilled water to 120 ml) or 63 ml per 450 ml blood of PCD (sodium citrate, 2·63 g; dextrose, 2·32 g; citric acid, 372 mg; sodium dihydrogen phosphate, 251 mg; distilled water to 100 ml).

(i) Centrifuge the blood (1000 g, 10 min), remove the plasma and buffy coat.

(ii) Wash the packed erythrocytes twice by centrifugation (1000 g, 10 min) in RPMI without serum.

(iii) Resuspend the cells to 50% (v/v) in medium (as above).

Preparation of infected cells from established culture

(i) Centrifuge the culture (750 g, 10 min), and discard the supernatant fluid.

(ii) Resuspend the packed cells in an equal volume of medium (see above).

(iii) Determine the parasite level from a thin smear by counting the numbers of infected cells per 10^4 erythrocytes.

(iv) Dilute the infected cells with freshly washed uninfected cells (see above) to give a final parasite level of 0·1 to 0·2% (the culture now contains equal volumes of cells and medium).

(v) To each ml of this suspension add 3 ml of medium to give a cell dilution of 1:8.

(vi) Dispense 1·5 ml aliquots of this suspension into 35 mm plastic Petri dishes (Falcon Plastics).

Preparation of infected cells from human or other primate infection

(i) Centrifuge the infected blood (1000 g, 10 min), remove the plasma and the buffy coat.

(ii) Wash the cells twice with RPMI 1640 minus serum.

(iii) Resuspend the infected cells in an equal volume of medium (see above).

(iv) Determine the parasite level and make dilutions with uninfected erythrocytes as above.

Incubation of cultures in candle jars

(i) Place Petri dishes in desiccator containing a white candle.

(ii) Light the candle and seal the lid of the desiccator with stopcock open until candle goes out.

(iii) Close stopcock and incubate desiccator at $37 \cdot 5 \pm 0 \cdot 5°C$. This is a simple but effective way of producing an atmosphere of low O_2 and high CO_2 content.

Changing culture medium

(i) Change the medium at least every 24 h by gently tilting the culture dish and aspirating the medium with a sterile Pasteur pipette. The disturbance of the settled cells should be kept to a minimum so that as much fluid as possible can be removed without removing cells.

(ii) Add 1·5 ml of RPMI 1640 + 10% serum to each culture dish and suspend the cells by gentle swirling.

(iii) Replace cultures in the candle jar and incubate at $37 \cdot 5 \pm 0 \cdot 5°C$.

Thin smears can be made at appropriate intervals by removing the clear fluid from the settled cells and spreading a small drop of concentrated cells on a glass slide.

Cultures can be safely continued without the addition of fresh cells for 6–8 d providing the parasite level does not exceed 5–6%; above this level the buffering system becomes inefficient and the culture may die out. If conditions are correct, a culture started at 0·1–0·2% parasitized erythrocytes will reach 5–6% after 96 h; thus fresh cells are added routinely every 4 d.

2° Continuous cultivation of *P. falciparum*
(Haynes *et al.*, 1976)

MEDIUM

TC Medium 199 (with Eagle's modified
salts; Taylor and Baker, 1968)... 100 ml
Glucose .... 200 mg
L glutamine.................. 2 mM
2-mercaptoethanol .... 3×10^5 M
DL-α-tocopherol (emulsified) ...... 3 mg
Gentamycin .... 2·5 mg
TES .... 10 mM

(One part TES acid to 1·4 parts TES sodium salt gives pH 7·35 at 37°C.)

TECHNIQUE

The method is similar to that described by Trager and Jensen (1976) (above).

(i) Thaw cryopreserved infected erythrocytes (Section II, A, 4).
(ii) Wash uninfected erythrocytes and mix with the thawed infected erythrocytes in a ratio of 25:1.
(iii) Dilute the cell suspension with growth medium to give an erythrocyte concentration of ca 5×10^7 ml^{-1} and a parasite density of ca 10^6 ml^{-1}.
(iv) Inoculate cultures and incubate at 37°C in nitrogen containing 3% (v/v) CO_2 and 6·6% (v/v) O_2. 3 ml of culture are added to each 5 cm^2 flask, or 0·2 ml per well of a plastic serological microtitration plate.

Subculture

(i) Every 2–4 d centrifuge the erythrocytes (400 g) and resuspend in a small volume of medium.
(ii) These samples are centrifuged (1300 g, 2 min) in sterile capillary tubes.
(iii) The capillaries are scored and broken to collect the upper brown layer of schizonts (usually 3×10^7 cells), which are added to a new 15 ml subculture containing 3×10^8 human erythrocytes in a 22 cm^2 flask.
(iv) The medium is usually changed every 1–2 d with each culture.

2. Continuous Flow Techniques

The simple dilution systems described above are limited by the number of parasites that can be grown in the cultures before the buffering capacity of the medium becomes inadequate and by the number of times it is practicable to change the medium manually. An efficient method of maintaining constant conditions is urgently needed. Trager and Jernberg (1961) described a special flask and mechanical equipment which provided for the automatic removal of old culture fluid and the addition of fresh medium to cultures in which extracellular malaria parasites were attached as a scum to a thin plasma clot lining the wall of the flask. Such an apparatus could probably be adapted for the cultivation of intracellular parasites. Another approach to maintaining constant conditions for the parasite is to provide the cells with a constant flow of medium throughout the culture period. Continuous flow techniques have been devised that employ cellulose acetate membranes (Trigg, 1968, 1969a) and a tidal flow system (Tiner, 1969). Although these methods showed slight improvements in the growth and multiplication as compared to the dilution systems, they do not appear as efficient as the continuous flow method of Trager (1971) which operates without membranes separating the cells from the medium and has been used subsequently for the long-term culture of *P. falciparum* (Trager, 1976; Trager and Jensen, 1976).

1° Continuous flow technique
(Trager, 1971; Trager and Jensen, 1976).

APPARATUS

The flow apparatus (Fig. 1) is assembled from standard laboratory equipment except for the culture vessel itself (A). This is a vial (58 mm high,

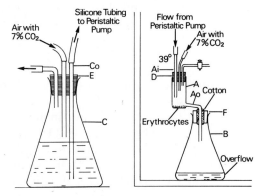

FIG. 1. Diagram of continuous flow apparatus (Trager, 1971; reproduced by kind permission of the author and the Society of Protozoologists).

24 mm outside diameter) with an outflow tube (6 mm outside diameter) that slopes downwards and then turns down vertically (Ao). The height of the outlet above the flat-bottomed vial is critical and should be 2–3 mm. The vial is fitted with a silicone rubber stopper D, bearing a straight tube (Ai) of 3–4 mm outside diameter with a narrowed outlet about 10 mm from the bottom of the vial for delivery of culture medium. The stopper also holds 2 similar tubes plugged with cotton wool for delivery and outlet of the gas mixture. The vertical portion of the outlet tube (Ao) is inserted through cotton wool in a silicone rubber stopper (F) fitted to a 125 ml Ehrlenmeyer flask (B) to collect the overflow. The vial and overflow flask are incubated at 37°C. The inflow tube (Ai) is connected by silicone rubber tubing to a reservoir flask outside the incubator. The tubing passes through a peristaltic pump capable of varying flow. The reservoir flask (C) is a 250 ml Ehrlenmeyer flask with a silicone rubber stopper (E) bearing one tube (Co) of 3–4 mm outside diameter reaching to the bottom of the flask and gas delivery and outlet tubes plugged with cotton wool.

For sterilization, the culture vial A, without its stopper but plugged with cotton wool is assembled together with overflow flask B. The tops of both vial and flask are covered with aluminium foil and the whole autoclaved together. The silicone rubber stopper assembly (D) with the upper end of the tube (Ai), capped with aluminium foil, are autoclaved together. The silicone rubber stopper assembly (E) for the reservoir flask, together with the required length of silicone tubing attached to tube Co which is capped at the distal end with aluminium foil, are also autoclaved together.

TECHNIQUE

(i) 120 ml of medium plus 12 ml of heparinated autologous serum are added to the flask C which is then sealed with sterile assembly A.

(ii) The silicone rubber tubing is inserted into the peristaltic pump and then, still capped with aluminium foil, passed into the incubator.

(iii) The flask C is connected to a supply of $7\% \ CO_2 \ 1\% \ O_2 \ 92\% \ N_2$ gas mixture.

(iv) $1\cdot5$–$2\cdot0$ ml of freshly collected heparinated infected blood are introduced into vial A; the blood must not enter the overflow tube Ao.

(v) The cotton wool plug in vial A is replaced by the stopper assembly D, and the overflow flask B with attached vial A is placed in the incubator.

(vi) The aluminium foil caps are quickly removed from the end of the silicone rubber tubing and from the top of the glass tube Ai and the tubing attached to the inflow tube.

(vii) The gas inlet tubing to the vial A is connected to the gas mixture source and the outlet clamped shut to force the gas to leave the vial by the overflow tube Ao.

(viii) The peristaltic pump is switched on at maximum rate to fill the tubing and to deliver a few drops of medium on top of the blood.

(ix) The pump is then slowed to give a flow rate of about 50 ml d^{-1}, which permits the erythrocytes to settle before the overflow begins. Thus, only a small proportion of the cells should be lost in the initial overflow and none thereafter.

3. *Isolation of Merozoites*

Dennis *et al.* (1975) designed a culture chamber with a polycarbonate sieve to isolate *P. knowlesi* merozoites as they are released from schizonts developing *in vitro*. This provides uncontaminated merozoites in high yield (ca 5×10^{10} merozoites ml^{-1} schizonts). Although merozoite viability diminishes rapidly during the 30 min after isolation, these preparations are valuable for biochemical, physiological and immunological studies.

1° Collection of merozoites
(Dennis *et al.*, 1975)

APPARATUS

The cell sieve apparatus (Fig. 2) consists of a Perspex chamber (97·5 cm diameter) containing a magnetically operated stirrer. The base of the

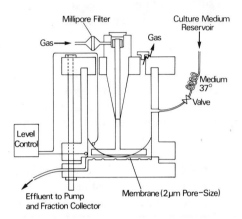

FIG. 2. Diagram of cell sieve apparatus for isolation of *Plasmodium knowlesi* merozoites (Dennis *et al.*, 1975, reproduced by kind permission of the authors and the Cambridge University Press).

chamber is formed by a 2–3 μm pore size polycarbonate sieve (General Electric) stretched by a rubber O-ring in the top section of the chamber which engages with a groove in the supporting base. The membrane is fixed about 1 mm below the under surface of the stirrer blade and forms the upper boundary of a space about 0·5 mm deep connected to a peristaltic pump (MHRE 22 Flow Inducer; Watson Marlow). The cell has an inlet and outlet for the gas mixture. The medium is prewarmed to 37°C and its volume inside the chamber (about 20 ml) is kept constant by a make and break electrical contact between two platinum probes which operate a valve controlling the entry of the medium. The optimum stirrer speed is about 200 rpm and the pumping rate through the sieve about 3 ml min^{-1}. The effluent from the chamber is collected in 10 ml aliquots using a fraction collector.

TECHNIQUE

(i) Bleed monkey when parasitaemia is between 10–30% and the parasites are predominantly schizonts with 6 or more nuclei (10 ml of blood + 1 ml 0·2 M sodium chloride, buffered with 0·1 M phosphate pH 7·4 containing heparin 16 iu ml^{-1} to prevent clotting).

(ii) Centrifuge blood suspension at 350 g for 8 min at 25°C.

(iii) Remove brown schizont layer and suspend in medium 199 (Wellcome Reagents) buffered with 5% (v/v) 0·52 M sodium bicarbonate and containing added glucose (2 mg ml^{-1}) and penicillin (5 iu ml^{-1}).

(iv) Recentrifuge at 350 g for 8 min at 25°C.

(v) Resuspend schizont layer in medium 199 as above (1:20).

(vi) Add approximately 20 ml of the suspension to the cell sieve apparatus and incubate at 37°C whilst passing through the cell a gas mixture of 5% CO_2 95% air.

(vii) Pump medium through the sieve at 3 ml min^{-1}.

If the culture has been inoculated with schizonts having 6 or more nuclei, merozoites start passing through the sieve 30–60 min after initiating the cultures; maximum release occurs after 1–3 h. During maximum merozoite release the effluent contains less than 0·5 ml parasitized erythrocytes provided the chamber is fitted with a 2 μm sieve; with a 3 μm sieve contamination is variable but may be as high as 10%.

4. Cryopreservation of Infected Blood for Cultivation in vitro

The successful long term cultivation of P. falciparum means that it is now possible for certain studies to be made in laboratories which do not have access either to the human infections or to primate hosts which are used as laboratory sources of this parasite. Methods for the cryopreservation of parasitized erythrocytes which allow retrieval and direct use in a culture will aid such studies. In addition, biological material can be preserved and greater flexibility in design of experiments can be obtained. Pavanand et al. (1974) have shown that P. falciparum in erythrocytes frozen in 12% (v/v) dimethylsulphoxide, stored for up to 24 months in liquid nitrogen and thawed by immersion in isotonic saline containing 8% glucose, grew in vitro as well as unfrozen control samples. The major problem is the prevention of haemolysis upon thawing; Diggs et al. (1975) overcame this by slow thawing at room temperature followed by the stepwise reconstitution of isotonicity using sorbitol solutions of varying strengths after freezing in glycerol. Parasites thus frozen could multiply in vitro for 22 days. The viability of the cryopreserved parasite inoculum varied from 20–50% as judged by the numbers of mature trophozoites after 24 h in vitro; apparently only ring stages survived thawing (Haynes et al., 1976). A similar and equally efficient technique has been developed by Wilson et al. (unpublished results) in which haemolysis is prevented by more gradual reconstitution in a longer series of sorbitol solutions of increasing strengths.

1° Cryopreservation and retrieval of parasites
(Diggs *et al.*, 1975)

CRYOPRESERVATION

(i) Centrifuge infected blood (800 g, 10 min) and discard the supernatant fluid.

(ii) Measure packed cell volume (PCV) since generally not more than 2 ml of cells should be processed per tube.

(iii) Suspend packed cells in 1–5 × PCV of cold medium 199 without isoleucine but containing HEPES (Grand Island Biological) and with 2 mg ml^{-1} additional glucose.

(iv) Warm the cell suspension to room temperature for 5 min.

(v) Add 0·4 × PCV of a solution, containing 6·3 M glycerol, 0·14 M sodium lactate, 5 × 10^{-3} M–KCl with sufficient NaH_2PO_4 to adjust the pH to 7·4 ± 0·1, at 0·75 ml min^{-1} with continuous mixing on a mechanical mixer (Vortex: Fisons MSE Scientific Instruments).

(vi) Equilibrate at room temperature for 5 min.

(vii) Add an additional 1·2 × PCV of the freezing mixture at a similar rate to above and allow a further 5 min equilibration.

(viii) Add a further 2·4 × PCV of the freezing mixture and divide into 2 ml aliquots.

(ix) Cool to −70°C overnight.

(x) Transfer to liquid nitrogen for long-term storage.

RETRIEVAL

(i) Thaw at room temperature over a period of 10–20 min.

(ii) Add 2 volumes of 27% sorbitol in 0·01 M phosphate buffer pH 7·4 ± 0·1 at room temperature with continual mixing on a mechanical mixer.

(iii) Equilibrate for 5 min.

(iv) Add 2 volumes of 5% (w/v) sorbitol and repeat 2 and 3.

(v) Centrifuge (300 g, 10 min) and discard supernatant.

(vi) Wash cells with a volume of 5% (w/v) sorbitol equal to that added previously.

(vii) Wash in medium 199 and incubate *in vitro* as described above under the appropriate technique.

2° Cryopreservation and reconstitution of ring-stage parasites of
P. falciparum for culture
(Wilson *et al.*, unpublished data)

CRYOPRESERVATION

(i) Separate erythrocytes from heparinated infected human blood by centrifugation (600 g, 10 min).

(ii) Wash cells in phosphate buffered saline (PBS, pH 7·4) and reconstitute to 40% haematocrit in human AB or another suitable serum.

(iii) Distribute 0·25 ml aliquots in plastic (Sterilin) screwtop ampoules.

(iv) Add an equal volume of either of the two following sterile cryoprotectant solutions dropwise with continuous agitation over one minute at 4°C: either 38 g glycerol, 2·9 g sorbitol, 0·63 g NaCl, water to 100 ml; or 20% (v/v) dimethylsulphoxide (DMSO) in PBS.

(v) Freeze cells in DMSO as soon as possible by plunging into liquid N_2. Glycerolized cells should be held at 4°C for 10 min before plunging in liquid N_2.

RETRIEVAL AND RECONSTITUTION

(i) Transfer ampoule rapidly to waterbath at 37°C and shake gently for 1 min.

(ii) Centrifuge ampoule at 500 g for 3 min.

(iii) Replace supernatant by dropwise addition of 1 ml cold 17·5% (w/v) sorbitol in PBS, with agitation to resuspend cells.

(iv) Transfer specimen to glass tube and add in a similar way 2 ml 10% (w/v) sorbitol, then 2 ml 7·5% (w/v) sorbitol.

(v) Centrifuge (500 g, 3 min) and discard supernatant.

(vi) For minimal lysis, the washing procedure should be continued similarly with the following three additional solutions (all concentrations w/v): 1 ml 7·5% sorbitol + 2 ml 5% sorbitol + 2 ml 2·5% sorbitol; 1 ml 5% sorbitol + 2 ml 2·5% sorbitol + 2 ml culture medium; 1 ml 2·5% sorbitol + 2 ml culture medium.

(vii) Reconstitute final cell pellet to 40% haematocrit in serum appropriate for culture.

5. Tissue Culture Techniques

Recently Speer et al. (1976a,b) have reported multiplication of asexual erythrocytic stages of P. berghei in primary mouse and hamster bone marrow cells and also Leydig testicular tumour cell cultures. Cell line cultures were initiated with 2×10^5 cells in 0·5 ml of appropriate growth medium in Leighton tubes containing coverslips. Either immediately or 24–48 h later, the cell cultures were inoculated with 0·5 ml of cell suspension containing 10^6 parasitized cells. In the Leydig cell cultures intracellular parasites were detected from 15–96 h after inoculation and were most numerous at 36 h. All intracellular stages of parasite development were observed, although gametocytes were few and found only at 36 and 48 h. There was an increase in the number of merozoites between 24–48 h relative to the number of cells. Intracellular parasites were normal in morphology and staining characteristics up to 48 h, but were abnormal after 72 h *in vitro*. It appears that the merozoites were not phagocytosed but actively penetrated the cultured cells, since these cells did not phagocytose uninfected erythrocytes, heat-inactivated parasites or polystyrene spheres. Infected tissue cells showed progressive nuclear and cytoplasmic degeneration which ultimately resulted in their death, but uninfected cells in the same culture appeared normal. Cultures containing cells infected with *P. berghei* were infective to mice after 48 h *in vitro*. In mouse and hamster bone marrow cultures, intracellular parasites were most numerous 7 and 8 days after inoculation, but surprisingly the cultures were noninfective to mice after 3 days *in vitro*.

B. GAMETOCYTOGENESIS

Gametocytogenesis in malaria parasites is poorly understood. There is no special technique as yet for producing gametocytes *in vitro* although their development has been observed in cultures primarily set up for studies on asexual stages. Phillips *et al.* (1976) observed that gametocytes differentiated from ring-stage parasites in cultures of *P. falciparum* isolated from Gambian patients and incubated in medium 199 supplemented (litre^{-1}) with 2 g glucose, 1·98 g sodium bicarbonate, 100 iu streptomycin and 100 iu penicillin. Gametocytes were first morphologically recognizable after 40–50 h in culture and immature crescent forms were observed by 55–65 h. Gametocytes appeared to differentiate from ring stages which were the products of schizogony *in vitro*. Similar observations were made by Mitchell *et al.* (1976), on cultures of *P. falciparum* isolated from *Aotus* monkeys. Smalley (1976), working with human isolates of *P. falciparum*, found that gametocyte development to the crescent form took up to 10 days *in vitro* although even after this time functionally mature gametocytes were not obtained. His technique differed from that of Phillips *et al.* (1976) only by supplementing the medium with 25% (v/v) foetal calf serum and 30 mg% glutamine. However, 59% of the cultures started with blood from patients with only asexual parasites in the peripheral blood produced developing gametocytes compared with 100% of those from patients in whom mature gametocyte production had already started *in vivo*. Clearly, gametocyte development can occur in cultures similar to those used for asexual stages but the factors which trigger it are not understood. (See note on p. 109.)

III. CULTIVATION OF SPOROGONIC STAGES

Very little work apart from that of Ball and his colleagues has been performed on the cultivation of sporogonic stages of malaria parasites (see Taylor and Baker, 1968; Ball, 1972; Vanderberg *et al.*, 1977; Chapter 7, Section VI, of this book). Early work showed that it is possible to maintain different stages of the sporogonic cycle of the avian *P. relictum* in the mosquito stomach, but it has not been possible to obtain the complete cycle of a single culture isolate *in vitro*. The techniques involve tedious microsurgical procedures, and maintenance of sterility is a major problem. More recent techniques have shown that the mosquito stomach is not essential for continued development of the oocyst. Schneider (1968a), starting with 8–9 d old oocysts of *P. gallinaceum*, obtained sporozoites in modified Grace's medium without mosquito cells; later she cultured the same species with Grace's "*Aedes aegypti*" cell line (now shown to be of lepidopteran origin; Greene *et al.*, 1972) and found that oocysts developed more rapidly (Schneider, 1968b). This method using Grace's cell line was adopted by Ball and Chao (1971) who approximately doubled the extent of development of a single isolate of *P. relictum in vitro*.

Little work has been performed on the sporogonic stages of mammalian malarial parasites. Alger (1968) used a simple technique of incubating blood, taken from infected mice and containing male and female gametocytes, in capillary tubes at room temperature, to obtain ookinete formation by *P. berghei*. Yoeli and Upmanis (1968) did not get ookinete formation using this system unless an aqueous extract of *Anopheles stephensi* midgut was added to the cultures. Ookinete formation without additions could be obtained if the blood was first ingested by a mosquito from an infected host and then rapidly ejected during the act of engorgement. Ookinete formation of *P. berghei* has been reported in primary cultures of *Anopheles stephensi* and epithelial cell lines from fat-head minnow (Rosales-Ronquillo and Silverman, 1974); they were grown intracellularly in fat-head minnow cells by Rosales-Ronquillo *et al.* (1974). Shapiro *et al.* (1975) were unable to obtain high yields of ookinetes by this method. It is not clear if ookinete formation of other species can be obtained by such procedures.

IV. CULTIVATION OF EXOERYTHROCYTIC STAGES

Although the exoerythrocytic stages of avian malarial parasites have been grown routinely in tissue culture for several years (Taylor and Baker, 1968), little work has been performed on developing systems for mammalian malaria parasites. These, unlike the avian species, grow within liver parenchyma cells, which are notoriously difficult to maintain *in vitro*. Recently, research in the field of mammalian liver tissue culture has advanced a great deal (Seglen, 1976). Long-term cultures started from isolated adult liver parenchymal cells of rodents have been shown to maintain some of the functions of the cells of origin, including the induction of tyrosine aminotransferase (Gerschenson *et al.*, 1970), retention of specific cell antigens (Iype *et al.*, 1972), synthesis of albumin (Kaighn and Prince, 1971) and the retention of certain cytological characters (Williams *et al.*, 1971). However, since these cells are dividing, unlike adult liver cells *in vivo*, they lack many of the functions of liver parenchymal cells *in vivo* (Lambiotte *et al.*, 1972). Waymouth *et al.* (1971) have established permanent cell lines from liver of mouse foetuses of 16 d gestation in a completely defined medium and one of these, FL83B, exhibits striking morphological similarities to liver parenchymal cells and produces 12 mouse serum proteins, including albumin and high density lipoprotein.

Beaudoin *et al.* (1974) have used Waymouth's technique to infect mouse embryo liver cells with merozoites of exoerythrocytic stages of *P. lophurae* and *P. fallax*. These mammalian cell cultures were not characterized and were not completely homogeneous, but development of the merozoites into mature schizonts was obtained after 48 h in the parenchyma-like cells, and to a lesser extent in fibroblastic cells. These results obtained with avian parasites in cells of mammalian origin show that the growth of the exoerythrocytic stages of mammalian malaria parasites *in vitro* may be possible. The emphasis must be placed now on developing techniques that will provide sporozoites of a mammalian parasite with which to infect liver cells grown *in vitro*. Cultures could be inoculated with sporozoites taken either directly from the salivary glands of the mosquito by sterile dissection, or from blood of the host inoculated 30 min previously with large numbers of sporozoites. Both methods have technical problems but the production of viable sporozoites may be aided by the addition of 1% albumin, which has a stimulatory effect on motility and infectivity of *P. berghei* sporozoites (Vanderberg, 1974).

An alternative to using sporozoites as the inoculum is the production of infected liver cell cultures from heavily infected livers by the methods developed for normal adult cultures.

Organ culture techniques have also been used successfully for the growth of exoerythrocytic stages *in vitro* (Yoeli *et al.*, 1968). Exoerythrocytic stages of *P. berghei* were grown to mature schizonts *in vitro* in isolated perfused tree-rat (*Thamnomys surdaster*) liver. Mature exoerythrocytic schizonts were found in the liver parenchymal cells and showed all the morphological characteristics of parasites grown *in vivo*.

REFERENCES

Alger, N. E. (1968). *In vitro* development of *Plasmodium berghei* ookinetes. *Nature, London* **218**, 774.

Anderson, C. R. (1953). Continuous propagation of *Plasmodium gallinaceum* in chicken erythrocytes. *American Journal of Tropical Medicine and Hygiene* **2**, 234–242.

Anfinsen, C. B., Geiman, Q. M., McKee, R. W., Ormsbee, R. A. and Ball, E. G. (1946). Studies on malaria parasites VIII: Factors affecting the growth of *Plasmodium knowlesi in vitro. Journal of Experimental Medicine* **84**, 607–621.

Ball, G. H. (1972). Use of invertebrate tissue culture for the study of plasmodia. *In* "Invertebrate Tissue Culture" (C. Vago, ed.), Vol. 2, pp. 321–342. Academic Press, New York and London.

Ball, G. H. and Chao, J. (1971). The cultivation of *Plasmodium relictum* in mosquito cell lines. *Journal of Parasitology* **57**, 391–395.

Bass, C. C. and Johns, J. M. (1912). The cultivation of malaria plasmodia (*Plasmodium vivax* and *Plasmodium falciparum*) *in vitro. Journal of Experimental Medicine* **16**, 567–579.

Beaudoin, R. L., Strome, C. P. A. and Clutter, W. G. (1974). Cultivation of avian malaria parasites in mammalian liver cells. *Experimental Parasitology* **36**, 355–359.

Bertagna, P., Cohen, S., Geiman, Q. M., Haworth, J., Königk, E., Richards, W. H. G. and Trigg, P. I. (1972). Cultivation techniques for the erythrocytic stages of malaria parasites. *Bulletin of the World Health Organization* **47**, 357–373.

Butcher, G. A. and Cohen, S. (1971). Short-term culture of *Plasmodium knowlesi. Parasitology* **62**, 309–320.

Butcher, G. A., Mitchell, G. H. and Cohen, S. (1973). Mechanism of host specificity in malaria infection. *Nature, London* **244**, 40–42.

Coombs, G. H. and Gutteridge, W. E. (1975). Growth *in vitro* and metabolism of *Plasmodium vinckei chabaudi. Journal of Protozoology* **22**, 555–560.

Dennis, E. D., Mitchell, G. H., Butcher, G. A. and Cohen, S. (1975). *In vitro* isolation of *Plasmodium knowlesi* merozoites using polycarbonate sieves. *Parasitology* **71**, 475–481.

Diggs, C. L., Joseph, K., Flemmins, B., Snodgrass, R. and Hines, F. (1975). Protein synthesis *in vitro* by cryopreserved *Plasmodium falciparum. American Journal of Tropical Medicine and Hygiene* **24**, 760–763.

Geiman, Q. M., Siddiqui, W. A. and Schnell, J. V. (1966). *In vitro* studies on erythrocytic stages of Plasmodia: Medium improvement and results with several species of malaria parasites. *Military Medicine* **131** (Supplement), 1015–1025.

Gerschensen, L. E., Andersson, M., Molson, J. and Okigaki, T. (1970). Tyrosine transaminase induction by dexamethasone in a new rat liver cell line. *Science, New York* **170**, 859–861.

Greene, A. E., Chaney, J., Nichols, W. W. and Coriell, L. L. (1972). Species identity of insect cell lines. *In Vitro* **7**, 313–322.

Haynes, J. D., Diggs, C. L., Hines, F. A. and Desjardins, R. E. (1976). Culture of human malaria parasites, *Plasmodium falciparum. Nature (London)* **263**, 767–769.

Iype, P. T., Baldwin, R. W. and Glaves, D. (1972). Culture from adult rat liver cells II. Demonstration of organ-specific cell surface antigens on cultured cells from normal liver. *British Journal of Cancer* **26**, 6–9.

Kaighn, M. E. and Prince, A. M. (1971). Production of albumin and other serum proteins by clonal cultures of normal human liver. *Proceedings of the National Academy of Sciences (USA)* **68**, 2396–2400.

Lambiotte, M., Susor, W. A. and Cohn, R. D. (1972). Morphological and biochemical observations on mammalian liver cells in culture. Isolation of a clonal strain from rat liver. *Biochimie* **54**, 1179–1187.

Mitchell, G. H., Butcher, G. A. and Cohen, S. (1976). A note on the rapid development of *Plasmodium falciparum* gametocytes *in vitro*. *Transactions of the Royal Society of Tropical Medicine and Hygiene* **70**, 12–13.

Pavanand, K., Permanich, B., Chuanak, P. and Sookto, P. (1974). Preservation of *Plasmodium falciparum*-infected erythrocytes for *in vitro* cultures. *Journal of Parasitology* **60**, 537–539.

Phillips, R. S., Trigg, P. I., Scott-Finnigan, T. J. and Bartholomew, R. K. (1972). Culture of *Plasmodium falciparum in vitro*: a subculture technique used for demonstrating antiplasmodial activity in serum from some Gambians, resident in an endemic malarious area. *Parasitology* **65**, 525–535.

Phillips, R. S., Wilson, R. J. M. and Pasvol, G. (1976). Differentiation of gametocytes of *Plasmodium falciparum in vitro*. *Transactions of the Royal Society of Tropical Medicine and Hygiene* **70**, 286.

Richards, W. H. G. and Williams, S. G. (1973). The removal of leucocytes from malaria infected blood. *Annals of Tropical Medicine and Parasitology* **67**, 249–250.

Rosales-Ronquillo, M. C. and Silverman, P. H. (1974). *Plasmodium berghei* ookinete formation in vector and non-vector cell substrates. *Proceedings of the 3rd International Congress of Parasitology, München* **1**, 124–125.

Rosales-Ronquillo, M. C., Nienaber, G. and Silverman, P. H. (1974). *Plasmodium berghei* ookinete formation in a non-vector cell line. *Journal of Parasitology* **60**, 1039–1040.

Schneider, I. (1968a). Cultivation *in vitro* of *Plasmodium gallinaceum* oocysts. *Experimental Parasitology* **22**, 178–186.

Schneider, I. (1968b). Cultivation of *Plasmodium gallinaceum* oocysts in Grace's cell strain of *Aedes aegypti* (L.). *Proceedings of the 2nd International Colloquium on Invertebrate Tissue Culture, Tremezzo* 247–253.

Seglen, P. O. (1976). Preparation of isolated rat liver cells. *In* "Methods in Cell Biology" (D. M. Prescott, ed.), Vol. 13, pp. 29–83. Academic Press, New York and London.

Shapiro, M., Espinal-Tejada, C. and Nussenzweig, R. S. (1975). Evaluation of a method for *in vitro* ookinete development of the rodent malarial parasite. *Journal of Parasitology* **61**, 1105–1106.

Siddiqui, W. A. and Schnell, J. V. (1973). Use of various buffers for *in vitro* cultivation of malarial parasites. *Journal of Parasitology* **59**, 516–519.

Siddiqui, W. A., Schnell, J. V. and Geiman, Q. M. (1970). *In vitro* cultivation of *Plasmodium falciparum*. *American Journal of Tropical Medicine and Hygiene* **19**, 586–591.

Smalley, M. E. (1976). *Plasmodium falciparum* gametocytogenesis *in vitro*. *Nature, London* **264**, 271–272.

Smalley, M. E. and Butcher, G. A. (1975). The *in vitro* culture of the blood stages of *Plasmodium berghei*. *International Journal for Parasitology* **5**, 131–132.

Speer, C. A., Silverman, P. H. and Schiewe, S. G. (1976a). Cultivation of the erythrocytic stages of *Plasmodium berghei* in primary bone marrow cells. *Journal of Parasitology* **62**, 657–663.

Speer, C. A., Silverman, P. H. and Schiewe, S. G. (1976b). Cultivation of the erythrocytic stages of *Plasmodium berghei* in Leydig cell tumor cultures. *Zeitschrift für Parasitenkunde*, **50**, 237–244.

Taylor, A. E. R. and Baker, J. R. (1968). "The Cultivation of Parasites *in vitro*", pp. 53–76. Blackwell, Oxford and Edinburgh.

Tiner, J. D. (1969). Growth of *Plasmodium berghei* in parallel perfused blood cell suspensions. *Military Medicine* **134**, 945–953.

Trager, W. (1971). A new method for intraerythrocytic cultivation of malaria parasites (*Plasmodium coatneyi* and *P. falciparum*). *Journal of Protozoology* **18**, 239–242.

Trager, W. (1976). Prolonged cultivation of malaria parasites (*Plasmodium coatneyi* and *P. falciparum*). *In* "Biochemistry of Parasites and Host-Parasite Relationships" (H. Van den Bossche, Ed.), pp. 427–434. North-Holland Publishing Company, Amsterdam, New York and Oxford.

Trager, W. and Jensen, J. B. (1976). Human malaria parasites in continuous culture. *Science, New York* 193, 673–675.

Trager, W. and Jernberg, N. A. (1961). Apparatus for change of medium in extracellular maintenance *in vitro* of an intracellular parasite (malaria). *Proceedings of the Society for Experimental Biology and Medicine* 108, 175–178.

Trigg, P. I. (1968). A new continuous perfusion technique for the cultivation of malaria parasites *in vitro*. *Transactions of the Royal Society of Tropical Medicine and Hygiene* 62, 371–378.

Trigg, P. I. (1969a). The use of proprietary tissue culture media for the cultivation *in vitro* of the erythrocytic stages of *Plasmodium knowlesi*. *Parasitology* 59, 925–935.

Trigg, P. I. (1969b). Some factors affecting the cultivation *in vitro* of the erythrocytic stages of *Plasmodium knowlesi*. *Parasitology* 59, 915–924.

Trigg, P. I. (1975). Invasion of erythrocytes by *Plasmodium falciparum in vitro*. *Parasitology* 71, 433–436.

Trigg, P. I. and Gutteridge, W. E. (1971). A minimal medium for the growth of *Plasmodium knowlesi in vitro*. *Parasitology* 62, 113–123.

Trigg, P. I. and McColm, A. A. (1976). Cultivation of the erythrocytic stages of *Plasmodium berghei berghei in vitro*. *Parasitology* 73, xxxiii–xxxiv.

Trigg, P. I. and Shakespeare, P. G. (1976a). The effect of incubation *in vitro* on the susceptibility of monkey erythrocytes to invasion by *Plasmodium knowlesi*. *Parasitology* 73, 149–160.

Trigg, P. I. and Shakespeare, P. G. (1976b). Factors affecting the long-term cultivation of the erythrocytic stages of *Plasmodium knowlesi*. *In* "Biochemistry of Parasites and Host-Parasite Relationships" (H. Van den Bossche, Ed.), pp. 435–440. North-Holland Publishing Company, Amsterdam, New York and Oxford.

Vanderberg, J. (1974). Studies on the motility of *Plasmodium* sporozoites. *Journal of Protozoology* 21, 527–537.

Vanderberg, J. P., Weiss, M. M. and Mack, S. R. (1977). *In vitro* cultivation of sporogonic stages: a review. *Bulletin of the World Health Organization* 55, 377–392.

Waymouth, C., Chen, H. W. and Wood, B. G. (1971). Characteristics of mouse liver parenchyma cells in chemically defined media. *In Vitro* 6, 371.

Williams, C. M., Weisburger, E. K. and Weisburger, J. H. (1971). Isolation and long term culture of epithelial-like cells from rat liver. *Experimental Cell Research* 69, 106–112.

Yoeli, M and Upmanis, R. S. (1968). *Plasmodium berghei* ookinete formation *in vitro*. *Experimental Parasitology* 22, 122–128.

Yoeli, M., Upmanis, R. S., Kronman, B. and Schoenfeld, C. (1968). Growth of the pre-erythrocytic tissue stages of *Plasmodium berghei* in isolated, perfused tree rat liver. *Nature, London* 216, 1016–1017.

Note added in proof

Carter and Beach (1977: Gametogenesis in culture by gametocytes of *Plasmodium falciparum*. *Nature, London* 270, 240–241) have reported that *P. falciparum* gametocytes grown from parasites maintained by continuous culture *in vitro* can be stimulated to undergo gametogenesis (exflagellation). Immature gametocytes in a culture maintained in RPMI 1640 (Section II, A, 2, 1°) without the usual dilution, matured in a further 8 d when 50% v/v human AB serum was substituted for the usual 10% serum. Mature gametocytes were also grown in 2 of 3 cultures containing only 10% serum. Prolonged maintenance of cultures for at least 2 weeks was essential; the success may also be related to the fact that the parasites had been recently isolated from man. Exflagellation was induced by suspending the mature gametocytes in foetal calf serum adjusted to pH 8·0 with 1·5% w/v NaHCO$_3$ or in human serum similarly adjusted to pH 8·2.

Chapter 6

Cell and Tissue Culture

J. F. Ryley and R. G. Wilson

Imperial Chemical Industries Ltd., Pharmaceuticals Division, Alderley Park, Macclesfield, Cheshire, England

I. INTRODUCTION

Why grow parasites in tissue culture? Some parasites—e.g. coccidia—are naturally intracellular and cannot be grown in the absence of suitable host cells, while others, though extracellular—e.g. some trypanosomes—have so far only been grown *in vitro* in the presence of cultured cells. Cultivation of protozoa in tissue culture is not an end in itself, although some published studies may give the impression that it is, but a powerful tool in investigations of various aspects of parasitology. Several different potential uses of the tissue culture system will be illustrated by examples from the coccidia, although analogous studies have been or could be carried out with other organisms.

Roberts *et al.* (1971) used light and electron microscopy to study the penetration of *Eimeria larimerensis* sporozoites into cultured bovine kidney (MDBK) cells, while Jensen and Hammond (1975) appear to have demonstrated that invasion of MDBK cells by *E. magna* sporozoites involves invagination rather than penetration of the host cell membrane. Cinephotomicrography using a warm stage and a suitable culture chamber allows study of a process which normally takes place in the hidden depths of the host. Again, the cell culture model is useful to study subsequent stages of the intracellular life cycle of the parasite, including problems associated with sexual differentiation. Whole parasites can be observed in stained preparations rather than sectioned material from infected tissues, while extended observation of individual parasites can be carried out in cultures using phase contrast or interference contrast optics and recorded by time-lapse cinephotomicrography. Thus Speer and Hammond (1972) reported a study on gametogony and oocyst formation of *E. magna* in cultured cells. The tissue culture system is particularly useful for studies of this type involving a change from asexual to sexual reproduction since it is possible to isolate parasites from many of the host-mediated stimuli present in the natural situation (e.g. Bedrník, 1970, using *E. tenella*). Parasite development in tissue culture may however differ in timing, size and form from that in the natural situation, and for parasites with complicated developmental patterns, *in vitro* studies may only increase the confusion (e.g. Speer and Hammond, 1971, and Ryley and Robinson, 1976, for *in vitro* and *in vivo* studies on the life cycle of *E. magna*).

Immunity in coccidiosis is a very involved topic, with much to be learned concerning the significance of localized and disseminated immune responses. Certain aspects of the interaction between parasite and individual immune factors can be conveniently studied in cell culture in the absence of complications from the host; thus Long and Rose (1972) report an investigation of the effect of serum antibodies on cell invasion by sporozoites of *E. tenella*, while their later studies (Rose and Long, 1976) deal with interactions between parasites and phagocytic cells.

The biochemistry of parasites in general and of intracellular parasites in particular is a topic where information is somewhat fragmentary. A cell culture system has been used by Gutteridge *et al.* (1969) for bulk production of amastigotes of *Trypanosoma cruzi* for biochemical studies, and hopefully development of techniques will make possible the production of other parasite material in quantity. Nevertheless, with systems currently available it is possible to study various aspects of parasite biochemistry. Thus the cell culture model is eminently suited for histochemical studies at the light and electron microscope level; Fayer and Thompson (1975) studied *Sarcocystis* in cultures of embryonic bovine tracheal cells. The use of radioactive metabolites makes possible the study of various anabolic processes; for example, Ouellette *et al.* (1973) followed the incorporation of radioactive pyrimidine nucleosides into DNA and RNA of *E. tenella* in cell culture. Investigation of the nutritional requirements of intracellular parasites is difficult at the best of times, but simplification of cell culture techniques and media should allow more meaningful studies to be made. Strout (1975) cultivated *E. tenella* in embryonic chick kidney cells in a chemically defined medium, while Sofield and Strout (1974) attempted to study the amino acid requirements of *E. tenella* in cell culture. Studies of metabolite antagonism in cell culture (e.g. Ryley and Wilson, 1972a) are also relevant to nutritional studies.

Drug screening and evaluation can readily be carried out in cell cultures. Drug screening is a process of investigating very large numbers of compounds on a regular basis (e.g. weekly) to see if they interact with the parasite. A simple "yes/no" answer is all that is initially required, to indicate whether more detailed investigation is warranted. With such a programme, time, effort, animal accommodation, labour and particularly the amount of compound available for test are important factors to consider, and a cell culture screen or pre-screen has many attractions. The relevance of results obtained in such a system to the behaviour of a compound in the natural host is however open to question and Ryley and Wilson (1976) have discussed *in vitro/in vivo* correlations in drug screening with *E. tenella*. Drug evaluation studies may be enhanced by observations made in cell culture. Morphological observations may be made on whole parasites under the influence of drugs, the point of action in the life cycle can often be identified, and the possibility of recovery following drug withdrawal studied. Where a number of different species of the same genus may be involved in the natural situation, cell culture studies make possible the separation of differences in species susceptibilities to the drug from differences due to uneven distribution and metabolism of the drug in the tissues of the host; see Ryley and Wilson (1972b) for such a study with nine drugs and three species of coccidia.

We will make preliminary remarks about tissue culture, describe in full techniques which we use successfully with coccidia—which will also serve to illustrate the general principles involved—and then refer briefly to other parasites to which the tissue culture technique is being applied. In many of these instances, the systems are far from established, and some are of but limited interest. It is hoped that the references given plus the techniques illustrated more fully with the coccidia will be of help to the interested reader.

II. TISSUE CULTURE METHODS

General principles and techniques of cell and tissue culture are discussed in detail by Paul (1970) and Whitaker (1972). Descriptions of a wide range of techniques and applications are given by Kruse and Patterson (1973).

A. HARDWARE

Cells are capable of growing on a variety of surfaces including glass, polystyrene, cellulose and metals. In recent years glass vessels have largely been replaced by disposable polystyrene vessels, specially treated to provide a suitable surface. The purpose of an experiment will to some extent determine the hardware to be employed; a selection is shown in Fig. 1. Many workers have grown cells on glass coverslips in stationary Leighton tubes. To facilitate handling of large numbers, special racks need to be employed. Where coverslips are to be used, some of the disposable multiwell plastic plates shown are equally satisfactory. Indeed in certain circumstances they are preferable, since replicate coverslips can be cultured within a single container under identical cultural conditions. Such plates are admirable for large throughput drug screening programmes, where the cells can be grown directly on the bottom of the wells and the monolayer stained for examination using an inverted microscope.

For continuous observation of parasite development, possibly involving time lapse cinephotomicrography, a specialized chamber with adequate optical properties and facilities for gas and medium exchange will be required. The Sykes–Moore tissue culture chamber has been used for some years for this purpose, although there are now available from Sterilin disposable polystyrene chambers based on that originally described by Cruickshank *et al.* (1959).

B. HOST CELLS

Primary cultures may be derived from a large number of organs from almost any species. The choice of tissue is often decided by the requirement of the experiment and the age, size and availability of the tissue. When cultured, cells become undifferentiated, losing their specialized functions, and therefore cannot be expected to display the properties of intact tissues. In general, the younger the tissues, the better they will grow *in vitro*. Embryonic tissues grow particularly well, but the organs are small. Organs from post-natal animals are larger but grow less well with increasing age,

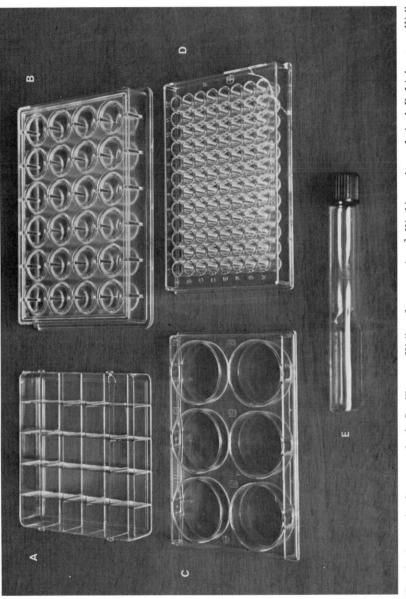

FIG. 1. Assorted culture vessels. A. Sterilin tray. Well surface area, 4 cm². Working volume, 1–4 ml. B. Linbro tray. Well surface area, 1·8 cm². Working volume, 1–2·5 ml. C. Mini Petri plate. Well surface area, 9·6 cm². Working volume, 2–5 ml. D. Microtiter plate. Well surface area, 0·38 cm². Working volume, 0·1–0·25 ml. E. Leighton tube. Coverslip area, 3–5 cm². Working volume, 2 ml. (Reproduced by permission of the Cambridge University Press from *Parasitology*, volume 73, part 2.)

probably because of the increasing amounts of connective tissue. Hence, kidneys are often chosen for primary cultures because they contain relatively small amounts of connective tissue.

The tissue is disaggregated by physical methods and/or enzymic digestion, usually with trypsin. The removal of primary culture cells from the surface of the vessel in which they are grown and their transfer into new vessels marks the beginning of a cell line. Such stripping of cells from the culture vessel is usually accomplished by trypsin or versene (ethylene diamine tetra-acetic acid, disodium salt; EDTA) alone or in combination. Whether further subcultures are feasible depends on the age of the donor organ or tissue and is characteristic of the species. Once a cell line demonstrates the potential for unlimited subculture it is said to have become "established".

C. MEDIUM

The uninitiated may be confounded by the extensive range of commercially available media, many of which have been formulated for specific cells or cell types. (For the origins and modifications of commercially available media see review by Morton, 1970.) All media contain a balanced salt solution (BSS) to maintain osmotic pressure and hydrogen ion concentration (see Chapter 1). The two most commonly used are Earle's (Chapter 1: IV, B, 1, f), which is strongly buffered with sodium bicarbonate, and Hanks's (Chapter 1: IV, B, 1, g) which contains less bicarbonate, being designed to equilibrate with air. Glucose is normally added as a carbohydrate source. Serum is an essential addition to most media, although serum-free chemically defined media have been developed (e.g. by Gorham and Waymouth, 1965). Foetal calf serum (FCS) generally has greater growth promoting properties than that from older animals, but is considerably more expensive. A variety of supplements may be added to this basic medium to improve cell development and function, e.g. specific amino acids, vitamins, lactalbumin hydrolysate or tryptose phosphate broth. A BSS with lactalbumin hydrolysate (LAH) and calf serum is a simple, cheap and effective medium for growing primary cultures. Established cell lines are most often grown in Eagle's minimum essential medium (MEM).

Comprehensive details of the composition and storage life of commercially available culture media are contained in the Gibco Bio-Cult catalogue and the Flow Manual. The latter also describes the characteristics and culture conditions for cell lines supplied by the American Type Culture Collection (ATCC) and available from Flow Laboratories (see Appendix).

D. USE OF DRUGS

Several workers have described the use of cell cultures for studying the effects of drugs on various parasites and have employed a variety of methods for drug formulation. In our experience, it is not essential to use a solution; a fine suspension is normally adequate. If small numbers of drugs are being tested, it may be possible or advantageous to compare the solubility of the drugs in a number of solvents. When a wide variety of compounds is being tested, as in a large throughput random screen, it is more convenient to use a single solvent. We have found dimethylsulphoxide (DMSO) to be satisfactory; provided its final concentration is less than 1%, no interference with host cell or parasite development has been observed. Other commonly used solvents are Tween 20 or 80, ethanol and Dispersol OG (a dispersing agent based on polyglyceryl ricinoleate; ICI). For insoluble compounds, physical methods of dispersion often employed are ball-milling, shaking with glass beads, grinding in a tissue disintegrator with a Teflon pestle (Tri-R; Camlab), or sonication.

III. COCCIDIA

Since Patton's (1965) initial observation of incomplete development of *E. tenella* in primary bovine kidney cells, numerous workers have attempted to obtain the development of *Eimeria* and *Isospora* in cell culture. These investigations, involving a wide variety of primary cell cultures and established cell lines, were summarized comprehensively by Doran (1973).

Thus far, only *E. tenella* has completed the endogenous cycle from sporozoite to oocyst in cell culture (Doran, 1970; Strout and Ouellette, 1970), although gametocytes and oocysts have been observed in cultures inoculated with merozoites obtained from *in vivo* infections of *E. bovis* (Speer and Hammond, 1973), *E. brunetti* (Shibalova, 1970) and *E. magna* (Speer and Hammond, 1972). It is interesting to note that Doran (1974) was unable to obtain any further development when merozoites produced *in vitro* were inoculated into fresh cultures of chick kidney cells. Several species, including *E. brunetti*, *E. necatrix*, *Isospora lacazei* and/or *I. chloridis*, will develop to mature second generation schizonts following sporozoite inoculation. Although it is possible to obtain some development of coccidia in cell line cultures from animals other than the natural host, numerous species have progressed farther into the life cycle when cultures from the natural host were used (see, e.g., Doran and Augustine, 1973). A comparative study by Doran *et al.* (1974) showed that different strains of the same species (*E. tenella*) differ in their capacity to infect, develop and produce oocysts under the same cultural conditions. This observation suggests that for a specific type of study a particular strain might be most suitable.

A. EIMERIA TENELLA

Although *E. tenella* will grow in embryonic and post-embryonic primary cells as well as many cell lines, complete development has most readily been accomplished in primary cultures. The preparation of the latter and the growth of *E. tenella* will be described.

1° Preparation of primary chick kidney cell cultures

(i) Kill several 2–3 week old chickens with chloroform.

(ii) Swab ventral surfaces with 70% ethanol and, using sterile scissors and forceps, open to reveal the kidneys.

(iii) Again using sterile instruments, remove kidneys to sterile beaker containing about 50 ml sterile phosphate buffered saline (PBS; Chapter 1: IV, B, 1, b).

(iv) Transfer kidneys to sterile glass Petri dish, and with sterile forceps and scissors, remove adhering fat, membranes and blood vessels.

(v) Transfer "cleaned" kidneys to another sterile glass Petri dish and mince finely with two scalpels.

(vi) Transfer minced tissue to a sterile 50 ml flask and wash with 25 ml PBS; allow cells to settle and remove supernatant. Repeat washing until supernatant is clear.

(vii) Wash once with 10 ml 0.25% trypsin solution in PBS and then trypsinize with stirring (magnetic) at 37°C for periods of approximately 15, 10, 3 and 3 min using 4 lots of 5–10 ml trypsin in PBS (T_1, T_2, T_3, and T_4).

(viii) Retain the supernatants from T_3 and T_4, which should contain mainly cell aggregates, in a tube containing 1 ml FCS as trypsin inhibitor.

(ix) Centrifuge to deposit cells, wash by centrifugation with PBS and suspend cells in 20 ml growth medium (Hanks's BSS with 0.125% LAH, 5% FCS and 0.15% $NaHCO_3$—or 0.035% $NaHCO_3$ for closed vessels with air rather than CO_2-air).

(x) Filter suspension through sterilized wire mesh (or muslin) into a beaker or flask to remove any large clumps of undegraded tissue.

(xi) Prepare a 1:3 dilution of the suspension in 0.5% trypan blue in PBS and count viable (unstained) aggregates using a haemocytometer.

(xii) Prepare a suspension in growth medium and dispense into culture vessels using an automatic pipetting syringe (Horwell or Becton, Dickinson). The suspension should be stirred throughout dispensing to ensure even distribution of the cells. The cell concentration and volume inoculated will vary according to the culture vessel being used. In general, $1\cdot5 \times 10^3$ viable cell aggregates cm^{-2} should produce a semiconfluent monolayer within 3 d.

(xiii) Incubate culture plates at 41°C in an atmosphere of $5\% CO_2$–95% air for 2–3 d until cells have formed a semi-confluent monolayer. A CO_2 incubator can be used, but a convenient alternative is to put the culture plates in a plastic "sandwich" box which can be gassed and sealed with tape and then placed in an ordinary incubator.

2° Preparation of sporozoites for inoculation of cell cultures

(i) Suspend sporulated oocysts (not more than 3 ml packed volume) in 17 ml $0\cdot85\%$ saline and 10 ml sodium hypochlorite solution (30% NaOCl; not less than 14% available chlorine). Leave at room temperature for 20–30 min, mixing at intervals.

(ii) Centrifuge (5 min, 500 g); sterilized oocysts float on the surface, while unwanted degraded debris is sedimented.

(iii) Remove oocyst layer with a wide-bore Pasteur pipette and wash with sterile water by centrifugation until free from hypochlorite.

(iv) Suspend oocysts in saline and rupture to release the sporocysts. This can be accomplished by shaking with grade 7 ballotini beads ($0\cdot5$ mm diam.), by use of a Teflon coated tissue grinder (Tri-R; Camlab) or with an ultrasonic disintegrator. The breakage has to be carried out under sterile conditions, and we find mechanical shaking with glass beads the quickest and most convenient method. Shaking machines are difficult to obtain; in the absence of anything else, a Mickle disintegrator is satisfactory, although the volume which can be used is small. As a last resort, a Whirlimixer (Fisons) will suffice. The efficiency of a tissue homogenizer will depend on the speed of rotation of the pestle and the tightness of its fit. Whatever system is used, the resultant suspension will contain unbroken oocysts, intact sporocysts and completely smashed parasites; the particular system will have to be evaluated by the operator to obtain optimal production of free sporocysts. We find 20 s shaking with glass beads gives a satisfactory level of breakage.

(v) Spin to deposit sporocysts, resuspend in saline or PBS containing $0\cdot25\%$ trypsin and 5% chick bile (or 1% bile salt), and adjust pH to $7\cdot6$. Incubate at 41°C until satisfactory excystation is achieved (30–90 min with *E. tenella*).

(vi) Wash by centrifugation and resuspend in pH 8 buffer (Wagenbach, 1969) and separate sporozoites by passage through a column of glass beads (200 μm dia, Superbrite 100–5005 obtainable from Minnesota Mining and Manufacturing Company). The special buffer contains in 1 litre: NaCl 6·0 g, KCl 0·28 g, CaCl$_2$ 0·16 g, KHCO$_3$ 0·14 g and Tris 2·02 g; pH is adjusted to 8·0 with HCl. (Some workers consider differential centrifugation to remove debris and intact oocysts and sporocysts to be adequate.)

Column sterilization can be achieved by autoclaving, but in practice we have found that cleaning of beads and adequate sterile conditions are produced by treating the packed column with concentrated nitric acid overnight and then washing through with sterile distilled or deionized water to remove the acid. The column is then washed with sterile pH 8 buffer until the effluent attains pH 8, at which time the column is ready for use. After use the debris should be eluted from the top of the column and the beads washed *in situ* with tap water. Nitric acid is then used for final cleaning and resterilization. If the column shows any signs of blockage, it should be dismantled, the glass wool replaced and the beads repacked.

(vii) Wash by centrifugation and resuspend sporozoites in growth medium.

(viii) When repeated culture work is envisaged, it is convenient to prepare a large batch of sporozoites and store in liquid nitrogen, using one or two ampoules from the stock for each experiment. DMSO is added to a suspension of sporozoites in growth medium (with 5% FCS) to give a final concentration of 10%. We use up to 5×10^7 sporozoites ml^{-1}, dispensing 1 ml amounts into screw-capped polypropylene vials. These are frozen at a cooling rate of 1°C min^{-1} over the range of 0 to -30°C and stored in liquid nitrogen.

(ix) To inoculate cultures, remove original growth medium and replace with sporozoite suspension (using an automatic pipetting syringe if large numbers of vessels are used). To prevent overgrowth of cells, it is advisable to reduce the level of FCS in the medium to 2–3% at this stage. An inoculum sufficient to give 25–50,000 sporozoites cm^{-2} surface area should be adequate for *E. tenella*. Incubate at 41°C. If the inoculum has been purified by passing through a glass bead column, there should be no need to remove the suspending medium. However, if a dirty or a particularly large inoculum has been used, it may be preferable to remove the medium after 24 h and replace with fresh. This will provide a more favourable environment and allow the intracellular sporozoites to develop free from toxic factors due to degenerate extracellular parasites and debris.

Using primary chick kidney cultures, mature first generation schizonts develop between 48 and 72 h after inoculation with *E. tenella* sporozoites and mature second generation schizonts can be found from 90 h onward. If it is desired to obtain oocyst production, less confluent monolayers should be used at the time of inoculation. It is also preferable to use a larger quantity of medium to maintain pH and nutritional requirements without having to change the medium during cultivation.

B. OTHER COCCIDIA

Similar methods can be used for other species of coccidia or coccidia-like parasites, although details of host cell and method of producing inocula will vary. Sporozoites of other species of *Eimeria* or *Isospora* may be obtained by the methods described. For some purposes it may be desirable to inoculate cultures with merozoites derived from infections in the host. Speer and Danforth (1976) found it satisfactory in the case of *E. magna* to scrape the relevant part of the intestinal mucosa of the rabbit into saline containing 4000 iu ml^{-1} penicillin G + 5000 μg ml^{-1} streptomycin. After stirring gently for an unspecified period, the suspension is centrifuged (200 g, 1 min) to sediment grosser debris and the supernatant containing merozoites used to inoculate cell cultures. *Toxoplasma gondii* is best obtained from peritoneal exudate of experimentally infected mice. *Besnoitia jellisoni* can be obtained by removing cysts from the peritoneum of infected mice and crushing them in a tissue grinder under sterile conditions, while zoites of *Sarcocystis* can be obtained by crushing cysts dissected from the muscles of the host (Fayer, 1973). Antibiotics need be used only when there is doubt about the sterility of the inoculum.

Toxoplasma will grow in a variety of host cells (see Doran, 1973); Fayer *et al.* (1972) used cultures of Vero (African green monkey kidney) cells in Eagle's MEM containing Earle's BSS and 5% FCS. The same authors used embryonic bovine trachea (EBTr) cells for *Besnoitia* and embryonic bovine kidney (EBK) cells for *Sarcocystis*, both in Eagle's basal medium with Hanks's BSS and 10% FCS.

IV. KINETOPLASTIDA

A. FLAGELLATED FORMS

Although stages of many trypanosomes corresponding to some of those found in the insect vector can be grown in a variety of diphasic or monophasic media (see chapter 4), tissue cultures have proved useful in certain situations.

1. *Trypanosoma vivax* does not grow in media usually used for trypanosomes, but cultures have been established by Trager (1959, 1975) using organ cultures of *Glossina morsitans* or *G. tachinoides*. Fly pupae were surface sterilized and allowed to hatch in sterile sand in sterile vials. Under CO_2 anaesthesia, salivary glands, anterior midgut and hindgut except for the rectum were removed and cultured in modified Eagle's medium with 20% sheep serum at 28°C in Falcon plastic culture dishes. The inoculum was obtained by brief fractional centrifugation of infected blood. When growth occurred, it did so in layers attached to tissues or membranes, rather like the natural situation in the proboscis of the fly; subculture was achieved by transferring fragments of gut coated with attached trypanosomes to a new dish of fly organ culture. Organisms multiplied first as elongate spiral trypomastigotes and later transformed to epimastigotes. In older cultures, transitional forms appeared, including some approaching in appearance the morphology of metacyclic trypomastigotes. The technique, which is only in a developmental stage, was fully described by Trager (1975); it seems applicable also to *T. congolense* and *T. brucei*, for which Cunningham (1973) gave additional details. (See also Chapter 4, Section V, F.)

2. Trypanosome cultures usually grow only at 25–28°C and produce only morphological forms similar to those found in the vector; they die if the temperature is raised to blood heat. Using cultures of bovine fibroblasts, mouse L-cell line or cell line monkey kidney in RPMI 1640 medium with 25 mM HEPES buffer (see Chapter 1: IV, B, 2) and 20% FCS at 37°C, Hirumi *et al.* (1977) obtained multiplication of blood forms in the fluid overlying the cell sheet (see Chapter 4: V, E).

Dougherty *et al.* (1972) have similarly grown *T. lewisi* in the presence of LBN rat kidney cells using RPMI 1640 medium with 20% FCS. Cultures maintained at room temperature produced predominantly epimastigotes, non-infective to rats; adaptation of these cultures to 35°C gave trypomastigotes which produced benign infections when injected into newborn rats; fatal infections in similar animals developed after 3 *in vivo* passages.

B. *TRYPANOSOMA CRUZI* AMASTIGOTES

The intracellular forms of *T. cruzi* have been grown *in vitro* in a variety of tissues; Pipkin (1960) reviewed some of the earlier systems used. Probably the most satisfactory current method is that of Gutteridge *et al.* (1969). Flagellated forms of *T. cruzi* were grown at 28°C in Yaeger's LIT medium (liver infusion/tryptose; Castellani *et al.*, 1967), modified by replacing the 10% haemoglobin solution with 25 mg haemin in 5 ml 50% aqueous triethanolamine per litre of medium, with weekly subculture. Cultures were grown using 20 ml medium in a 120 ml "medical flat" bottle with screw cap; a gas phase of 5% CO_2–95% air was used. Cultures destined for inoculation of tissue cultures were incubated at 33°C. Under these conditions about 10% of the population at the end of exponential growth were metacyclic trypomastigotes. Girardi human heart cells (Bio-Cult or Flow) were grown at 37°C as monolayers in Roux bottles using medium 199 supplemented with 10% calf serum and 0·07% $NaHCO_3$. Cells were subcultured every 3–4 d using 0·2% EDTA to disperse the cell sheet. Aliquots of heart cell suspension (2 ml; 5–6 × 10^5 cells ml^{-1}) in medium 199 were dispensed into Leighton tubes with coverslips (or appropriate quantities into disposable multiwell plastic dishes) and incubated for 2 d at 37°C to establish a confluent monolayer. The medium was then replaced with 2 ml amounts of medium 199 containing 10% calf serum, 0·2% $NaHCO_3$ and 10% by volume of *T. cruzi* in LIT medium, diluted to give 10^7 flagellates ml^{-1}, about 10% of which should be metacyclics. Incubation temperature following inoculation was reduced to 33°C. Drugs for screening or evaluation can be added when the medium (199 containing 10% calf serum, 0·2% $NaHCO_3$ and 10% LIT medium) is renewed 4 d after inoculation, and monolayers fixed in Bouin's fluid and stained with Giemsa's stain for examination 4 d later. Using Roux bottles and an inoculum of 2 × 10^6 flagellates ml^{-1} with an incubation period of 7 d at 33°C, the system has been scaled up to give amastigote forms in quantity for biochemical studies.

C. INTRACELLULAR *LEISHMANIA*

Early attempts to cultivate *Leishmania* involved the use of tissue explants in plasma clots. Since then, *Leishmania* spp. have been found to infect several types of cell culture including monkey kidney and human amnion. In 1964, Lamy *et al.* described the serial passage of intracellular *L. donovani* by repeated subculture of infected dog sarcoma cells. Frothingham and Lehtimaki (1969), using five different strains of *Leishmania*, obtained more consistent results by inoculating infected sarcoma cells into fresh cultures of uninfected sarcoma cells. This particular system has been adapted recently by Mattock and Peters (1975a,b,c) to investigate the antileishmanial activity of a variety of drugs and synthetic compounds. Dog sarcoma cells (Samso series 503; DS) supplied by Professor Lamy were grown at 37°C in medium 199 with 10% FCS, 0·088% $NaHCO_3$, 2·0 mM glutamine and antibiotics.

Cells were inoculated with promastigotes maintained on standard 4N blood agar slopes (Chapter 3: III, C, 1°) with medium 199 as overlay or amastigotes prepared from lesions of infected hamsters, and incubated at a lower temperature. Mattock and Peters found this cell type to be better than a hamster peritoneal cell line (HP) maintained in Puck's medium. In general, HP cells were more delicate than DS cells, their rate of growth was less constant once infected with *L. mexicana mexicana*, they could not readily be passaged, and they did not survive cryopreservation. *L. m. mexicana* developed readily at 32°C, while *L. tropica major* preferred incubation at 33°C. Promastigotes of *L. donovani* produced poor infections in cell culture at 35°C; a more satisfactory procedure was to inoculate dog sarcoma cultures with amastigotes prepared from hamster liver or spleen. The major environmental factor affecting the morphology of the parasites appeared to be temperature (Frothingham and Lehtimaki, 1969). Thus for growth of amastigotes cultures of *L. donovani* were incubated at 36°C and cultures of the "cutaneous" strains at 33°C; lower temperatures favoured extracellular growth of flagellated parasites.

Using DS cells and *L. m. mexicana* parasites supplied by Professor W. Peters, we have investigated the possibility of using such a system for the random screening of compounds for antileishmanial activity. The parasites were found to develop readily in the variety of multiwell culture vessels employed, but unfortunately we were unable to obtain adequate parasite control by any of the "standard" drugs investigated.

V. TICK-BORNE PARASITES

Only one attempt seems to have been made to culture the rickettsial parasite *Anaplasma marginale*, and only indirect evidence based on animal infectivity of serial cultures rather than direct visual observation has been obtained for parasite growth. Marble and Hanks (1972) used primary cultures of rabbit bone marrow and washed infected erythrocytes as inoculum.

More success has been obtained with *Theileria*. Lymph node or spleen from animals infected with *T. annulata* was cut into small pieces, fastened with plasma clots to small coverslips and cultured in medium 199 containing 2·5% chick embryo extract and 40% calf serum (Hawking, 1958). Spleen tissue gave better growth of host cells, but lymph nodes had higher initial parasite burdens. Growth of both parasites and host cells was variable, but it was possible with adequate replication to evaluate 40 drugs in the system. Prolonged culture or serial sub-culture was not achieved.

Hulliger (1965) described a somewhat more successful system in which

T. parva, *T. annulata* and *T. lawrencei* were grown in cultures containing bovine lymphocytes in suspension associated with feeder monolayers of baby hamster kidney cells. Not all attempts at isolation gave viable cultures, but successful isolates were maintained for up to 205 d with twice weekly sub-cultures. *T. parva* was more difficult to grow than the other two species. Parasites were all of the "theilerial body" type—previously known as Koch's blue bodies, agamonts or macroschizonts. The component chromatin particles replicated by binary fission and the composite parasites divided at the time of host cell mitosis in close association with the spindle fibres; there was no evidence of an extracellular form of the parasite capable of reinfecting new cells *in vitro*. Further experimentation by Malmquist *et al.* (1970) led to the establishment of three spleen cell lines from calves experimentally infected with *T. parva* in which the infected lymphoblasts could be readily subcultured in Eagle's MEM with 20% FCS without the necessity of having a feeder layer of hamster cells.

VI. MALARIAL OOCYSTS

Ball and Chao (1971) reviewed the attempts which have been made to obtain *in vitro* development of the malaria parasite from zygote to sporozoite and described a system which has given encouraging results with *Plasmodium relictum*. Whole organ cultures of stomachs of the mosquito *Culex tarsalis* which had been fed on infected canaries or pigeons were cultivated in a modified Grace's insect culture medium in the presence of a feeder layer of Grace's cell line of *Aedes aegypti*. The whole development from zygote to sporozoite could not be obtained in a single culture, but sporozoites for instance could be obtained in 6–8 d if cultures were started with stomachs containing 5 d old oocysts. (See Chapter 5: Section III.)

REFERENCES

Ball, G. H. and Chao, J. (1971). The cultivation of *Plasmodium relictum* in mosquito cell lines. *Journal of Parasitology* **57**, 391–395.

Bedrník, P. (1970). Cultivation of *Eimeria tenella* in tissue cultures. II. Factors influencing a further development of second generation merozoites in tissue cultures. *Acta Protozoologica Warszawa* **7**, 253–262.

Castellani, O., Ribeiro, L. V. and Fernandes, J. F. (1967). Differentiation of *Trypanosoma cruzi* in culture. *Journal of Protozoology* **14**, 447–451.

Cunningham, I. (1973). Quantitative studies on trypanosomes in tsetse tissue culture. *Experimental Parasitology* **33**, 34–45.

Cruickshank, C. N. D., Cooper, J. R. and Conran, M. B. (1959). A new tissue culture chamber. *Experimental Cell Research* **16**, 695–698.

Dougherty, J., Rabson, A. S. and Tyrrell, S. A. (1972). *Trypanosoma lewisi: in vitro* growth in mammalian cell culture media. *Experimental Parasitology* 31, 225–231.

Doran, D. J. (1970) *Eimeria tenella:* from sporozoites to oocysts in cell culture. *Proceedings of the Helminthological Society of Washington* 37, 84–92.

Doran, D. J. (1973). Cultivation of coccidia in avian embryos and cell culture. *In* "The Coccidia" (D. M. Hammond & P. L. Long, eds), pp. 183–252. University Park Press, Baltimore.

Doran, D. J. (1974). *Eimeria tenella:* merozoite production in cultured cells and attempts to obtain development of culture-produced merozoites. *Proceedings of the Helminthological Society of Washington* 41, 169–173.

Doran, D. J. and Augustine, P. C. (1973). Comparative development of *Eimeria tenella* from sporozoites to oocysts in primary kidney cell cultures from gallinaceous birds. *Journal of Protozoology* 20, 658–661.

Doran, D. J., Vetterling, J. M. and Augustine, P. C. (1974). *Eimeria tenella:* an *in vivo* and *in vitro* comparison of the Wisconsin, Weybridge and Beltsville strains. *Proceedings of the Helminthological Society of Washington* 41, 77–80.

Fayer, R. (1973). Preparing *Sarcocystis* organisms for cell cultures. *In* "The Coccidia" (D. M. Hammond & P. L. Long, eds), pp. 451–452. University Park Press, Baltimore.

Fayer, R. and Thompson, D. E. (1975). Cytochemical and cytological observations on *Sarcocystis* sp. propagated in cell culture. *Journal of Parasitology* 61, 466–475.

Fayer, R., Melton, M. L. and Sheffield, H. G. (1972). Quinine inhibition of host cell penetration by *Toxoplasma gondii, Besnoitia jellisoni*, and *Sarcocystis* sp. *in vitro*. *Journal of Parasitology* 58, 595–599.

Frothingham, T. E. and Lehtimaki, E. (1969). Prolonged growth of *Leishmania* species in cell culture. *Journal of Parasitology* 55, 196–199.

Gorham, L. W. and Waymouth, C. (1965). Differentiation *in vitro* of embryonic cartilage and bone in a chemically-defined medium. *Proceedings of the Society for Experimental Biology and Medicine, New York* 119, 287–290.

Gutteridge, W. E., Knowler, J. and Coombes, J. D. (1969). Growth of *Trypanosoma cruzi* in human heart tissue cells and effects of aminonucleoside of puromycin, trypacidin and aminopterin. *Journal of Protozoology* 16, 521–525.

Hawking, F. (1958). Chemotherapeutic screening of compounds against *Theileria annulata* in tissue culture. *British Journal of Pharmacology* 13, 458–460.

Hirumi, H., Doyle, J. J. and Hirumi, K. (1977). African trypanosomes: *in vitro* cultivation of animal-infective *Trypanosoma brucei*. *Science, New York* 196, 992–994.

Hulliger, L. (1965). Cultivation of three species of *Theileria* in lymphoid cells *in vitro*. *Journal of Protozoology* 12, 649–655.

Jensen, J. B. and Hammond, D. M. (1975). Ultrastructure of the invasion of *Eimeria magna* sporozoites into cultured cells. *Journal of Protozoology* 22, 411–415.

Kruse, P. F. and Patterson, M. K. (eds) (1973). "Tissue Culture: Methods and Applications". Academic Press, New York etc.

Lamy, L. H., Samso, A. and Lamy, H. (1964). Instillation, multiplication et entretien d'une souche de *Leishmania donovani* en culture cellulaire. *Bulletin de la Société de Pathologie Exotique* 57, 16–21.

Long, P. L. and Rose, E. M. (1972). Immunity to coccidiosis: effect of serum antibodies on cell invasion by sporozoites of *Eimeria in vitro*. *Parasitology* 65, 437–445.

Malmquist, W. A., Nyindo, M. B. A. and Brown, C. G. D. (1970). East Coast Fever: cultivation *in vitro* of bovine spleen cell lines infected and transformed by *Theileria parva*. *Tropical Animal Health and Production* 2, 139–145.

Marble, D. W. and Hanks, M. A. (1972). A tissue culture method for *Anaplasma marginale*. *Cornell Veterinarian* **62**, 196–205.

Mattock, N. M. and Peters, W. (1975a). The experimental chemotherapy of leishmaniasis. I. Techniques for the study of drug action in tissue culture. *Annals of Tropical Medicine and Parasitology* **69**, 349–357.

Mattock, N. M. and Peters, W. (1975b). The experimental chemotherapy of leishmaniasis. II. The activity in tissue culture of some antiparasitic and antimicrobial compounds in clinical use. *Annals of Tropical Medicine and Parasitology* **69**, 359–371.

Mattock, N. M. and Peters, W. (1975c). The experimental chemotherapy of leishmaniasis. III. Detection of antileishmanial activity in some new synthetic compounds in a tissue culture model. *Annals of Tropical Medicine and Parasitology* **69**, 449–462.

Morton, H. J. (1970). A survey of commercially available tissue culture media. *In Vitro* **6**, 89–108.

Ouellette, C. A., Strout, R. G. and McDougald, L. R. (1973). Incorporation of radioactive pyrimidine nucleosides into DNA and RNA of *Eimeria tenella* (Coccidia) cultured *in vitro*. *Journal of Protozoology* **20**, 150–153.

Patton, W. H. (1965). *Eimeria tenella*: cultivation of the asexual stages in cultured animal cells. *Science, New York* **150**, 767–769.

Paul, J. (1970). "Cell and Tissue Culture", ed. 4. Livingstone, Edinburgh.

Pipkin, A. C. (1960). Avian embryos and tissue culture in the study of parasitic protozoa. II. Protozoa other than *Plasmodium*. *Experimental Parasitology* **9**, 167–203.

Roberts, W. L., Speer, C. A. and Hammond, D. M. (1971). Penetration of *Eimeria larimerensis* sporozoites into cultured cells as observed with the light and electron microscopes. *Journal of Parasitology* **57**, 615–625.

Rose, M. E. and Long, P. L. (1976). Immunity to coccidiosis: interactions *in vitro* between *Eimeria tenella* and chicken phagocytic cells. *In*: "Biochemistry of Parasites and Host-Parasite Relationships" (H. van den Bosche, ed.), pp. 449–455. North-Holland Publishing Company, Amsterdam.

Ryley, J. F. and Robinson, T. E. (1976). Life cycle studies with *Eimeria magna* Pérard, 1925. *Zeitschrift für Parasitenkunde* **50**, 257–275.

Ryley, J. F. and Wilson, R. G. (1972a). Growth factor antagonism studies with coccidia in tissue culture. *Zeitschrift für Parasitenkunde* **40**, 31–34.

Ryley, J. F. and Wilson, R. G. (1972b). Comparative studies with anticoccidials and three species of chicken coccidia *in vivo* and *in vitro*. *Journal of Parasitology* **58**, 664–668.

Ryley, J. F. and Wilson, R. G. (1976). Drug screening in cell culture for the detection of anticoccidial activity. *Parasitology* **73**, 137–148.

Shibalova, T. A. (1970). Cultivation of the endogenous stages of chicken coccidia in embryos and tissue culture. *Journal of Parasitology* **56** (Section II), 315–316.

Sofield, W. L. and Strout, R. G. (1974). Amino acids essential for the *in vitro* cultivation of *Eimeria tenella*. *Journal of Protozoology* **21**, 434.

Speer, C. A. and Danforth, H. D. (1976). Fine-structural aspects of microgametogenesis of *Eimeria magna* in rabbits and in kidney cell cultures. *Journal of Protozoology* **23**, 109–115.

Speer, C. A. and Hammond, D. M. (1971). Development of first- and second-generation schizonts of *Eimeria magna* from rabbits in cell culture. *Zeitschrift für Parasitenkunde* **37**, 336–353.

Speer, C. A. and Hammond, D. M. (1972). Development of gametocytes and oocysts of *Eimeria magna* from rabbits in cell culture. *Proceedings of the Helminthological Society of Washington* **39**, 114–118.

Speer, C. A. and Hammond, D. M. (1973). Development of second-generation schizonts, gamonts and oocysts of *Eimeria bovis* in bovine kidney cells. *Zeitschrift für Parasitenkunde* **42**, 105–113.

Strout, R. G. (1975). *Eimeria tenella* (Coccidia): development in cells cultured in a chemically defined medium. *Abstract of the American Society of Parasitologists New Orleans meeting* p. 103, No. 250.

Strout, R. G. and Ouellette, C. A. (1970). Schizogony and gametogony of *Eimeria tenella* in cell cultures. *American Journal of Veterinary Research* **31**, 911–918.

Trager, W. (1959). Tsetse-fly tissue culture and the development of trypanosomes to the infective stage. *Annals of Tropical Medicine and Parasitology* **53**, 473–491.

Trager, W. (1975). On the cultivation of *Trypanosoma vivax*: a tale of two visits to Nigeria. *Journal of Parasitology* **61**, 3–11.

Wagenbach, G. E. (1969). Purification of *Eimeria tenella* sporozoites with glass bead columns. *Journal of Parasitology* **55**, 833–838.

Whitaker, A. (1972). "Tissue and Cell Culture". Baillière Tindall, London.

Chapter 7

Chicken Embryos

P. L. LONG

Houghton Poultry Research Station, Houghton, Huntingdon, Cambridgeshire, England

I. USE OF CHICKEN EMBRYOS IN MICROBIOLOGICAL RESEARCH

It is now well established that chicken embryos provide a highly convenient, inexpensive and attractive means by which micro-organisms may be studied. Because many organisms develop within host cells, chicken embryos have been widely used for the diagnosis and propagation of a number of viruses as well as the growth of extracellular organisms. Levaditi (1906) was the first to use the chicken embryo for the growth of an infectious agent when he cultivated spirochaetes. This observation was soon confirmed and followed up by a number of workers during the next 20 years.

Research workers have made use of embryos and tissue culture techniques because they provided easier ways of cultivating a range of organisms. The embryo, protected by membrane and shell, is a naturally sterile environment and provides living cells in various stages of development. Avian embryos, depending upon their age, are relatively immunologically incompetent and are, therefore, very receptive and often easier to infect than adult animals. Cross-infection, a constant problem when working with disease of mature animals, is avoided.

Rous and Murphy (1911) propagated sarcoma from the fowl in the chorioallantois (CAM). Woodruff and Goodpasture (1931) grew fowlpox virus on the CAM and Goodpasture *et al.* (1931) reported the successful growth of vaccinia and *Herpes simplex* viruses. Successful vaccines for smallpox, yellow fever, equine encephalomyelitis, pigeon and fowlpox and infectious laryngotracheitis of fowls have been produced using chicken embryo infections. Rickettsiae were cultivated by inoculating the yolk sac of the embryo and this route of inoculation has subsequently proved most useful for the cultivation of a number of viruses.

Beveridge and Burnet (1946) thought that chicken embryos could not produce antibodies. This is now known to be incorrect. Chicken embryos have reduced immunological competence to a range of antigens; the older the embryo the more competent it is to produce antibodies. In addition, cellular responses, including graft versus host reactions, occur in embryos.

The cellular reactions which occur in response to developing organisms are of particular interest. Parasitic infections studied by the author appear to stimulate at first heterophil polymorphonuclear cells (avian equivalent of the neutrophil polymorphonuclear leucocytes of mammals) and later, between the 14th and 18th days, lymphocytes. Bacteria and fungi appear to stimulate this type of response while virus infections appear to stimulate lymphocytic cells.

McCoy (1934) was able to obtain the development of the nematode *Trichinella spiralis* in chicken embryos. Fried (1969, 1973) found that a number of trematodes would grow on the chorioallantoic membranes of developing chicken embryos. Fried has used the method for studies on the behaviour, regeneration and repair process of trematodes, as well as studies of drug action and temperature tolerance; implantations of metacercariae were made onto the chorioallantois on d 9–10. This method gives an 8–10 d observation period before transfer to new embryos becomes necessary.

Holbrook and Stauber (1972, 1973) were able to grow the amastigotes of a protozoon, *Leishmania donovani*, in the liver of embryos incubated at 33°C but not at 37°C. Chicken embryos are poikilothermic and in my experience 11–18 d old embryos can survive at temperatures up to 43°C. It may well be that better growth of many viruses, bacteria, protozoa and helminths would occur if embryos were incubated at subnormal temperature or at temperatures near to that of the normal host. In any event the chicken embryo system provides the opportunity to study the growth of animal parasites at different temperatures.

Chicken embryo infections also enable studies to be made of infections (e.g. of viruses and protozoa) in a closed system. Recently Herman *et al.* (1973), using chicken embryos, have demonstrated that infection with chikungunya virus reduced the number of exoerythrocytic stages of

TABLE 1. *Summary of growth of Protozoa in chicken embryos*

Species	Route of inoculation[a]	Site(s) of development	Notes
Trichomonas foetus, *T. vaginalis, T. hepatica*	i/a	Allantois, alimentary tract and gall bladder	Serial passage of *T. foetus* possible
Hexamita meleagridis	i/a	Allantois	Present in intestine of hatched chick
Entamoeba histolytica	i/a	Allantois	Poor growth, enteric bacteria need to be inoculated
Trypanosoma spp. including *T. brucei,* *T. equiperdum, T. evansi,* *T. lewisi, T. cruzi*	i/a, i/v, i/yolk	Blood, allantoic fluid	Infections are frequently fatal
Leishmania donovani, *L. braziliensis* and *L. tropica*	On chorion, i/a	CAM, allantoic fluid, yolk membrane	Passage possible. Infections frequently fatal
Plasmodium gallinaceum, *P. lophurae,* *P. cathemerium,* *P. elongatum*	On chorion, i/v, i/yolk	Blood, various tissues within embryo and brain	Passage possible
Toxoplasma gondii	On chorion, i/a, i/v, i/yolk	CAM, various tissues within embryo and blood	Passage possible
Eimeria tenella, *E. necatrix,* *E. acervulina,* *E. brunetti*	i/a, i/v, i/yolk	CAM, liver when i/v route is used	Passage of *E. tenella* and *E. acervulina* possible

[a] i/a = intra-allantoic cavity; i/v = intra-venous inoculation; i/yolk = intra-yolk sac.

Plasmodium gallinaceum and that interferon may have been responsible for the protective effect. It is clear that research using chicken embryos for the study of parasitic infections has not developed as much as other test systems (e.g. tissue culture) despite the fact that growth of many protozoa (e.g. coccidia and malarial parasites) is greater in chicken embryos. The use of chicken embryos for the cultivation of protozoa is obviously most useful for those protozoa which do not grow in cell free media or do not grow readily in cells of tissue cultures. Embryo infections are therefore of great value for members of the Sporozoa, particularly coccidia and malarial parasites. Accordingly, special attention has been directed to the conditions needed to cultivate these parasites in embryos. The cultivation of protozoa in chicken embryos is summarized in Table 1.

II. USE OF CHICKEN EMBRYOS FOR THE GROWTH OF EXTRACELLULAR PROTOZOA

A. FLAGELLATES OF THE ALIMENTARY AND REPRODUCTIVE TRACTS

Nelson (1938) and Levine *et al.* (1939) obtained growth of extracellular flagellates when they cultivated *Trichomonas foetus* and *T. muris* in the allantoic cavity in the absence of bacteria. Nelson was able to stabilize the parasite by serial passage through 14 batches of embryos. McNutt and Trussell (1941) grew *T. vaginalis*, a parasite of the human reproductive tract, which developed in the amniotic and allantoic cavities but not in the albumen, yolk sac or blood stream, or in the embryo itself. Hughes and Zander (1954) cultivated another flagellate, *Hexamita meleagridis* from the turkey intestine, in chicken embryos. The author (unpublished observations) was able to confirm this work and carry the parasite by passage of allantoic fluid to five successive batches of embryos by the allantoic route. Eleven d old embryos were used and the parasites were present from the 12th–16th d; they were not seen subsequently. However, in 3 embryos allowed to hatch, the parasites were seen in the intestinal lumen. The chick could have become infected during "pipping" when some allantoic fluid could have been ingested as the embryo broke through the allantoic cavity and the shell membranes. McGhee (1949) was able to cultivate a flagellate symbiont (*Euryophthalmus davisi*) from a bug by lowering the incubator temperature. Inoculation of the intraallantoic cavity or the chorioallantoic membrane of 11–12 d old embryos resulted in growth in the allantoic cavity. Flagellates were found in the alimentary tract and gall bladder and also in the intestine of the hatched chick. Since it is likely that material from the alimentary and reproductive tracts will contain bacteria it is necessary to free the flagellates from associated bacteria as far as practicable. This can be attempted by centrifuging the parasites in suitable media at a relatively slow speed (ca 200–300 g) several times. The embryos should also be inoculated with penicillin and streptomycin (ca 2000 iu of each).

B. HAEMOFLAGELLATES

Various haemoflagellates have been grown in chick embryos (see review by Pipkin, 1960), including various species of *Leishmania* and *Trypanosoma* which developed mainly in the blood although there are reports of successful infections occurring when parasites were inoculated into the yolk sac or allantoic cavity. The flagellates were sometimes found in the allantoic fluid. *Trypanosoma brucei* sspp., *T. cruzi*, *T. evansi*, *T. "hippicum"* (= *T. evansi*), *T. lewisi*, *T. equinum*, *T. congolense*, and *T. equiperdum* have all been grown. With *T. cruzi*, infectious organisms were found in various organs including the chorioallantoic membrane. With most of the parasites serial passage was possible. However, trypanosomes of the subgenus *Trypanozoon* may frequently be fatal to embryos. Embryos aged between 6–14 d have been used and yolk sac inoculation is usually more reliable when young (6 d old) embryos are used. There are indications that when trypanosomes are inoculated late (beyond 10 d), the organisms are more likely to be found in the blood of the hatched chickens. The addition of hyaluronidase and glucose apparently causes a shortening of the incubation period with an increase in parasitism.

T. cruzi causes lesions on the chorioallantois, associated with the amastigotes. Tissue forms may be produced by lowering the incubation temperature to 25–28°C, which causes the death of the embryos although parasites will still replicate. Mello and Deane (1976) inoculated blood forms of four strains of *T. cruzi* onto the chorioallantois or the yolk sac of 9 d old embryos. Irrespective of the strain used the yolk sac route proved most successful. Transformation from the blood trypomastigotes to culture forms did not occur in embryos incubated at 37°C. However, stages of the invertebrate cycle were found in embryos incubated at 32–34°C. These stages were found in the embryo itself and its membranes.

No problem should arise from bacterial contamination as a result of inoculating blood or tissue forms into the embryo. It is of course preferable to free such stages from host cells as far as possible, by centrifugation, before their inoculation and to remove as much foreign protein from the inoculum as possible. For intravenous inoculation mortality can be reduced by removing clumps of cells and reducing the volume inoculated to 0·1 ml. Embryos are best inoculated at 10–12 d.

C. AMOEBAE

Attempts to cultivate *Entamoeba histolytica* by inoculating embryos by different routes have not been very successful. Everett *et al.* (1953) reported limited success when the parasites were inoculated along with their associated enteric bacteria and with antibiotics. These workers tried several routes of inoculation and the allantoic cavity and amniotic cavity were the most successful. High mortality occurred which was thought to be caused by the presence of bacteria in the inoculum. It appears that embryo infections are not particularly suitable for the cultivation of *E. histolytica* and that other more simple methods are available. Meerovitch (1956) inoculated *E. invadens*, a parasite of reptiles, intravenously and produced infections of the liver in embryos incubated at 30°C.

III. USE OF CHICKEN EMBRYOS FOR THE GROWTH OF INTRACELLULAR PROTOZOA

Chicken embryo infections are particularly valuable for the growth of intracellular parasites which show predilection for development in different organs and within different types of cell within these sites. In this connection it is interesting to note that the chorioallantoic membrane consists of three separate layers of cells; ectoderm, mesoderm and endoderm.

A. *LEISHMANIA*

Leishmania tropica grows within macrophages or in tissues containing cells of the reticuloendothelial system. Inoculation of embryos *via* the chorioallantois appears to be the best route, *Leishmania* being subsequently present in the CAM and in embryonic heart. Promastigotes are found in the allantoic fluid following inoculation of the allantoic cavity. *L. donovani*, *L. braziliensis* and *L. tropica* appear to grow best when the incubator temperature is 23–28°C. Although embryonic mortality is very high it is possible to passage these parasites because the embryonic membranes remain fairly intact despite the fact that the embryo itself is moribund. Intracellular stages develop near the point of introduction in the chorioallantoic membrane. This membrane contains phagocytic cells including macrophages. Promastigotes are commonly found in the yolk membrane which contains many blood cell precursors.

B. *PLASMODIUM*

The growth of malaria parasites in embryos has been reviewed by Pipkin and Jensen (1956). Adult chickens are extremely resistant to *P. gallinaceum* infection and McGhee (1949) found it necessary to inoculate massive numbers of sporozoites to obtain infection. Embryos are susceptible and the inoculation of sporozoites onto the chorion (ectoderm) of developing chicken embryos produces infection. However, if sporozoites are inoculated into the allantoic cavity, so that only the endothelial cells of the allantois are readily available, they fail to develop (McGhee, 1949).

Studies on the growth of *Plasmodium* in chicken embryos have been concentrated on those parasites affecting birds. Embryos have been used for the cultivation of avian malarial parasites introduced intravenously, *via* the yolk sac or onto the chorion. These studies were led by Wolfson (1940) and Stauber and Van Dyke (1945) who grew species of *Plasmodium* in duck embryos where development of the exoerythrocytic stages of the parasites occurred. Haas and Ewing (1945) inoculated embryos with sporozoites of *Plasmodium gallinaceum* by inducing mosquitoes to feed through the shell membrane upon the blood vessels of the chorioallantois. Similarly, it is possible to infect mosquitoes by allowing them to feed for 1–1·5 h on blood from the chorioallantoic blood vessels at a time of high parasitaemia. Infections can be obtained by placing salivary glands from infected *Aedes aegypti* on the chorion of 13 d old embryos or by placing sporozoites directly onto the membrane. Huff *et al.* (1960) used tissues from chick and turkey embryos infected with *P. gallinaceum* or *P. fallax* to make observations on the morphology and behaviour of exoerythrocytic stages. Huff succeeded in maintaining cultures for 5–7 months by making serial transfers.

It is possible to passage *P. gallinaceum* by weekly transfer of infected blood to the chorion. It is also possible to passage the parasite by inoculating intravenously about 0·1 ml of parasitized blood. Infection can also be induced by inoculation of the embryo itself or by inoculation of the yolk sac.

C. COCCIDIA

Coccidia do not grow in cell-free media and growth in cultured cells is limited to a few species. Even with these species growth is often poor and the completion of the endogenous life cycle is a rare event. Accordingly, growth of coccidia in embryos is an attractive method.

Avian embryos have been used for the cultivation of the asexual cycle of *Toxoplasma gondii*. The schizogonous stages of *T. gondii* appear to develop readily in chicken embryos in a wide variety of cells and tissues, the infections being initiated by different routes (Wolfson, 1942). The sexual stages (gamonts and oocysts) have not yet been grown in embryos. Although embryos are not as sensitive as mice to infection with *Toxoplasma* the prepatent period may be shorter in embryos. Growth can be obtained by inoculating macerated pieces of tissue (Jacobs, 1956). Furthermore, a strain of *T. gondii* has been maintained by weekly serial passage in chicken embryos without loss of virulence (Jacobs and Melton, 1954). Embryos are usually inoculated at 7 days of age with a dose of between 1×10^2–1×10^5; the survival time is shorter when high doses are inoculated. Embryos over 14 d old at the time of inoculation do not die as readily as younger ones and infections are not so severe.

Avian species of *Eimeria* have been grown in chicken embryos. Long (1965) obtained the entire endogenous cycle of *E. tenella* in epithelial cells of the chorioallantoic membrane following the inoculation of sporozoites into the allantoic cavity. Several species of *Eimeria* were shown to develop in this site.

Although parasites of the genus *Eimeria* belong to the same class as *Plasmodium*, they normally develop in epithelial cells. Inoculation of sporozoites of *Eimeria* onto the chorion of chicken embryos did not induce infections (Long, 1973). Some development of *E. tenella* occurred when sporozoites were introduced into the yolk sac. In this site the parasites appeared to develop within cells lining islands of haemopoietic tissue.

Species of *Eimeria* from chickens have been grown in the liver of chicken embryos after intravenous inoculation of sporozoites. These infections were initially limited to epithelial cells of the small bile ducts and were eventually found in the larger bile ducts and the gall bladder (Long, 1973). It was interesting that in these infections the parasites did not migrate to other sites, in contrast to infections with *Toxoplasma gondii* where development of the asexual forms occurs in a wide variety of sites following inoculation by different routes. The limitation of the development of *Eimeria* to specific sites and cell types distinguishes *Eimeria* from *Toxoplasma*.

IV. FACTORS AFFECTING THE SUCCESSFUL INFECTION OF EMBRYOS WITH *EIMERIA*

A. THE EFFECT OF TEMPERATURE AND DOSE

In the mature host *E. tenella* infection follows a well defined course culminating in oocyst production, the peak of which occurs 6–7 d after inoculation. When chicken embryos were incubated at 39°C, second generation schizonts occurred at the expected time (100–140 h), but the peak of oocyst production was delayed until the 9th d (Long, 1965). When chicken embryos were incubated at 41°C (the average body temperature of the mature host), *E. tenella* developed much more rapidly and was more pathogenic. The magnitude of infection with *E. tenella* in embryos is dose dependent (Long, 1970a).

B. THE EFFECT OF ROUTE OF INOCULATION

Work on the growth of *Eimeria* in chicken embryos has been carried out with embryos inoculated *via* the allantoic cavity. Initial attempts to infect them by inoculating sporozoites into the amniotic cavity, *via* the yolk sac, on the CAM or intravenously (i.v.) failed.

The introduction of sporozoites into the allantoic cavity did not give rise to infection in sites other than in the CAM. Inoculation of sporozoites onto the chorion did not produce infection in the allantoic membrane unless some damage of the chorion took place as a result of the inoculation procedure, when sporozoites entered the allantois *via* the damaged site. *Eimeria* sporozoites normally invaded epithelia and with few exceptions completed their development within epithelial cells. This is in complete contrast to the behaviour of *Plasmodium gallinaceum* in which sporozoites fail to infect embryos when they are inoculated *via* the allantoic cavity (McGhee, 1949), but readily infect them when sporozoites are placed onto the chorion (Fonseca *et al.*, 1946). Initial attempts to achieve infection in chick embryos in sites other than the CAM by inoculating sporozoites i.v. were unsuccessful. However, if large numbers of cleaned sporozoites are used schizogony and gametogony occur in the bile duct epithelium and in cells lining the sinusoids of the liver. The gametocytes and oocysts appear to develop in hepatocytes; this observation was subsequently confirmed by electron microscopy (Lee and Long, 1972).

V. APPLICATIONS OF *EIMERIA* INFECTIONS IN EMBRYOS

A. CHEMOTHERAPEUTIC STUDIES

Ryley (1968) examined several anticoccidial drugs for activity against *E. tenella* in chick embryos and found that not all of those active in chickens were equally effective in chick embryos. Only one route of drug administration was used and mortality caused by *E. tenella* was the only criterion considered. A more critical method for assessing infections in chick embryos is the use of focal lesion counts (Long, 1970b). These focal lesions are produced by the development of colonies of second generation schizonts in the CAM. Long and Millard (1973) compared the activity of nine anticoccidial drugs against *E. tenella* and *E. acervulina* var. *mivati* and found that inhibition of CAM focal lesions and oocyst production were more critical parameters than mortality. Protection against *E. tenella* could be effected by inoculating anticoccidials *via* the yolk sac.

It appears that infections in embryos offer some advantages over chicken infections for the evaluation of antiparasitic effects: (a) only small amounts of drug are required, (b) definite amounts can be given and dose response information obtained and (c) there is no danger of cross infection which is of particular value for drug resistance studies. The disadvantages are that only 2–3 species of *Eimeria* can be used.

B. SERIAL PASSAGE OF *EIMERIA*

Since unsporulated oocysts are produced in CAM cells these may be collected, cleaned and sporulated. The resulting sporozoites can then be used to inoculate other embryos. In this way *E. tenella* and *E. acervulina* var. *mivati* have been serially passaged in embryos (Long, 1972). *E. tenella* lost its pathogenicity to both embryos and chickens by the 62nd embryo passage and some changes in schizogony occurred. The reproduction, judged by oocyst production, of *E. tenella* and *E. acervulina* var. *mivati* appeared to increase gradually with each embryo passage.

When endogenous stages of *E. tenella* occur in the liver of chick embryos given sporozoites i.v. the schizonts develop within cells lining the sinusoids and gametocytes develop in parenchyma cells (hepatocytes). *E. tenella* appears to be consistent in the types of cell it parasitizes whether development occurs in the chicken intestine, embryo, CAM or chick embryo liver. The second generation schizonts develop in cells of mesodermal origin and gametocytes in endothelial cells.

With infections in embryos it is necessary to standardize the age and strain of the host, incubation temperature and strain and dose of sporozoites. It is necessary to ensure that hens producing fertile eggs are not given anticoccidial drugs as these may be maternally transferred.

VI. TECHNIQUES

A. SOURCE, STORAGE AND INCUBATION OF EMBRYOS

It is probably best to produce fertile eggs from breeding birds under the control of the laboratory but most workers have used eggs from a commercial hatchery. The source of eggs should be kept constant as different strains of embryos may vary in susceptibility to parasitism. It is necessary to candle eggs to determine inoculation sites and eggs with white shells are preferable; they should weigh between 50 and 60 g. Before inoculation the fertile eggs should be kept at an even temperature (10–15°C) as embryonic development begins above 27°C.

Ordinary bacteriological incubators fitted with water trays are suitable for incubating embryos at 37°C for a few days but for longer periods egg incubators with controlled ventilation and humidity are preferable. Suitable machines for small scale experiments are table-top observation incubators with egg capacities of 50, 100 or 130 (Curfew Appliances).

B. EMBRYO INOCULATION

The structure of an 11–12 d old developing chicken embryo is shown in Fig. 1. The various routes by which infective material may be introduced are shown in Fig. 2.

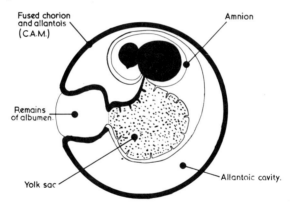

FIG. 1. Diagram of a chicken embryo about the 10th day of incubation showing location of some internal structures. Some commonly used sites of inoculation are also indicated.

Embryos are best incubated at 38°C until the 9th–12th d when they are usually inoculated and then incubated at 39–41°C. The egg is candled and the position of the air sac noted. The point of inoculation is marked in pencil and the area swabbed with 70% alcohol. Holes in the shell for allantoic and yolk sac inoculations may be made with a triangular needle but triangular cuts for intravenous inoculations are made with a carborundum disc fitted to a dental drill (C. Ash). A hole in the shell, but not the shell membrane, for

P. L. LONG

suprachorioallantoic inoculation is best made with a No 6 dental burr (C. Ash). As a precaution against bacterial infection they are inoculated with 2000 iu each of penicillin and streptomycin before being inoculated with parasites.

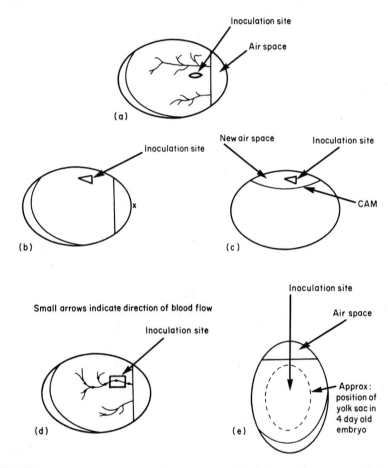

FIG. 2. Diagram showing different routes of inoculating embryos (for details, see text). (a) Intra-allantoic cavity; (b), (c) Suprachorioallantoic (on chorion); (d) Intravenous; (e) Intrayolk sac.

1. *Inoculation into the Allantoic Cavity* (Fig. 2a)

The inoculation site is chosen avoiding large blood vessels and a small hole made with a triangular needle. Inoculum (0·05–0·2 ml) is introduced 1–1·5 cm below the hole.

2. Inoculation of the Chorioallantoic Membrane (CAM) (Fig. 2b and c)

A piece of shell is cut out as in Fig. 2b, taking care not to damage the shell membrane underneath, and removed. A hole is then made in the air space (x, Fig. 2b) and negative pressure applied with a rubber teat to create a new air space at the inoculation site (Fig. 2c). Infective material (0·1–0·3 ml) is then dropped onto the CAM.

3. Intravenous Inoculation (Fig. 2d)

A vein of suitable size is chosen at the time of candling and the direction of blood flow noted. A rectangular or triangular hole is cut with a cutting wheel attached to a dental drill or by cutting with a serrated scalpel blade, taking care to avoid damage to the shell membrane underneath. The piece of shell is removed and a drop of liquid paraffin applied to make the shell membrane transparent. The vein chosen is then inoculated using a very fine hypodermic needle (gauge 30). The volume inoculated should be 0·05–0·1 ml and free from any clumps of cells.

4. Inoculation of the Yolk Sac (Fig. 2e)

Three to five d old embryos are most suitable for this route of inoculation. The eggs are placed in an upright position and the inoculum (0·1–0·3 ml) is introduced into the centre of the egg by using a fine needle (gauge 30) 3–4 cm long.

5. Cultivation of Embryos in vitro

The tough egg shell makes inoculation of the amniotic cavity and various parts of the embryo itself difficult, as is subsequent observation of various parts of the embryo. A new method of maintaining embryos was recently described in which embryos were grown in plastic Petri dishes (Auerbach et al., 1974). Fertile eggs incubated for 3–4 d (38°C) were surface sterilized with Iodicide (Rohm and Haas) and allowed to drain. The underside of the egg was cracked open and its contents placed in a 100×20 mm diameter dish (it is possible to achieve 80% success with this procedure). Five ml of sterile Ringer's solution (Chapter I: IV, B, 1, d) containing 750 units each of gentamycin (Flow Laboratories) and neomycin (E. R. Squibb) were added. The dish (without lid) was placed inside a 150×25 mm plastic dish containing 10 ml sterile distilled water and incubated at 37°C. Many deaths occurred but some embryos survived for 19 d. Mortality may be reduced by incubating the embryos first in an atmosphere of 5% CO_2. Shirley and

Sampson (1976) were able to inoculate sporozoites of *Eimeria* into various parts of the embryo without difficulty and even obtained development of *E. tenella* in the skin tissues.

C. PREPARATION OF INFECTIVE MATERIAL

1. *Recovery and Cleaning of Sporulated Oocysts*

Oocysts can be obtained from faeces by a variety of methods (see Davis, 1973). Faecal material is broken up in water using a mechanical mixer or by shaking up with glass beads. The suspension is passed through a wire mesh sieve (pore size 100 µm) and centrifuged (200 g, 3–5 min). The supernatant fluid is discarded and oocysts, present in the deposit, are mixed with saturated NaCl solution; the tube is filled to within 1 cm of the top with this solution. The tubes are centrifuged again for 3 min; the oocysts are removed from the surface film and placed in water. The salt solution is then removed by further centrifugation. The oocysts are allowed to sporulate suspended in 2% potassium dichromate at 28°C for 1–2 d. The potassium dichromate must be removed by centrifuging repeatedly in water before they are used.

2. *Sporozoites*

For *in vitro* work it is necessary to clean sporulated oocysts further; they should be freed from debris and bacteria using the method described by Wagenbach *et al.* (1966) or by modifications of it. The oocysts are mixed with about 12 ml of a saturated solution of NaCl in a 15 ml centrifuge tube. Distilled water (1–2 ml) is carefully run onto the surface of the salt solution and the tube centrifuged (200 g, 3 min). The oocysts are then removed from the area just below the water layer. These oocysts are then treated with 10% of a technical grade of sodium hypochlorite solution (final concentration 1·5% w/v sodium hypochlorite) for 10–15 min. The suspension is then diluted 1:10 (v/v) with water and the hypochlorite removed by repeated centrifugation.

Sporozoites are obtained from the cleaned oocysts by the following method:

(i) Oocysts are placed in 3 ml of phosphate buffered 0·9% NaCl solution (w/v) (PBS) (pH 7·6) in a centrifuge tube with 1 ml of glass balls about 0·2 mm in diameter (Jencons No. 8).

The tube is mechanically shaken using a Whirlimixer (Fisons) for 20–60 s to release sporocysts from the oocysts.

(ii) The sporocyst suspension is removed and the glass balls rinsed with more PBS into a 100 ml conical flask. To a 20–30 ml suspension is added sufficient trypsin and bile salts to give final concentrations of 0.25% and 0.5% (w/v) respectively. A stock solution of 2.5% (w/v) trypsin (1:250 powder; Difco) is made up in PBS at pH 7.6 and kept in small quantities at −20°C. The 5% (w/v) stock solution of bile salts (Difco) in PBS is kept similarly.

(iii) The suspension is placed in a water bath at 39°C until the sporozoites hatch. For *E. tenella* this may take 30–120 min. Other species hatch more rapidly and lower concentrations of trypsin (0.01–0.1% w/v) are used.

(iv) The sporozoites are then washed free of the excystation fluid by centrifuging in PBS (pH 7.0). They should then be used without delay.

N.B. Sporozoites of mammalian species of *Eimeria* can be obtained by this method, but mechanical fracture (step i) is not necessary as CO_2 is an essential first stimulus, assisted by anaerobic conditions and reducing agents such as cysteine and sodium dithionite. The oocysts are incubated overnight in 0.02 M cysteine hydrochloride in an atmosphere containing 5% (v/v) CO_2. After this treatment the pH is adjusted to 7.0 with 0.2 M-Na_2HPO_4 and then to 7.6 with 1.0 N-KOH. Trypsin (1:250 Difco) is added to a final concentration of 0.25% w/v. Fresh bile (4–10% w/v final concentration) or bile salts (Difco), final concentration 0.5% w/v, are added and the oocysts incubated at 37°C. Sporozoites emerge from the oocyst micropyle in 15–60 min. The sporozoites are then freed from the excystation medium by centrifugation in PBS as previously described. Sporozoites may be further purified using a glass ball column (Wagenbach, 1969). This procedure is necessary for the preparation of sporozoites for tissue culture work but is not essential for embryo infections.

D. RECOVERY OF SPOROZOITES OF *PLASMODIUM* FROM MOSQUITOES

The following details were kindly supplied by Dr. R. L. Beaudoin. Intact *Anopheles stephensi* mosquitoes, infected at least 18 d previously with the ANKA strain of *Plasmodium berghei berghei*, were lightly anaesthetized with chloroform and ground with a mortar and pestle in 0.5 ml serum for 3 min. The ground mosquitoes suspended in serum were mixed with medium 199 (Taylor and Baker, 1968) and transferred to a conical centrifuge tube. Medium 199 was added to make a volume of 15 ml and this was centrifuged (5 min, 700 g, 4°C) to remove coarse particles. The supernatant fluid containing the sporozoites was removed and stored at 4°C while the residue was mixed with an additional 0.5 ml mouse serum, transferred back to the mortar and pestle, and treated as in the initial step. The supernatant fluid from the second spin was added to that of the first and the mixture was recentrifuged (20 min, 10,000 g, 4°C). All but 2 ml of supernatant fluid was then aspirated and discarded. The residue was thoroughly mixed in the remaining 2 ml of supernatant and carefully overlaid onto a two phase gradient. The bottom phase was a mixture of 3 ml sodium diatrizoate 60% w/v, 1 ml mouse serum, and 3 ml medium 199. The top phase was a mixture of 2 ml sodium diatrizoate 60% w/v, 1 ml mouse serum and 4 ml medium 199. The gradient was centrifuged for 30 min at 10,000 g, 4°C. Sporozoites were found at the interface of the bottom layer and the adjacent layer overlying it and were carefully removed with a Pasteur pipette, resuspended in 30 ml of medium 199 and centrifuged (25 min, 10,000 g, 4°C). The supernatant fluid was decanted leaving the sporozoites in a residual button at the bottom of the tube; they were resuspended in 1 ml of medium 199 and thoroughly mixed. Aliquots were counted by standard haemocytometry and the concentration of the suspension adjusted to deliver a desired inoculum of sporozoites in 0·1 ml of medium 199.

E. PASSAGE OF *PLASMODIUM GALLINACEUM* IN CHICKEN EMBRYOS

The following information was kindly supplied by Dr. R. B. McGhee. Blood cells were collected from infected embryos in 0·027% sodium heparin in 0·78% saline at a ratio of one part heparin to nine parts of blood. Red blood cells were washed twice in 0·78% saline and reconstituted to the original volume.

Plasmodium gallinaceum (8a strain) was serially passaged in chicken embryos. White Leghorn embryos were infected when 10–11 d old by injecting parasitized cells from an infected embryo into a large allantoic vein. Blood films were prepared 3–4 d later by locating a small allantoic vein, drilling through the shell only far enough to abrade the shell membrane, and withdrawing a small amount of blood by means of a capillary pipette. After fixation with methanol, the films were stained with Giemsa's stain (1 part stain to 20 parts distilled water buffered at pH 7·2). Fifty parasites or 1000 erythrocytes, whichever came first, were counted and the parasitaemias expressed as parasites per 10,000 erythrocytes.

Approximately 10% of the parasites observed on blood films were schizonts containing four or more nuclei. The average number of merozoites per mature schizont of *P. gallinaceum* in chicken embryos was 25·1 ± 3·8; the normal duration of the cycle was 36 h.

F. METHODS FOR THE EXAMINATION OF LIVING EMBRYOS FOR PARASITES

Samples of allantoic fluid for examination can be removed by inserting a hypodermic needle into the allantoic cavity and withdrawing a few ml of fluid. In order to avoid puncturing blood vessels of the CAM, the shell over the air space should be carefully removed and part of the shell membrane "cleared" by placing a drop of liquid paraffin onto it. A hypodermic needle, fitted to a syringe, can then be introduced and fluid removed. The air space can then be covered with a piece of filter paper fixed in position with adhesive tape. Alternatively, allantoic fluid may be removed from other points on the shell surface after the embryo has been "candled".

REFERENCES

Auerbach, R., Kubai, L., Knighton, D. and Falkman, J. (1974). A simple procedure for the long-term cultivation of chicken embryos. *Developmental Biology* **4**, 391–394.

Beveridge, W. I. B. and Burnet, F. M. (1946). "The Cultivation of Viruses and Rickettsiae in the Chick Embryo". His Majesty's Stationery Office, London. (Special Report series No. 256.)

Davis, L. R. (1973). Techniques. *In* "The Coccidia" (D. M. Hammond and P. L. Long, eds), pp. 421–458. University Park Press, Baltimore.

Everett, M. G., Sadun, E. H. and Carrera, G. M. (1953). Infection of chick embryos with *Entamoeba histolytica. Experimental Parasitology* **2**, 141–146.

Fonseca, F., Cambournac, F. J. C., Pinto, M. R., Pereira, J. M. and Cunha, A. (1946). Studies on the exoerythrocytic cycle of malaria. *Parasitology* **37**, 113–117.

Fried, B. (1969). Transplantation of trematodes to the chick chorioallantois. *Proceedings of the National Academy of Science, Washington* **43**, 232–234.

Fried, B. (1973). The use of 3-day-old chick embryos for the cultivation of *Leucochloridium variae* McIntosh, 1932 (Trematoda). *Journal of Parasitology* **59**, 591–592.

Goodpasture, E. W., Woodruff, A. M. and Buddingh, G. J. (1931). The cultivation of vaccinia and other viruses in the chorioallantoic membrane of chick embryos. *Science, New York* **74**, 371–372.

Haas, V. H. and Ewing, F. M. (1945). Inoculation of chick embryos with sporozoites of *Plasmodium gallinaceum* by inducing mosquitoes to feed through the shell membrane. *United States Public Health Department Report* No. 60, 185–188.

Herman, R., Shiroishi, T. and Buckler, C. E. (1973). Viral interference with exoerythrocytic forms of malaria (*Plasmodium gallinaceum*) in ovo. *Journal of Infectious Diseases* **128**, 148–155.

Holbrook, T. W. and Stauber, L. A. (1972). *Leishmania* in the chick embryo. I. Multiplication of amastigotes of *L. donovani* in the liver. *Journal of Protozoology* **19**, 490–493.

Holbrook, T. W. and Stauber, L. A. (1973). *Leishmania* in the chick embryo. II. Effects of inoculum size, strain size, strain origin and stage injected and infectivity of embryo-derived parasites for hamsters. *Journal of Protozoology* **20**, 431–436.

Huff, C. G., Pipkin, A. C., Weathersby, A. B. and Jensen, D. (1960). The morphology and behavior of living exoerythrocytic stages of *Plasmodium gallinaceum* and *P. fallax* and their host cells. *Journal of Biophysical and Biochemical Cytology* **7**, 93–102.

Hughes, W. and Zander, D. V. (1954). Isolation and culture of *Hexamita* free of bacteria. *Poultry Science* **33**, 810–815.

Jacobs, L. (1956). Propagation, morphology and biology of *Toxoplasma*. *Annals of the New York Academy of Sciences* **64**, 154–179.

Jacobs, L. and Melton, M. L. (1954). Modification in virulence of a strain of *Toxoplasma gondii* by passage in various hosts. *American Journal of Tropical Medicine and Hygiene* **3**, 447–457.

Lee, D. L. and Long, P. L. (1972). An electron microscopal study of *Eimeria tenella* grown in the liver of the chick embryo. *International Journal for Parasitology* **2**, 55–58.

Levaditi, C. (1906). La spirillose des embryons de poulet. *Annales de l'Institut Pasteur* **20**, 924–938.

Levine, N. D., Brandley, C. A. and Graham, R. (1939). The cultivation of *Tritrichomonas foetus* in developing chicken eggs. *Science, New York* **89**, 160–161.

Long, P. L. (1965). Development of *Eimeria tenella* in avian embryos. *Nature, London* **208**, 509–510.

Long, P. L. (1970a). Some factors affecting the severity of infections with *Eimeria tenella* in chicken embryos. *Parasitology* **60**, 435–447.

Long, P. L. (1970b). *Eimeria tenella*: chemotherapeutic studies in chick embryos with a description of a new method (chorioallantoic membrane foci counts) for evaluating infection. *Zeitschrift für ParasitenKunde* **33**, 329–338.

Long, P. L. (1972). *Eimeria tenella* in chick embryos. Reproduction, pathogenicity and immunogenicity of a strain maintained in embryos by serial passage. *Journal of Comparative Pathology and Therapeutics* **82**, 429–437.

Long, P. L. (1973). The growth of *Eimeria* in culture cells and in chicken embryos. *In* "The Coccidia and Related Organisms, a Symposium" (B. M. McGraw, ed.), pp. 57–82. University of Guelph, Guelph, Ontario.

Long, P. L. and Millard, B. J. (1973). *Eimeria* infection of chicken embryos: the effect of known anticoccidial drugs against *E. tenella* and *E. mivati*. *Avian Pathology* **2**, 111–125.

McCoy, O. R. (1934). The development of adult Trichinae in chick and rat embryos. *Journal of Parasitology* **20**, 333.

McGhee, R. B. (1949). The course of infection of *Plasmodium gallinaceum* in chick embryos. *Journal of Infectious Diseases* **84**, 105–110.

McNutt, S. H. and Trussell, R. E. (1941). Comparison of growth of *Trichomonas foetus* and *Trichomonas vaginalis* in chick embryos. *Proceedings of the Society for Experimental Biology and Medicine* **46**, 489–492.

Meerovitch, E. (1956). Experimental infection of chick embryos with *Entamoeba invadens*. *Canadian Journal of Microbiology* **2**, 1–5.

Mello, Maria N. and Deane, Maria P. (1976). Patterns of development of *Trypanosoma cruzi* in the embryonated chicken egg. *Annals of Tropical Medicine and Parasitology* **70**, 381–388.

Nelson, P. M. (1938). Cultivation of *Trichomonas foetus* in chick embryos. *Proceedings of the Society for Experimental Biology and Medicine* **39**, 258.

Pipkin, A. C. (1960). Avian embryos and tissue culture in the study of parasitic protozoa. II. Protozoa other than *Plasmodium*. *Experimental Parasitology* **9**, 167–203.

Pipkin, A. C. and Jensen, D. V. (1956). Avian embryos and tissue culture in the study of parasitic protozoa. I. Malarial Parasites. *Experimental Parasitology* **7**, 491–530.

Rous, P. and Murphy, J. B. (1911). Tumor implantations in the developing embryo: experiments with a transmissible sarcoma of the fowl. *Journal of the American Medical Association* **56**, 741–742.

Ryley, J. F. (1968). Chick embryo infections for the evaluation of anticoccidial drugs. *Parasitology* **58**, 215–220.

Shirley, M. W. and Sampson, R. (1976). *In vitro* cultivation of chicken embryos: applications in microbiological research. *British Journal of Photography* (25th June 1976), 534–544.

Stauber, L. A. and Van Dyke, H. B. (1945). Malarial infections in the duck embryo. *Proceedings of the Society for Experimental Biology and Medicine* **58**, 125–126.

Taylor, A. E. R. and Baker, J. R. (1968). "The Cultivation of Parasites *in vitro*". Blackwell Scientific Publications, Oxford and Edinburgh.

Wagenbach, G. E. (1969). Purification of *Eimeria tenella* sporozoites with glass bead columns. *Journal of Parasitology* **55**, 412–417.

Wagenbach, G. E., Challey, J. R. and Burns, W. C. (1966). A method for purifying coccidian oocysts employing clorox and sulphuric acid–dichromate solution. *Journal of Parasitology* **52**, 1222.

Woodruff, A. M. and Goodpasture, E. W. (1931). The susceptibility of the chorioallantoic membrane of chick embryos to infection with the fowl pox virus. *American Journal of Pathology* **7**, 209–232

Wolfson, F. (1940). Successful cultivation of avian *Plasmodium* in duck embryos. *American Journal of Hygiene* **32**, 60–61.

Wolfson, F. (1942). Maintenance of human *Toxoplasma* in chicken embryos. *Journal of Parasitology* **28** (Supplement), 16–17.

HELMINTHS

Chapter 8

Trematoda*

BERNARD FRIED

Biology Department, Lafayette College, Easton, Pennsylvania, U.S.A.

* Detailed culture techniques are given under the relevant organism except for the schistosomes which are in Section I, F.

I. SCHISTOSOMATIDAE

A. *SCHISTOSOMA MANSONI*

Considerable advances have been made in the cultivation of *Schistosoma mansoni* since the review by Taylor and Baker (1968).

1. *Eggs*

For a discussion of attempts to maintain schistosome eggs *in vitro* see Taylor and Baker (1968). Although schistosome eggs are usually obtained from the livers of experimentally infected mammalian hosts, Jacqueline and Biguet (1973) have described a technique whereby aseptic infective eggs of *S. mansoni* can be isolated from the small intestine of hamsters (see original paper for details). Simple and effective procedures for obtaining schistosome eggs aseptically from the livers of mice (Voge and Seidel, 1972; Section I, F, 1°) or hamsters (Basch and DiConza, 1974) also have been described.

Muftic (1969) claimed to have cultivated miracidia of *S. mansoni* to sporocyst and cercarial stages in sterile haemolymph obtained from *Biomphalaria glabrata* snails. He reported that crystalline extracts obtained from snail haemolymph had ecdysone-like properties which influenced larval metamorphosis. Because of the excitement produced by this report, various schistosome specialists attempted to verify his work and an account of their failure is given by Bayne (1972). In brief, no ecdysone-like substance was isolated from *B. glabrata* and most of Muftic's findings are questioned.

Voge and Seidel (1972) successfully cultivated *S. mansoni* and *S. japonicum* miracidia to stages resembling young sporocysts in both axenic and monoaxenic cultures; these were maintained for 3 and 5 weeks respectively. The axenic cultures, in a defined medium plus 20% horse serum, and monoxenic cultures, which contained monolayers of either mouse fibroblasts or monkey kidney cells, gave essentially similar results; therefore only details of the axenic technique are provided (see Section I, F,

1° and 2°). Voge and Seidel reported cessation of miracidial swimming in solutions of osmolarity similar to or higher than that of *B. glabrata* haemolymph, but shedding of cilia occurred only in media suitable for continued development. During cultivation body size doubled, the numbers of large nuclei increased, the terebratorium was lost, and the nervous system dedifferentiated. Differences in behaviour and nuclear characteristics of cultivated miracidia of *S. mansoni* and *S. japonicum* were also documented.

Basch and DiConza (1974) also reported *in vitro* transformation of *S. mansoni* miracidia into sporocysts in the defined culture medium of DiConza and Basch (1974; see Section I, F, 3°) plus 20% human serum. This medium differed from that of Voge and Seidel in sugar concentration, pH, and buffer system. Only 2–3 ml of unchanged medium were needed to support the development of 100–300 miracidia for up to 10 d with essentially similar development to that reported by Voge and Seidel (1972; see Section I, F, 2°). However, Basch and DiConza (1974) then found that cultured mother sporocysts (up to 10 d *in vitro*) would continue to develop and produce cercariae, if inoculated into snails (*B. glabrata*).

3. Sporocysts*

Early studies on the *in vitro* maintenance of schistosome sporocysts (notably Chernin's work) have been described in Taylor and Baker (1968). Subsequently, successful monoxenic cultivation of *S. mansoni* daughter sporocysts has been reported by DiConza and Hansen (1973) who used insect cell lines, one of which was "considered to be from a mosquito line" and the other from *Aedes albopictus* (mosquito). The insect cells were grown in 50 ml capped flasks with 10 ml of medium (D and H sporocyst medium) consisting of 0·65% lactalbumen hydrolysate in Hanks's saline supplemented with 10% inactivated foetal bovine serum, 10% whole egg ultrafiltrate, 0·01% bovine serum fraction A, and 200 iu ml^{-1} penicillin and 200 μg ml^{-1} streptomycin (pH 6·8–7·0). When cells covered the bottom of the flask, the medium was changed; 35–65 daughter sporocysts from *B. glabrata*, rinsed in Chernin's BSS (see Taylor and Baker, 1968, p. 167) containing 25% foetal bovine serum, were immediately inoculated into each culture. Cultures were maintained at 28°C and the medium was changed twice a week by withdrawing and replacing 2·5 ml. Sporocysts doubled their length within 7 d, developed "hair-like" tegumentary processes (microvilli) and germ balls, and survived up to 35 d.

Hansen *et al.* (1973) later devised a sporocyst culture chamber constructed from a Nuclepore membrane to separate the sporocysts from

* Taylor and Baker (1968) incorrectly referred to these as "rediae".

the underlying *A. albopictus* cells in the culture. This facilitated observation of sporocysts and permitted free interchange of D and H sporocyst
medium between sporocysts and the cell culture. Cultures were maintained
for 21 d at 27·4°C and secondary daughter sporocysts ("progeny-
daughters") emerged from primary sporocysts by the ninth day. The length
of the primary daughter sporocysts increased 5-fold after 18 d. Hansen *et al.*
(1974a) reported observations on axenic culture using various "conditioned
media"; one, the D and H sporocyst medium, was conditioned for 7 to 10 d
with *A. albopictus* cells, passed through a 0·3 µm Millipore membrane to
remove cells, diluted with water and used. As expected, in unconditioned
medium sporocysts did not grow or develop, whereas in conditioned
medium sporocysts increased in length and developed tegumentary
processes, and small embryos. Subsequently Buecher *et al.* (1974) and
Hansen *et al.* (1974b) reported the emergence of progeny daughter
sporocysts in various fresh (unconditioned) media containing very exact
concentrations of sulphydryl compounds (reducing substances consisting of
mixtures of cysteine and glutathione bases) and a strictly controlled gas
mixture of O_2, CO_2 and N_2, obtained by using Berntzen's (1966) micro gas
apparatus. (Space does not allow a detailed discussion of the various media
employed by Hansen and her colleagues.) In the absence of either the
correct combination of reductants or gas mixture, cultivation in fresh
medium was not successful.

Hansen (1975) discussed the occurrence of numerous progeny daughter
sporocysts (up to 4 generations) as part of the usual life cycle of *S. mansoni*
(the original paper should be consulted for details). However, attempts to
culture "progeny daughter" sporocysts axenically in various media were
unsuccessful (Hansen, 1975).

DiConza and Basch (1974) described a simple and effective procedure for
the axenic cultivation of *S. mansoni* daughter sporocysts and interestingly,
the medium (basic sporocyst culture medium plus 20% human serum) was
the same as the one they had used for axenic cultivation of miracidia to
mother sporocysts (see Section I, F, 3°).

4. *Cercariae*

Numerous studies have reported cercarial transformation into schistosomules since the initial work of Jensen and his colleagues using Rose multi-
purpose dialysis chambers (reviewed by Taylor and Baker, 1968).

Stirewalt *et al.* (1966) established criteria to monitor cercaria to
schistosomule transformation, i.e. loss of cercarial tail, evacuation of
acetabular gland secretions, saline or serum tolerance (the cercaria is water
tolerant and saline or serum intolerant) but water intolerance of the

schistosomule, inability of the schistosomule to react positively to the
Cercarien-Hüllen Reacktion (CHR) test, and changes in the tegument during
transformation. Stirewalt *et al.* (1966) also prepared a biological membrane
by scraping, drying and plucking the excised skin from a rat's abdomen.
Passage of cercariae through this membrane produced schistosomules
which within 3 h differed significantly from cercariae in various metabolic
parameters (Bruce *et al.*, 1969). Stirewalt and Uy (1969) provided details of
an artificial skin membrane and collection chamber for obtaining schistosomules following penetration of the membrane. (Artificial skin membranes
are difficult to prepare; details are given by Stirewalt *et al.*, 1966 and
Stirewalt and Uy, 1969.) A disadvantage is that the artificial skin
preparation can be used only once or twice. Clegg and Smithers (1972)
reported an artificial membrane procedure that yielded only about 20–30%
schistosomules. Eveland and Morse (1975) commented that use of an
artificial membrane does not allow for the analysis of precise steps occurring
during cercarial-schistosomule transformation.

Eveland (1972) obtained cercarial transformation *in vivo*, using Millipore
chambers containing cercariae implanted into the peritoneal cavities of
mice, indicating that tissue contact was not a prerequisite for conversion.
Furthermore, Gilbert *et al.* (1972) showed that transformation occurred *in
vitro* without a tissue stimulus and that such a change was initiated by a
phospholipid fraction containing a crude lecithin extract. Eveland and
Morse (1975) however found that lecithin was highly toxic to cercariae.

Various techniques have been used to induce cercarial transformation *in
vitro*, e.g. high speed centrifugation (Gazzinelli *et al.*, 1973), vortex mixing
(Tiba *et al.*, 1974; Ramalho-Pinto *et al.*, 1974) and the passage of cercariae
through a narrow bore hypodermic needle (Colley and Wikel, 1974; see
Section I, F, 4°). Although these techniques appear effective, they may
produce a traumatic physical stress and may damage cercariae (Eveland and
Morse, 1975). Moreover these procedures do not allow analysis, under
physiological conditions, of factors involved in the conversion.

Eveland and Morse (1975) described an *in vitro* system for converting
cercariae into schistosomules in various media. Best results were obtained
when cercariae were placed in a mixture of Eagle's saline and rat serum (1 : 1)
at 37°C for 3 h followed by cultivation in Clegg's Medium II for 14 d; post
schistosomule development was apparently normal (see Section I, F, 5°).

5. *Schistosomules*

Clegg's (1965) work on the cultivation of *S. mansoni* remains the major study
of postlarval development of schistosomules *in vitro*; it has been described
in detail by Taylor and Baker (1968). It should be apparent from the

previous discussion on cercariae (Section I, A, 4) that investigators studying cercarial transformation relied on Clegg's work to assess further development of their schistosomules (e.g. Eveland and Morse, 1975; see Section I, F, 5°).

Moreover, Clegg's (1965) work influenced recent studies on the cultivation of schistosomules of *Trichobilharzia ocellata* by Howell and Bourns (1974; see Section I, F, 8°) and schistosomules of *Schistosoma haematobium* by Smith *et al.* (1976, see pp. 11–12). Clegg and his colleagues have recently used these *in vitro* techniques to study the effects of immune sera on schistosomules of *S. mansoni* (Clegg and Smithers, 1972) and *S. haematobium* (Smith and Webbe, 1974a) and cross immunity between both in a single culture system (Smith and Webbe, 1974b).

6. *Adults*

Early studies on the cultivation of adult *S. mansoni* were reviewed by Taylor and Baker (1968). Of these studies, that of Senft and Senft (1962; see Section I, F, 6°) using a defined medium (NCTC 109) is still the one of choice, and provides worm survival for about 20 d with production of nonviable eggs. A method that should gain popularity is that of Schiller *et al.* (1975), who maintained *S. mansoni* adults for at least 12 d with the production of viable eggs by slightly modifying the diphasic culture medium that Schiller (1965) used for the successful cultivation of the rat tapeworm, *Hymenolepis diminuta*. Schiller *et al.* (1975) showed that worms maintained aerobically in the diphasic medium produced significantly more viable eggs than worms maintained similarly but under anaerobic conditions (see Section I, F, 7°).

Other recent methods include that of Lancastre and Golvan (1973) who kept adults alive for as long as 120 d when cultured on Kb or HeLa cells, with oviposition occurring on the 15th d (eggs were abnormal). For short term maintenance, Coles (1972) used Earle's lactalbumin and 10% newborn calf serum and reported that worm pairs produced about 3 eggs d^{-1} during a 3 d period of cultivation under aerobic conditions (no eggs were produced when worms were maintained anaerobically). Michaels and Prata (1968) cultured worm pairs in Medium 199 (90 parts) and calf serum (10 parts) and reported that worms laid about 24 eggs d^{-1} during the first 5 d of maintenance.

Cowper *et al.* (1972) described a continuous flow heart lung apparatus for the maintenance of adults in an active state for 4–5 d (see the original paper for construction details). Worm movement was not observed after d 6, and the authors were concerned with the possibility of toxicity from circulating silica particles derived from the silicone rubber tubing used in the apparatus.

B. *SCHISTOSOMA HAEMATOBIUM*

Recently, Smith *et al.* (1976) cultivated *Schistosoma haematobium* in essentially the medium used by Clegg (1965) for *S. mansoni*. The rationale for the study was to provide one culture system for both *S. mansoni* and *S. haematobium* that would be of value in analysing cross-immunity. The authors provided detailed information on growth and development of *S. haematobium* in the golden hamster, *Mesocricetus auratus*, which is invaluable in evaluating the results of *in vitro* studies.

Cercariae of *S. haematobium* were obtained from *Bulinus truncatus* snails. Schistosomules used to initiate *in vitro* cultivation were obtained by allowing cercariae to penetrate isolated male mouse skin (Clegg and Smithers, 1972). Smith *et al.* (1976) reported that 40–55% of the *S. haematobium* cercariae, compared with 20–30% of *S. mansoni* cercariae (Clegg and Smithers, 1972), penetrated the mouse skin in 4 h and were collected as schistosomules.

Using aseptic procedures, the schistosomules were washed in Hanks's BSS plus antibiotics and then transferred to Leighton tubes containing 2 ml of the culture medium. Tubes were gassed with 8% CO_2 in air, closed with gas-tight screw tops and incubated at 37°C. The medium contained 50% serum (usually human) and 50% Earle's BSS containing 0·5% lactalbumin hydrolysate, penicillin (100 iu ml^{-1}) and streptomycin (100 µg ml^{-1}). The medium was replaced every 5 d and after the first replacement packed red cells (human, Type O) were added to give a final concentration of 1%. Of 8 selected donors used to obtain the human serum, only one (K.G.) had a serum producing consistently good growth of worms *in vitro*. Baboon, rhesus monkey or foetal calf sera were mostly ineffective in stimulating worm growth.

In the best cultures, *S. haematobium* developed at about the same rate as in the hamster up to 31 d when pairing first occurs *in vivo* and males produce sperm. Pairing never occurred *in vitro* and although some males produced sperm, females did not complete sexual maturation (did not develop vitellaria or eggs). Interestingly, somatic growth continued and at 70 d, some males had achieved almost the same length as in the hamster (approximately 7 mm).

Whereas *S. mansoni* can be cultivated in a wide variety of human sera, *S. haematobium* cannot, indicating significantly different nutritional requirements in the two species (Smith *et al.*, 1976).

C. *SCHISTOSOMA JAPONICUM*

For early studies on the *in vitro* maintenance of *Schistosoma japonicum* adults consult Ito *et al.* (1955) and Mao and Lyu (1957) (reviewed by Taylor and Baker, 1968).

Voge and Seidel (1972) cultivated miracidia of *S. japonicum* to stages resembling young sporocysts (see Section I, F, 1° and 2°). Recently, Hsu Shih-Ê (1974) reported that *S. japonicum* eggs can be cultured *in vitro* without undergoing a tissue migration phase. Eggs embryonated and produced miracidia that hatched within 13 d in a medium containing serum and red blood cells. Females embraced by males, as well as isolated females, not only laid eggs that were already in the uterus but produced new ones. The peak of egg production occurred 2–7 d following transfer of worms from the host to the culture medium.

Chow and Chiu (1972) cultured *S. japonicum* adults and schistosomules (obtained from mice) in Earle's saline containing 0·5 to 1% dog red blood cells, 0·5% lactalbumin hydrolysate, 100 iu ml^{-1} penicillin, 100 μg ml^{-1} streptomycin and either human, dog, pig, calf or cattle serum. Adult worms survived longest in calf serum (av. 59·5 d) and shortest in human serum (av. 41·5 d). Schistosomules developed well in the presence of dog serum, with reproductive organs appearing in 2–3 weeks.

D. *SCHISTOSOMA BOVIS*

Magzoub (1973) obtained adult *S. bovis* from experimentally infected mice and maintained them in a defined medium for 14–18 d without egg production. When horse serum was added to the defined medium survival was extended to 20–24 d and worms produced nonviable eggs (about 30 eggs worm^{-1} during the first 2 weeks). Worms showed considerable body activity and gut peristalsis in both media during the survival period.

E. *TRICHOBILHARZIA OCELLATA*

Howell and Bourns (1974) were the first to obtain significant cultivation of the avian schistosome, *Trichobilharzia ocellata*. The worms achieved organogeny in a medium based on Earle's saline containing lactalbumin hydrolysate and duck or chicken serum with homologous red blood cells (a modified Clegg II medium; Clegg, 1965). Results were similar irrespective of whether cultures were initiated with cercariae or with schistosomules obtained following cercarial penetration of bird skin or gelatin membranes so only the easier cercarial cultivation is detailed here (Section I, F, 8°). Interestingly, although most studies on the cultivation of mammalian schistosomes begin with the schistosomule stage, a recent report (McCowen *et al.*, 1968, abstract only) indicates that *S. mansoni* have been cultivated to the gametogeny stage beginning with cercariae.

F. DETAILED TECHNIQUES

1° *Schistosoma mansoni* and *S. japonicum* miracidia collection
(Voge and Seidel, 1972)

Mice infected with either *S. mansoni* or *S. japonicum* for 6 weeks or longer are killed by cervical dislocation and washed in 70% ethanol. The livers are removed aseptically and homogenized with 20 ml of 0·8% NaCl in a Waring Blender. Only sterile glassware and solutions should be used. The homogenate is centrifuged (2000 rpm, 2 min), resuspended in fresh saline and centrifuged again. The pellet is suspended in sterile spring water, re-centrifuged and the washed pellet placed in a sidearm flask containing 1 litre of sterile spring water. The flask should be covered with black plastic, except for the sidearm which is exposed to overhead light. This procedure should yield a large number of aseptic free-swimming miracidia in 15–20 min concentrated in the sidearm region as they are phototactic.

2° *Schistosoma mansoni* and *S. japonicum* miracidia
(Voge and Seidel, 1972)

MEDIUM

BME (Basal Medium Eagle) vitamins and BME amino acids (Gibco) with the following additives in mg litre^{-1} distilled water (unless otherwise stated):

Additional amino acids: serine, 6·0; proline, 2·9; L-alanine, 2·4; aspartic acid, 2·8; glutamic acid, 4·7; glycine, 2·4; L-alanine, 2·4.
Organic acids: malic, 40; alpha-ketoglutaric, 30; succinic, 10; fumaric, 5; citric, 10.
Salts: $CaCl_2.2H_2O$, 530; KCl, 150; $MgSO_4.7H_2O$, 450; Na_2HPO_4 (anhydrous), 70; NaCl, 1500.
Sugars: galactose, 5·0 g and glucose, 0·5–1·0 g.

This solution is supplemented with 20% horse serum (Gibco) inactivated (56°C, 30 min). The freezing point depression of the medium is about 132 mosmol and pH should be adjusted to 6·7–7·0 with sodium bicarbonate.

TECHNIQUE

Miracidia, obtained as described previously (Section I, F, 1°) must first be immobilized: various solutions are suitable, e.g. 0·8% NaCl or Earle's salt solution (Chapter 1: IV, B, 1, f). In general, solutions with an osmolarity of about 120 mosmol litre^{-1} cause immobilization within 10–20 min. For best results, immobilize miracidia in the culture medium (defined medium plus 20% horse serum) in Stender dishes. Following immobilization (10–20 min) transfer 20–30 miracidia to 16 × 125 mm screw cap tubes containing 4 ml of the complete medium. Antibiotics are not obligatory for cultivation, but can be used (300 µg ml^{-1} streptomycin, 300 iu ml^{-1} penicillin). Tubes are sealed with parafilm and kept slanted at room temperature (23–24°C) or at 26°C. Half the medium is replaced with fresh medium three times a week.

Axenic cultures can be maintained for 3 weeks. Miracidia shed their cilia, double their body size, increase the number of their large nuclei, lose the terebatorium and dedifferentiate the nervous system.

3° *Schistosoma mansoni* Daughter sporocysts
(DiConza and Basch, 1974)

MEDIA

(i) Chernin's Balanced Salt Solution (BSS)

	g litre^{-1}
NaCl	2·8
KCl	0·15
Na$_2$HPO$_4$ (anh.)	0·07
MgSO$_4$.7H$_2$O	0·45
CaCl$_2$2H$_2$O	0·53
NaHCO$_3$	0·05
Glucose	1·0
Trehalose	1·0
Penicillin G	200 iu ml^{-1}
Streptomycin sulphate	200 µg ml^{-1}
Phenol red (0·4%)	5·0 ml litre^{-1}

Dissolve all the above ingredients *except* CaCl$_2$.2H$_2$O in 800 ml double distilled water.

Dissolve the CaCl$_2$.2H$_2$O in 200 ml double distilled water.

Combine above 2 parts whilst stirring.

Sterilize medium by filtration (Millipore "H" filters, 0·45 µm pore size were used by Chernin but would not have removed the smallest bacteria).

Store medium at 5°C in rubber stoppered vessels until used.

(ii) Eagle's (MEM) amino acids (IX No. 1, i.e. $\frac{1}{50}$ Gibco 50 × MEM amino acids) plus Eagle's (MEM) Vitamins (Gibco No. 2, 100 ×, diluted $\frac{1}{100}$) with the following additives, in mg litre^{-1} distilled water:

Additional amino acids*: L-serine, 3·9; O-phosphorylethanolamine, 1·3; L-aspartic acid, 1·0; L-threonine, 4·1; DL-O-phosphoserine, 2·2; L-glutamine, 7·5; L-glutamic acid, 5·5; L-citrulline, 1·9; L-glycine, 2·6; L-alanine, 1·2; L-valine, 1·1; L-cystine, 2·1; L-methionine, 0·9; L-isoleucine, 2·6; L-leucine, 4·4; L-tyrosine, 3·0; L-phenylalanine, 2·6; L-ornithine, 4·0; L-lysine, 4·3; L-histidine, 2·5; L-arginine, 2·7.

Organic acids†: Malic, 40; α-ketoglutaric, 30; succinic, 10; fumaric, 5; citric, 10.

Salts‡: NaCl, 1300; KCl, 97; $MgCl_2.6H_2O$, 325; Na_2HPO_4, 17; $CaCl_2.2H_2O$, 558.

Sugar‡: Glucose, 100.

Buffer: HEPES, 1500.

The basic medium is supplemented with 20% (v/v) human serum. Selected donors who have fasted for 6–12 h yield serum which apparently gives the best cultivation results (see original paper for discussion). Serum is inactivated at 56°C for 30 min and stored at −20°C until used. The complete medium is adjusted to pH 7·0–7·2 with 10N-NaOH. The osmolarity of the medium is 126 mosmol kg^{-1} H_2O as determined with a Fiske osmometer.

TECHNIQUE

Mother and daughter sporocysts are obtained from experimentally infected *B. glabrata* snails 14–18 d after miracidial exposure: donor snails are transferred to 50 mm glass Petri dishes, containing 5000 iu penicillin G, 5000 µg streptomycin sulphate, and 330 µg gentamicin in 10 ml of distilled water, for 1 h. The antibiotic mixture is renewed for a second hour and then replaced with distilled water. The snail antennae containing mother sporocysts are pinched off with sterile watchmaker's forceps. Daughter sporocysts are released by gentle teasing with sterile forceps and then transferred through 3 washes of Chernin's BSS (see Taylor and Baker, 1968, p. 167) containing 25% foetal bovine serum.

Sporocysts are cultured in 10 × 75 mm glass culture tubes each containing 2 ml of complete medium and sealed with size 00 silicone stoppers. About 50 to 150 daughter sporocysts should be inoculated into each tube. Cultures are maintained upright at 25 ± 1°C and the medium is left unchanged. To facilitate observation of sporocysts the bottom of the

---

* Additional amino acids were added to the medium based on analysis of haemolymph from laboratory stocks of *B. glabrata* maintained by DiConza and Basch.

† Based on Voge and Seidel (1972).

‡ Based on published analyses for salts and glucose (Basch and DiConza, 1973) except that $NaHCO_3$ is omitted from the medium. Determined empirically.

tube is immersed in mineral oil in a 35 mm plastic Petri dish and viewed with an inverted microscope.

After 14 d sporocysts should have increased their length 3–4 times and contain small germ balls. A dense covering of microvilli should develop on the sporocyst tegument. Such sporocysts, when implanted into uninfected *B. glabrata*, produced infective cercariae within 20–46 d (DiConza and Basch, 1974).

4° *Schistosoma mansoni* cercaria to schistosomule transformation
(Shearing technique of Colley and Wikel, 1974)

Cercariae are obtained from *Biomphalaria glabrata* snails infected with a Puerto Rican strain of *S. mansoni*. Suspensions of cercariae are concentrated to about 1000 ml^{-1} by means of an 8 μm pore size Millipore filter. Five to 10 ml volumes of such suspensions are subjected to shearing forces created by 10–14 passages through 22 gauge hypodermic needles (N.B. narrower gauge needles produced lower yields due to increased trauma and destruction of cercarial bodies). Organisms are then transferred to tissue culture medium RPMI 1640 (Flow Laboratories) with 2% pooled normal human serum, 300 iu ml^{-1} penicillin and 200 μg ml^{-1} streptomycin (BBL). Schistosomules are maintained in this medium in either 30 × 10 mm or 60 × 15 mm plastic culture dishes (Falcon Plastics) at 37°C in 5% CO_2, 95% air.

Colley and Wikel claimed that the shearing force separated most of the cercarial bodies and tails and gave a population in which only 3–5% of the original cercariae remained intact. Subsequent *in vitro* cultivation (the medium they used is not stipulated) produced, within 48 h, organisms with schistosomule characteristics, i.e. loss of acetabular secretory materials, serum and saline tolerance, inability to react in the CHR reaction, and change in tegumentary surface.

5° *Schistosoma mansoni* cercaria to schistosomule transformation
(Eveland and Morse, 1975)

MEDIA

(i) Best results were obtained in Eagle's balanced saline (BSS; not defined in original paper) plus 50% fresh or frozen rat serum. N.B.: inactivated serum (56°C, 30 min) was not effective in inducing cercarial tail loss.

(ii) Clegg's (1965) Medium II

Earle's saline (modified—see Chapter 1:
IV, B, 1, f, and note below) ... 1 part
Inactivated rabbit serum .... 1 part
Glucose .... 0·1% w/v
Lactalbumin hydrolysate (Nutritional
Biochemical Corporation. Enzyme
hydrolysate) ....0·5% w/v
Rabbit red cells .... 1% v/v
Penicillin .... 100 iu ml^{-1}
Streptomycin .... 100 μg ml^{-1}

N.B. Earle's saline: to maintain isotonicity the NaCl content of Earle's saline is reduced by 0·18 g for every gram of lactalbumin hydrolysate used in the medium (0·5% w/v aqueous lactalbumin hydrolysate is isotonic).

TECHNIQUE

S. mansoni cercariae are collected in deionized water and concentrated by the method of Lewert and Para (1966). Infected snails are placed in a beaker containing 200 ml of dechlorinated tap water. Snails are placed under light to allow cercariae to emerge. The snails are removed from the beaker, the water containing cercariae is decanted, snail faeces and macroscopic debris removed, and the cercariae concentrated to a small volume with a 47 mm Millipore filter of 5 μm pore size. The cercariae are washed and reconcentrated 2 or 3 times by this method, with 200 ml of dechlorinated tap water added for each washing. The final volume is adjusted so that 650–2000 cercariae are contained in 1·0 ml. Concentrated cercarial suspensions are diluted 1:10 in the test medium (Eagle's BSS plus 50% rat serum). The final concentration of cercariae is 65–200 ml^{-1} and the total volume about 2 ml. Tubes which can be stoppered tightly should be used. To control pH, tubes should be gassed with 10% CO_2 in air and then stoppered tightly; they are incubated at 37°C for 3 h.

After 3 h this procedure should produce about 96–100% live organisms, of which 44–60% should have lost their tails. Furthermore none of these organisms should be water tolerant. Also, the external glycocalyx (a feature of the cercarial tegument) should be lost (as determined by electron microscopy). Organisms preincubated in Eagle's BSS plus 50% rat serum and then cultivated in Clegg's (1965) Medium II should remain viable for at least 14 d, evacuate their acetabular gland contents within the first 3 d and have intestinal contents by 7–9 d.

6° *Schistosoma mansoni* adults
(Senft and Senft, 1962)

MEDIUM

NCTC 109 (Taylor and Baker, 1968, pp. 351–354).

TECHNIQUE

Individual worm pairs are hooked out of the mesenteric veins of the liver from an infected white mouse (using sterile procedures) and transferred to sterile NCTC 109. They are washed in this several times before being transferred to the culture test-tubes (1 or 2 pairs test-tube^{-1}) containing 2 ml of culture medium. The tubes are tightly stoppered and incubated in an upright position at 37°C. The medium should be changed every 2–3 d.

The Senfts found that schistosomes laid eggs and appeared normal for 20 d after which they deteriorated.

7° *Schistosoma mansoni* adults
(Schiller *et al.*, 1975 and Schiller, 1965)

MEDIUM

The medium is diphasic, consisting of Difco nutrient agar (70%) and fresh, defibrinated rabbit blood (30%), overlaid with Hanks's balanced salt solution (HBS, Chapter 1: IV, B, 1, g): 16 g nutrient agar (Difco) and 3·5 g NaCl dissolved in 700 ml distilled water. After autoclaving, this solution is mixed thoroughly with 300 ml sterile, defibrinated rabbit blood (inactivated 56°C, 30 min). The pH is adjusted with $NaHCO_3$ to 7·8 and 5·0 ml aliquots are dispensed to sterile, cotton stoppered 25-ml Ehrlenmeyer flasks. After gelation of the blood-agar, 5 ml of HBS, containing 200 iu ml^{-1} penicillin and 200 µg ml^{-1} streptomycin are added to each flask. This medium is then preincubated in air for 24 h at 34°C to enhance the diffusion of nutrients.

TECHNIQUE

Scanty details are given for *S. mansoni* and the reader is left to deduce what he can from the two papers. Worms were incubated in pairs under aerobic or anaerobic conditions for a maximum of 12 d. During the first 4–6 d egg-laying under aerobic conditions was equivalent to that occurring *in vivo*.

8° *Trichobilharzia ocellata*
(Howell and Bourns, 1974)

MEDIUM

Clegg's Medium II (1965: see Section I, F, 5°, ii) should be used except that inactivated serum of chicken or duck (Peking, Mallard or Black) and homologous red blood cells are substituted for rabbit blood components. For instance if commercial chicken serum (Difco) is used, prepare chicken red cells as described by Clegg (1965) or Chapter 1 (IV, A, 2, a) for rabbit red cells. Mycostatin (25 μg ml^{-1}), penicillin (100 iu ml^{-1}) and streptomycin (100 μg ml^{-1}) should be added to the complete medium. After gassing (5% CO_2 in air) and equilibration at 39–40°C the pH should be 7·4–7·6.

TECHNIQUE

Cercariae are concentrated by centrifugation (2000 rpm, 5 min in siliconed tubes), are rinsed several times in sterile Earle's BSS (Difco; Chapter 1: IV) infected snails should be coated with a silicone preparation (Dricote, Fisher Scientific). To facilitate cercarial emergence, snails are exposed to light. Cercariae are concentrated by centrifugation (2000 rpm, 5 min in siliconed tubes), are rinsed several times in sterile Earle's BSS (Difco; Chapter 1: IV, B, 1, f) containing penicillin (100 iu ml^{-1}), streptomycin (100 μg ml^{-1}) and mycostatin (50 μg ml^{-1}) and incubated at 40°C for 20 min. Cercariae are then transferred to sterile siliconed centrifuge tubes and rinsed several times in Earle's saline plus antibiotics before transfer of groups of 200–300 in small volumes of saline to culture vessels, either Falcon TC flasks (30 ml) or Leighton tubes (Bellco, 20 ml) containing 5 or 2·5 ml of medium, respectively. The medium is replaced every 48 h and the red cells are changed weekly. The culture vessels are held upright (static cultures) at 39–40°C; some worms are removed for examination at each change of medium.

A lipid coat should develop around cercarial bodies within 2 or 3 d of culture; the significance of this coat is not understood and it subsequently becomes dislodged from the worms. Organisms ingest red cells and the caeca darken with haematin-like material. Maximal development *in vitro* to the organogeny stage (but not to gametogeny) occurs within 12 d. Males should attain a length of about 2·1 mm and females about 1·4 mm (representing an approximate 8-fold increase compared with cercariae). Interestingly, when Howell and Bourns injected worms cultured for 7 and 9 d into the leg veins of ducks, they obtained patent infections.

II. FASCIOLIDAE

A. *FASCIOLA HEPATICA*

1. *Miracidia, Rediae and Cercariae*

Pullin (1973) described a complex defined basic culture medium (BCM) for *in vitro* cultivation of *F. hepatica* miracidia, rediae and cercariae. The medium was partly based on haemolymph analysis of *Lymnaea truncatula* and partly on formulations in NCTC 109 and Eagle's Medium (Pullin, 1971). Because only limited success has been achieved with BCM it is not described in detail. Those interested should also consult Pullin (1970) and Wilson *et al.* (1971).

BCM can be made from stock solutions and contains inorganic salts, glucose, amino acids, vitamins, antibiotics, but no protein or lipid. Wilson *et al.* (1971) achieved transformation of *F. hepatica* miracidia into mother sporocysts in BCM, but this was also achieved in *Lymnaea* sp. haemolymph. Mother sporocyst development beyond 1 d was not obtained in BCM.

Rediae survived in BCM up to 5 d without further development (Pullin, 1973). Contamination was a major problem although Pullin states that "all stock solutions were sterilized before division into aliquots for storage"; his method of sterilization was not given. Bacteria may have been introduced with rediae at time of transfer to BCM. Pullin (1973) claims that immature cercariae derived from dead/dying rediae appeared to undergo further development in BCM, and when these cercariae were later removed to water they encysted. Since cercariae were obtained from moribund rediae, his claim of *in vitro* cercarial development should be accepted with caution until verification can be made using cercariae derived from "healthy" rediae.

2. *Metacercariae*

A discussion of chemical excystation of *F. hepatica* was not included by Taylor and Baker (1968) and therefore brief mention is made of two early papers. Wikerhauser (1960) treated cysts with 0.5% acidified pepsin solution for 2–3 h, followed by a 1% $NaHCO_3$ wash and a final treatment of 0.4% trypsin plus 20% ox bile. He claimed 80% excystation within 2–3 h following treatment in trypsin-bile. Dixon (1966) described excystation of *F. hepatica* as an active process occurring in two phases, activation and emergence. Activation was initiated by a high CO_2 concentration, reducing conditions and a temperature of 39°C. The CO_2 need be applied for only 5 min but exposure to the reducing agent (a solution of sodium dithionite or cysteine) should be for about 30 min. The emergence phase is stimulated by

10% sheep bile. Readers who attempt to excyst *F. hepatica* metacercariae should consult Dixon (1966).

A simple and effective excystation procedure which eliminates the need for gassing with CO_2 or adding a reductant is that of Sewell and Purvis (1969, p. 28). Dr Caroline Davies (Imperial College of Science and Technology, London) informed me that the procedure works very well provided the excysting medium is *not* sterile. According to Dr Davies, Professor J. D. Smyth suggested that this may be because bacteria present in the non-sterile excysting medium provide advantageous reducing conditions.

3. Cultivation of Excysted Metacercariae

Surprisingly few papers are available on the cultivation of *F. hepatica* from the metacercarial stage. Wikerhauser and Cvetnić (1967) excysted *F. hepatica* according to Wikerhauser (1960) and rinsed the organisms in warm, sterile Hédon-Fleig's medium (Dawes, 1954) supplemented with antibiotics, 1000 iu ml^{-1} of penicillin and 1000 µg ml^{-1} of streptomycin, and maintained them in this medium at 37°C with a gas phase of air for only 2 d. They achieved survival up to 14 d but no postmetacercarial development when excysted metacercariae were transferred to monoxenic cultures containing a primoculture of bovine embryo kidney cells and their Medium A (5% inactivated horse serum plus 0·5% lactalbumin hydrolysate in Hanks's solution). In a later study, Wikerhauser et al. (1970) extended the survival time of excysted metacercariae to 29 d in monoxenic cultures, without postmetacercarial development. Survival was enhanced by transferring flukes to fresh cell cultures on the 7th d with a change of the fluid overlay on d 14.

To my knowledge the only published report of postmetacercarial development of *F. hepatica* is that of Osuna-Carillo and Guevara-Pozo (1974). They maintained excysted metacercariae in a medium of inactivated horse serum plus red cells in 10% CO_2 and 90% N_2 at 38°C for 54 d, and reported an increase in worm size, development of genital primordia and intestinal crura (see Section II, B).

4. Adults

The reader interested in maintaining adult *F. hepatica* should use Rohrbacher's (1957) technique (see Section II, A, 1) which allows for maintenance up to 4 weeks. A more recent maintenance study on adult *F. hepatica* is that of Wikerhauser and Cvetnić (1967). In a cell free medium (Hédon-Fleig's) with penicillin and streptomycin, they obtained adult survival for 5 d at 37°C, whereas in a bovine cell culture medium adult survival was only 2 d. Foster (1970) has maintained adult *F. hepatica* at 28°C for up to 16 d in an Earle-Eagle's medium which was changed daily. Because of the low temperature used by Foster one questions the physiological status of these worms.

Several studies have examined egg production of *F. hepatica in vitro*. For instance, Ractliffe *et al.* (1969) used egg laying of *F. hepatica* to assess the value of a culture medium. Their study involved a multi-variable approach to the *in vitro* cultivation of a trematode and should be of interest to the helminthologist with a good facility for statistics. Ractliffe *et al.* (1969) also contributed a sophisticated culture system in which media can be renewed automatically. They studied worms in a glucose-saline medium (which they admitted was an abnormal environment) and provided some evidence that worms can actually produce eggs in it. Supplements of 30% calf serum plus 10% calf blood were particularly favourable for egg laying. Unfavourable components included 130 units ml^{-1} penicillin and 1 g litre^{-1} of cholesterol. One questions the use of cholesterol in such high concentration in an aqueous medium! Locatelli and Paoletti (1970) also observed egg production during *in vitro* maintenance of *F. hepatica*. Whereas 1800 eggs were produced per adult on the first day, this number dropped to 300 by d 2, less than 100 by d 3, and was negligible by d 6.

1. Detailed Techniques

1° *Fasciola hepatica* chemical excystation of metacercariae
(Sewell and Purvis, 1969)

N.B.: The original technique has been modified slightly, based on Dr Caroline Davies' suggestions (personal communication).

Cercariae of *F. hepatica*, obtained from either naturally or experimentally infected *Lymnaea* sp., encyst on cellophane or the walls of the maintenance vessel. Cysts scraped from the walls, or cellophane bearing cysts, can be placed in dry vessels (that can be capped tightly). Add 0·05N-HCl at 39°C to the vessel and then immediately add an equal volume of a solution containing 1% NaHCO$_3$, 0·8% NaCl and 20% ox bile at 39°C. Cap the vessel immediately to avoid loss of CO$_2$ which is produced. Incubate at 39°C for 4–5 h by which time 70–80% should have excysted.

2° *Fasciola hepatica* excysted metacercariae
(Osuna-Carrillo and Guevara-Pozo, 1974)

MEDIUM

Inactivated horse serum (5·0 ml, 56°C, 30 min) plus 7 drops of sheep red blood cell suspension washed in Alsever's solution (the constituents were obtained from Instituto Llorente, S.A., Madrid).

TECHNIQUE

Metacercariae are obtained from experimentally infected *Lymnaea truncatula* snails and excysted according to a modified Dixon (1966) procedure. N.B.: I have been unable to translate their exact excystation procedure and subsequent aseptic treatment of excysted metacercariae. I suggest that the reader who uses their medium and procedures excysts metacercariae according to Sewell and Purvis (1969: see Section II, B, 1, 1°), then allows the excysted metacercariae to settle in vessels containing sterile Hédon-Fleig's or Locke's solution (Chapter 1: IV, B, 1, d) containing penicillin $(100–200 \text{ iu ml}^{-1})$ and streptomycin $(100–200 \text{ µg ml}^{-1})$.

Forty excysted metacercariae are pipetted into a Carrel flask containing 5 ml of medium; the flask is gassed with a $10\% \text{ CO}_2$ $90\% \text{ N}_2$ mixture and incubated at 38°C. The medium is changed every fourth day. Treatment and transfer steps are carried out in a sterile laminar air flow chamber. During medium change, some specimens are removed, examined microscopically and stained to evaluate growth and development.

The authors claim an increase in length of about four-fold during the 54 d period, with worms developing intestinal ramifications, genital primordia, testicular masses, the outline of the cirrus sac and vitellaria.

3° *Fasciola hepatica* adults
(Rohrbacher, 1957)

MEDIUM

Balanced Saline

	g litre^{-1}
NaCl	2·4
KCl	0·75
$CaCl_2$	0·55
$MgCl_2.6H_2O$	0·2
$Na_2HPO_4.2H_2O$	2·7
$NaHCO_3$	0·84
Sodium citrate	2·9
Glucose	3·6

(i) Dissolve all above compounds except $CaCl_2$, $MgCl_2.6H_2O$, $NaHCO_3$ and glucose, in 600 ml double distilled water.

(ii) Dissolve $CaCl_2$ and $MgCl_2.6H_2O$ in 200 ml double distilled water.

(iii) Add ii to i stirring all the time. Sterilize by filtration.
(iv) Dissolve NaHCO₃ in 100 ml double distilled water and sterilize by positive pressure filtration.
(v) Dissolve glucose in 100 ml double distilled water and autoclave.
(vi) Mix iii, iv and v thoroughly, maintaining sterility.

Autoclaved Liver Extract

Macerate bovine liver in a Waring blender with an equal volume of distilled water, autoclave to precipitate the protein, and filter. The filtrate is bottled, autoclaved and used in the proportion of 200 ml liver extract to 800 ml of saline.

Crude Liver Extract

Injection crude liver—MRT (manufactured by Marvin R. Thompson Inc., Stamford, Conn. U.S.A.) is dispensed in quantities of 10 ml litre^{-1} of medium.

TECHNIQUE

The parasites are removed from infected bovine livers immediately after slaughter of the animals. They are expressed from the bile ducts into sterile balanced saline without glucose and transported rapidly to the laboratory (50 flukes in 200 ml saline) at 37°C.

On arrival at the laboratory, the flukes are transferred to fresh sterile saline containing 2·5 mg ml^{-1} streptomycin and 1000 iu ml^{-1} penicillin G. After 18 h in this solution, at 37°C, any contaminated (i.e. discoloured) flukes should be discarded and the remainder cultivated in the full culture medium. Each fluke can be cultured in 50 ml of medium in a rubber-stoppered 4 oz oil sample bottle. The medium should be changed at least once a week.

Rohrbacher found that adult flukes survived under the above conditions for 3–4 weeks.

B. *FASCIOLOPSIS BUSKI*

Lo and Cross (1974) reported observations on the *in vitro* cultivation of the giant intestinal fluke, *Fasciolopsis buski*. Of various media tried, best success was achieved in Rohrbacher's (1957) saline (see Taylor and Baker, 1968, p.

174): egg laying was observed in culture (about 2500 eggs worm^{-1}) for the first 10 d and two adult worms survived for 20 d.

III. STRIGEIDAE

A. *COTYLURUS LUTZI*

Considerable success has been achieved in the cultivation of the strigeid *Cotylurus lutzi*, and in time this trematode may be the first whose entire life cycle is completed *in vitro*. The larval stages of *C. lutzi*, including the tetracotyle (an encysted metacercaria with characteristic muscular pseudo-suckers called cotylae), infect *Biomphalaria glabrata* snails. Tetracotyles of *C. lutzi*, fed to laboratory finches, develop into adults and produce eggs within 7 d in the intestine (Basch, 1969).

Voge and Jeong (1971) cultivated tetracotyles of *C. lutzi* to ovigerous adults within 7 d (about the same time as in the finch host) in an NCTC 135–50% chicken serum medium. Cultures were maintained at 39–41°C (pH 7·3–7·4) with a gas phase of air. Although eggs were laid in the medium, they did not possess an outer shell nor did they embryonate (see Section III, 1, 1°). Basch *et al.* (1973) modified the Voge and Jeong medium by adding a mucosal extract obtained from the upper intestine of adult white leghorn chickens (i.e. 40% NCTC 135, 40% chicken serum and 20% chicken mucosal extract), and cultivated *C. lutzi* tetracotyles to adults containing *normal* eggs. Miracidia from eggs obtained *in vitro* were infective to *B. glabrata* snails and eventually produced cercariae that encysted as tetracotyles in the snails. These tetracotyles, when fed to finches, produced normal adults (see Section III, 1, 2°).

Basch and DiConza (1975) recently reported the *in vitro* cultivation of pre-adult and adult stages of *C. lutzi*. Larval *C. lutzi* removed from *B. glabrata* snails one day after cercarial infection were cultivated for up to 48 d at 25 ± 1°C in the DiConza and Basch sporocyst culture medium (see Section I, F, 3°) and some of these forms developed into tetracotyles with rigid cyst walls. Such tetracotyles placed in the adult culture medium of Basch *et al.* (1973) excysted, developed towards the adult stage, but did not produce viable eggs.

1. *Detailed Techniques*

1° *Cotylurus lutzi* tetracotyles to ovigerous adults
(Voge and Jeong, 1971)

MEDIUM

Of various media tried the most successful is NCTC 135 (Gibco) with 50%

chicken serum (v/v) (inactivated; 56°C, 30 min). Fluid media contain 100 μg ml^{-1} streptomycin and 100 iu ml^{-1} penicillin.

TECHNIQUE

Tetracotyle larvae are removed from experimentally infected *B. glabrata* snails, and transferred through 3 successive 10 min changes of Earle's saline (Chapter 1: IV, B, f) containing 100 iu ml^{-1} penicillin and 100 μg ml^{-1} streptomycin after which they are incubated in prewarmed sterile Earle's saline at 40–41°C: activation occurs within 20 min without the need for added enzymes or bile salts. Worms are then placed, 10–15 per tube, in 10 × 100 mm screw cap tubes containing 4 ml of fluid medium (pH 7·3– 7·4). The caps of the tubes are sealed externally with double layers of parafilm and the tubes are incubated in an inclined position at 39 or 41°C. After the first 2 d of culture, one-half of the medium is replaced with fresh medium every 24 h.

This procedure should produce ovigerous adults with *abnormal* eggs in about 7 d.

2° *Cotylurus lutzi* tetracotyles to adults producing normal eggs
(Basch *et al.*, 1973)

MEDIUM

Of various media employed the one that is most successful consists of 40% NCTC 135 (Gibco) and 40% chicken serum (Gibco), with 20% mucosal extract from chicken upper small intestine: 100 iu ml^{-1} penicillin, 100 μg ml^{-1} streptomycin and 10 μg ml^{-1} gentamicin are added. The mucosal extract is made up as follows.

About 35 cm of upper intestine (including the duodenum and part of the upper ileum) is removed, from a freshly killed 2–3 kg adult white leghorn chicken, slit longitudinally and washed in several changes of cold (4°C) saline. The mucosal layer is scraped off with a glass microscope slide, collected, and homogenized in a glass tissue grinder. This homogenate is made up to 20 ml with normal saline and then centrifuged (60 min, 3000 rpm, 5°C). The supernatant fluid is filtered twice through paper, then through 0·65 and 0·45 μm pore size Millipore filters for sterilization. This extract can be stored at −10°C until use. N.B.: dialysis of this extract using cellophane sacs produces a dialysate that is incapable of supporting production of normal worm eggs when added to the basic medium.

TECHNIQUE

Procedures follow very closely those of Voge and Jeong (1971; see Section

III, A, 1, 1°) with some minor exceptions. Cultures are set up in 25 × 150 mm glass tubes containing 5 ml of medium. About 15 "well-encysted" tetracotyles are placed in each tube which is closed by a silicone stopper and then sealed with parafilm. Medium is changed on alternate days. Cultures are maintained at 40°C in a stationary water bath. (N.B.: Full experimental details omitted from original paper; trial and error may be needed for successful repetition!)

Cultures produce ovigerous worms as early as 5 d (even earlier than development in the finch). Eggs obtained from such worms, when incubated in distilled water, embryonate and produce miracidia which hatch and are capable of infecting *B. glabrata* snails.

IV. DIPLOSTOMATIDAE

A. *DIPLOSTOMUM* SPP.

Kannangara and Smyth (1974) contributed a significant study on the *in vitro* cultivation of *Diplostomum spathaceum* and *D. phoxini* metacercariae (Section IV, 1, 1°). These metacercariae lack genital rudiments and are apparently more difficult to culture than trematodes whose larvae contain well-developed genital anlage. Kannangara and Smyth explored various techniques, including truly *in vitro* cultivation and chick embryo cultivation, and used numerous media including liquid, semi-solid and diphasic types. They found that the larvae could be stored in NCTC 135 at 4°C for at least 2 months and still retain viability. At the end of their study, they commented on the impossibility of developing a "universal" medium for the cultivation of digenetic trematodes.

1. *Detailed Technique*

1° *Diplostomum spathaceum* and *D. phoxini*
(Kannangara and Smyth, 1974)

This paper should be referred to for a full account of the various media used and the development of both species *in vivo* and *in vitro*.

MEDIUM

Considerable success with both species was obtained with a semi-solid medium referred to as whole egg macerate, prepared as follows: using aseptic procedures the contents of 2 hens' eggs are coagulated by heating (1·5 h, 80°C). The coagulated material is transferred to a sterile glass homogenizer with an equal volume of NCTC 135 containing 2% yeast extract (Gibco), 1% glucose and 20% foetal calf serum. The mixture is then slowly homogenized for 15 sec in a sterile glass Waring blender and the homogenate stored in sterile bottles at 4°C.

TECHNIQUE

Metacercariae of *D. spathaceum* are removed aseptically from the lens capsule of the roach, *Rutilus rutilus*, and metacercariae of *D. phoxini* from the brain of minnows, *Phoxinus phoxinus*. Metacercariae are placed in sterile NCTC 135 (Gibco?) before inoculation into disposable plastic tubes containing whole egg macerate medium. The gas phase is air, pH is maintained at approximately 7·4 and cultures are continuously agitated in an agitating incubator (Gallenkamp) (40 rev min^{-1}, 41°C).

In whole egg macerate medium *D. spathaceum* produced eggs on d 12, compared with d 6 *in vivo* (ducks). This is the first study in which *D. spathaceum* has been cultivated *in vitro* to the egg-producing stage. *D. phoxini* has been cultivated previously (see Taylor and Baker, 1968, pp. 179–182). In Kannangara's and Smyth's study an improved vitelline and growth response was obtained in *D. phoxini* and worms produced eggs on d 4 compared with d 2 *in vivo* (ducks). Kannangara and Smyth discussed the probable importance of physical factors associated with the viscosity of their medium.

V. ECHINOSTOMATIDAE

A. *ECHINOPARYPHIUM SERRATUM*

1. *Metacercariae*

Howell (1968) devised a chemical procedure for excysting metacercariae of *Echinoparyphium serratum* (see Section V, A, 1, 1°) and achieved post-metacercarial development of this echinostome in various media. Best development, although suboptimal when compared with worms grown in ducklings, was achieved in a yolk–albumen–saline medium supplemented with yeast extract (see Section V, A, 1, 2°). Worms cultivated in this medium at 39°C with a gas phase of air and agitated continuously, developed to the vitellogenesis stage and some appeared to contain eggs. Worms grown *in vitro* were stunted and development was delayed when compared with those grown in ducklings. For instance, vitellogenesis was achieved in about 140 h *in vitro* compared with 60 h *in vivo*. Howell's (1968) chemical excystation procedure essentially involved an acid pepsin treatment, followed by treatment with the reductant, sodium dithionite, followed by excysting fluid (trypsin-sodium cholate solution). Later, Howell (1970) was mainly concerned with detailed factors involved in excystment of this species and found that acid pepsin treatment was not obligatory for excystation; he also emended some of the excystation procedures published in his earlier paper.

2. Detailed Techniques

1° *Echinoparyphium serratum* excystation
(Howell 1968, 1970)

N.B.: Procedural differences are apparent in the two papers and both should be consulted. The procedure presented below is based on information derived from both papers.

Cysts of *Echinoparyphium serratum* are obtained from the pericardial tissue of naturally infected fresh water snails, *Isidorella brazieri*. Pericardial tissue containing cysts is dissected from the snails in 0.7% saline. The following steps are conducted at 37–39°C in a water bath or in a thermostatically controlled cabinet equipped with a binocular microscope so that excystation can be observed if desired. Cysts are transferred to a McCartney bottle or Leighton tube containing 5 ml of 0.025% pepsin (3X crystalline, May and Baker) in Hanks's (Chapter 1: IV, B, g) saline adjusted to pH 2.0 with 0.1 N-HCl for 15 min. Cysts are then rinsed briefly in prewarmed (?37–39°C) 0.85% NaCl before transfer to a reducing agent [0.02 M sodium dithionite, May and Baker, made up by adding the crystalline salt to 10 ml of *pregassed* (29% CO_2 in N_2) Hanks's saline to avoid oxidation]. After 15 min in the reductant, cysts are briefly rinsed in prewarmed 0.85% NaCl and transferred to 10 ml of the excysting solution [0.3% trypsin (4X U.S.P. pancreatin, Nutritional Biochemicals), 0.05% sodium cholate (Nutritional Biochemicals) in Hanks's saline gassed to pH 7.3 with 9.8% CO_2 in N_2].

This procedure should result in about 75% excystation after 10 min in the trypsin–sodium cholate solution.

2° *Echinoparyphium serratum* metacercariae
(Howell, 1968)

MEDIUM

Of various media used by Howell, best development is obtained in his Medium 21 which consists of 10 parts hens' egg yolk, 2 parts hens' egg albumen, 2 parts Hanks's saline (Chapter 1: IV, B, g) and 4 mg ml^{-1} yeast extract (pH 6·5–6·7) containing 100 iu ml^{-1} penicillin and 100 µg ml^{-1} streptomycin. No special gas phase is used with Howell's yolk-containing media. The egg yolk and albumen are separately drained into sterile bottles after painting the shell with alcohol-iodine. The methods of preparing yolk and albumen are based on techniques in Bell and Smyth (1958) as follows: Yolk is removed by aseptically inserting a wide bore pipette into the centre of the yolk of a hen's egg. The liquid fraction of hen-egg albumen is most distinct in fresh eggs from which 5–10 ml can be sucked off with a fine Pasteur pipette. The yeast extract is prepared as a 6% aqueous stock solution and sterilized by heating at 100°C for 20 min.

TECHNIQUE

After 20 min in the trypsin–sodium cholate, excysting fluid (see Section V, A, 1, 1°) metacercariae are transferred to sterile Petri dishes containing Hanks's saline–antibiotic solution. Twenty excysted metacercariae are then transferred to a McCartney bottle containing Hanks's saline with 100 iu ml^{-1} penicillin and 100 µg ml^{-1} streptomycin. The metacercariae are rinsed several times with repeated changes of the Hanks's saline–antibiotic solution and finally resuspended in 10 ml of culture medium 21 per McCartney bottle. These are incubated in a shaking water bath (120 agitations min^{-1}) at 39 ± 1°C.

At least some of the organisms should attain the vitellogenesis stage in about 140 h.

B. *ECHINOSTOMA MALAYANUM*

1. *Adult*

Jaw and Lo (1974) cultured immature and mature adults of *Echinostoma malayanum* obtained from experimentally infected rats, *Rattus norvegicus*. Worms were maintained individually in 2 ml of various media in Leighton tubes kept in the dark at 37°C without agitation. Maximal survival of immature worms was 14 d in Medium 199 and 53 d for mature adults in Earle's saline containing 20% pig serum. Although eggs were laid, growth or development was not observed and certain organs degenerated with prolonged maintenance. Mature worms usually survived twice as long as immature ones, regardless of the medium used.

VI. PSILOSTOMATIDAE

A. *SPHAERIDIOTREMA GLOBULUS*

Psilostomes are echinostome-like trematodes which lack oral collar spines. Macy *et al.* (1968) successfully excysted the metacercariae of the psilostome, *Sphaeridiotrema globulus* with acid pepsin and alkaline trypsin solutions (see Section VI, A, 1, 1°); their paper provided considerable information on physicochemical properties of the avian gut (i.e. redox potential, pH values, emptying times, etc.). The *in vitro* excystation requirements of *S. globulus* were also correlated with conditions found in the avian intestinal tract, a study which provides important information on the host specificity of trematodes.

The cultivation of chemically excysted metacercariae of *S. globulus* to ovigerous adults in a defined medium [NCTC 109 plus 20% egg yolk (NCTC 109-20Y); pH 7·4–8·0; gas phase of 10% O_2, 10% CO_2 and 80% N_2; 42°C] by Berntzen and Macy (1969) was a landmark in the *in vitro* cultivation of an hermaphroditic digene. The eggs obtained were normal and capable of embryonation and hatching. Unfortunately, attempts to infect snails with miracidia obtained from these eggs were not made (see Section VI, A, 1, 2°).

Cultivationists in the nineteen-seventies frequently use the procedure of Berntzen and Macy to prepare egg yolk supplement. Egg yolk supplemented media were not successful in the cultivation of *Cotylurus lutzi* (Voge and Jeong, 1971; Basch *et al.*, 1973), or *Metagonimus yokogowai* (Yasuraoka and Kojima, 1970), but were successful for the cultivation of *Leucochloridiomorpha constantiae* (Fried and Contos, 1973). To date relatively little information is available on factors in egg yolk (physical, chemical or both) which contribute to successful cultivation.

1. *Detailed Techniques*

1° *Sphaeridiotrema globulus* **metacercarial excystation**
(Macy, Berntzen and Benz, 1968)

MEDIUM

Of various solutions used in the above study, the following are the most useful for optimal excystation in a physiologically useful time:

Sol. 1. 1% pepsin (1:10,000, Sigma Chemical) and 1% HCl in 0·85% saline.

Sol. 2. Earle's BSS (Chapter 1: IV, B, f) plus 1% trypsin (1:250, Difco) plus sufficient $NaHCO_3$ to give a pH of 8·0.

TECHNIQUE

Encysted metacercariae of *S. globulus* are obtained from naturally infected *Flumenicola virens* snails by one of two methods. In the direct method snails are crushed, tissues are dissected and cysts are removed with a Pasteur pipette to saline. Cysts freed of snail tissue are transferred to a second dish of saline. The second method (indirect) is useful when large numbers of cysts are required. This technique involves digestion and differential centrifugation and the original paper should be consulted for details.

For excystation using the direct method, transfer a known number of cysts to a culture tube containing 10 ml of pepsin (Solution 1) maintained at either 37 or 42°C for 30 min. Rinse the cysts briefly in a prewarmed saline solution and then transfer them to a culture tube containing 10 ml of alkaline trypsin (Solution 2) maintained at 37 or 42°C for 2 h.

This procedure should provide about 80% excystation.

2° *Sphaeridiotrema globulus* metacercariae to ovigerous adults
(Berntzen and Macy, 1969)

MEDIUM

Of 5 different media tried the most successful was NCTC 109 (Difco) with 20% egg yolk (NCTC 109-20Y) prepared as follows. Fresh chicken eggs are placed in 70% ethanol for 30 min then opened, under a UV lamp, and the yolk separated from the albumen and placed in a sterile beaker. The yolk is broken with a sterile glass rod and then transferred to a sterile graduated cylinder. To each 80 ml of NCTC 109, 20 ml of yolk are added and mixed with a magnetic stirrer. The medium is centrifuged in sterile tubes (2000 g 1 h, room temp.). Remove the supernatant to a sterile bottle and adjust the pH to 7·8 with 5% $NaHCO_3$ just before use.

N.B.: during cultivation Berntzen and Macy maintained the O_2 level in the medium at 10% for all cultures. However, the CO_2, N_2 and $NaHCO_3$ levels were varied in different experiments to give desired pH values within a range of 7·2–8·0. Adjustments of gas mixtures were accomplished by a specially designed micro gas machine. For the details of the design and how to use the apparatus see Berntzen (1966).

TECHNIQUE

The excystation procedure is essentially as described previously (see Section VI, A, 1, 1°) but following isolation of cysts from snails, aseptic techniques are used. Before acid pepsin treatment, cysts should be incubated at 37°C for 30 min in 0·85% NaCl containing 50 mg ml^{-1}

streptomycin and 24 mg neomycin ml^{-1}. The excysting medium, used just before *in vitro* cultivation, is 1% trypsin adjusted to pH 8·3 with $NaHCO_3$. Excysted metacercariae are transferred to Kimax 15 × 150 mm screw cap tubes (25 tube^{-1}) each containing 10 ml of NCTC 109-20Y. Cultures are maintained at either 37 or 42°C in an automatic roller drum set to rotate for 5 min at 2 h intervals. Five ml of medium are replaced with fresh, daily.

Under optimal conditions (i.e. NCTC 109-20Y, 42°C, pH 7·8, gas phase of 10% O_2, 10% CO_2, 80% N_2) worms should become ovigerous in about 126 h. This is still considerably slower than *in vivo* (duck), where ovigerous worms occur about 68 to 72 h after infection.

VII. BRACHYLAEMIDAE

A. *LEUCOCHLORIDIOMORPHA CONSTANTIAE*

Leucochloridiomorpha constantiae metacercariae have been cultivated *in vitro* to ovigerous adults within 4 days in NCTC 135 (Gibco) plus 20% egg yolk (NCTC 135-20Y) at 37·5–41°C, pH 7·2–8·0 in static cultures with a gas phase of air (Fried and Contos, 1973). Some eggs had shells, but their viability was not determined. Procedures and media preparation closely followed those of Berntzen and Macy (1969, see Section VI, A, 1, 2°). Subsequently Contos and Fried (1976) showed that tegument development of worms cultivated in NCTC 135-20Y was similar to that of worms grown in the domestic chick.

VIII. GYMNOPHALLIDAE

A. *PARVATREMA TIMONDAVIDI*

Yasuraoka *et al.* (1974) cultivated metacercariae of *Parvatrema timondavidi* to ovigerous adults in NCTC 109 supplemented with 20% inactivated chicken or bovine serum. They used roller tube cultures, a gas phase of 8% CO_2 in air and maintained their cultures at 37 or 41°C. Eggs obtained *in vitro* were capable of embryonation and miracidial formation.

1. *Detailed Technique*

1° *Parvatrema timondavidi*
(Yasuraoka *et al.*, 1974)

MEDIUM
(i) Krebs–Ringer–Tris-malate solution

This solution is most conveniently made as a 10 times stock solution of the saline and of Tris-malate, both of which may be stored. Appropriate quantities of each are then mixed, and brought to the desired volume.

$10 \times$ KREBS–RINGER'S SALINE

	g litre^{-1}
NaCl	70·140
KCl	3·579
CaCl$_2$	2·886
MgSO$_4$	2·958
MgSO$_4$.7H$_2$O	6·049
Distilled water	to one litre

$10 \times$ TRIS-MALATE BUFFER, pH $7 \cdot 2$ ($0 \cdot 25$ M)

Tris (hydroxymethylaminomethane) ("Trizma Base" Sigma Chemical)	30·275
Maleic acid	29·018

Dissolve in about 950 ml distilled water in a beaker. Adjust to pH 7·2 by adding about 5–10 ml of 12 M-NaOH drop by drop with mixing and make up to 1 litre.

When making up 1 × saline-buffer, check the pH before adjusting to final volume, add 0·1 N-HCl or NaOH as needed, then bring to volume.

(ii) NCTC 109 (Difco) supplemented with 20% bovine or chicken serum (inactivated; 56°C, 30 min) and 200 iu ml^{-1} penicillin and 100 µg ml^{-1} streptomycin.

TECHNIQUE

Metacercariae are obtained from the inner part of the umbo area of naturally infected *Tapes philippinarum* bivalves. Before cultivation the viscous material surrounding the metacercariae must be removed by incubating them in sterile 0·85% NaCl (pH 7·1) at 37°C for 8 h. This treatment should yield 70–80% excystation. Excysted metacercariae are then introduced into 12 × 100 mm tubes containing 5–10 ml sterile Krebs–Ringer–Tris-malate solution (pH 7·1) with double the antibiotic strength (400 iu ml^{-1} penicillin, 200 µg ml^{-1} streptomycin) listed under medium. Larvae are allowed to settle and most of the saline is removed. Worms are then transferred to another tube containing the salt solution and the washing process is repeated five times to get rid of debris. For cultivation 50 or 100 metacercariae are transferred to 14 × 150 mm culture tubes containing 2 ml of the medium. The tubes are loosely capped and incubated at either 37 or 41°C in a flowing atmosphere of 8% CO_2 in air. The tubes are rotated in drums at 12 rev. h^{-1}.

Development *in vitro* proceeded at about the same rate (36 h for the production of ovigerous worms) as in the experimental mouse host. However, the number of eggs laid *in vitro* was much lower. Although this species infects birds in nature, the authors were unable to infect baby chicks or ducklings and therefore were not able to compare their *in vitro* results with observations in an experimental avian host.

IX. HETEROPHYIDAE

A. *METAGONIMUS YOKOGAWAI*

Yasuraoka and Kojima (1970) excysted metacercariae of *Metagonimus yokogawai* in a trypsin solution following pre-treatment in acidified pepsin. Excysted metacercariae were then cultivated to ovigerous adults (but with imperfectly formed eggs) in a medium consisting of chick embryo extract, human serum, and NCTC 109 in a ratio of 4:3:3 at 37·5°C and a gas phase of 8% CO_2 in air (see Section IX, A, 1, 2°).

1. *Detailed Techniques*

1° *Metagonimus yokogawai* excystation
(Yasuraoka and Kojima, 1970)

Metacercariae of *M. yokogawai* are obtained from naturally infected "sweet fish", *Plecoglossus altivelis*. Cysts are located mainly on the scales. Scales are removed from the fish with a scalpel, rinsed in several changes of sterile 0·4% NaCl and then placed in 0·7% HCl-0·03% Difco Pepsin N.F. (1:3000) for 3 h at 39–40°C. (This removes cysts from scales, but does not induce excystation.) The cysts are rinsed twice in sterile 0·8% NaCl and then incubated (30 min, 37°C) in the excystation medium [Haemoglobin, 20,000 units ml^{-1} trypsin diluent, pH 7·1 (Mochida Pharmaceutical)]: possibly similar results could be obtained using a 0·5% solution of trypsin (4-X U.S.P. Pancreatin, Nutritional Biochemical) in Earle's BSS (see Chapter 1: IV, B, 1, f) adjusted to pH 7·1 with $NaHCO_3$.

2° *Metagonimus yokogawai* metacercariae
(Yasuraoka and Kojima, 1970)

The original paper should be consulted for details of the various media used and for a comparison of *in vivo* and *in vitro* development.

MEDIUM

Of various media tested the most successful is Medium C which consists of 4 parts of CEE_{100} (Weinstein and Jones, 1956, 1959, see Chapter 1: IV, A, 1) 3 parts of human serum (from a blood bank; inactivated 56°C, 30 min), and 3 parts of NCTC 109 (Difco) and contains penicillin, 200 iu ml^{-1}, and streptomycin, 100 μg ml^{-1}; it is adjusted to pH 7·2 with 5% $NaHCO_3$ and should be replaced daily or at least every other day with the usual sterility precautions.

TECHNIQUE

Following excystation, metacercariae are washed several times in Krebs–Ringer–Tris solution (Section VIII, A, 1, 1°, i) containing 200 units ml^{-1} penicillin and 100 μg ml^{-1} streptomycin. Fifty or 100 metacercariae are placed aseptically in culture tubes (14 × 150 mm with Morton Closure) containing 2 ml of the medium and rotated in drums (12 rev. h^{-1}) in a CO_2 incubator (37·5°C) gassed with 8% CO_2 in air.

This procedure should result in ovigerous adults (but with imperfectly formed eggs) within 12 d. The inadequacy of the medium is apparent when one compares development of this species in the mouse where gravid worms with viable eggs are obtained in 5 d. Moreover, worms grown *in vitro* were considerably smaller than those obtained from mice.

Attempts to cultivate *M. yokogowai* in defined medium plus egg yolk (successful for *Spaeridiotrema globulus*, Berntzen and Macy, 1969) or in defined medium plus serum (successful for *Cotylurus lutzi*, Voge and Jeong, 1971) were unsuccessful, indicating that the physicochemical needs of different trematodes vary considerably.

X. MICROPHALLIDAE

A. *MICROPHALLUS PYGMAEUS*

The cultivation of the microphallid, *Gynaecotyle adunca*, has been discussed by Taylor and Baker (1968). Subsequent cultivation studies on microphallids involve the maintenance of daughter sporocysts of *Microphallus pygmaeus* obtained from the marine gastropod, *Littorina saxatillis tenebrosa*. Pascoe *et al.* (1970) and Richards *et al.* (1972) showed that a modified Medium 199 was more suitable for long term maintenance of *M. pygmaeus* daughter sporocysts than either artificial or natural sea water (see Section X, A, 1, 1°).

1. *Detailed Technique*

1° *Microphallus pygmaeus* daughter sporocysts
(Richards *et al.*, 1972)

MEDIUM

(i) Medium 199 (Glaxo) containing a modified Earle's salt solution plus the following ingredients, added to simulate the internal environment of the marine molluscan host (mg litre^{-1}): sodium chloride, 17,181; potassium chloride, 342; calcium chloride, 935; magnesium chloride, 5102; sodium sulphate, 4012; sodium bromide, 85; strontium chloride, 11; boric acid, 27; maltose, 2000; L-arabinose, 500. (The maltose and arabinose are added because they are suspected of being present in small amounts in *Littorina saxatilis*.) The final solution, pH 6·8–7·4, is stored at 2°C and used within 10 d.

(ii) Natural Sea Water. Sea water is filtered and sodium penicillin G 170 iu ml^{-1} and streptomycin sulphate 170 µg ml^{-1} are added (Pascoe *et al.* 1968).

(iii) Artificial Sea Water (Barnes, 1954).

	g litre^{-1}		
NaCl	23·991		
KCl	0·742		
$CaCl_2$	1·135	($CaCl_2.6H_2O$	2·240)
$MgCl_2$	5·102	($MgCl_2.6H_2O$	10·893)
Na_2SO_4	4·012	($Na_2SO_4.10H_2O$	9·100)
$NaHCO_3$	0·197		
NaBr	0·085	(NaBr$.2H_2O$	0·115)
$SrCl_2$	0·011	($SrCl_2.6H_2O$	0·018)
H_3BO_3	0·027		

Dissolve in distilled water and make up to 1 litre. Antibiotics added as for natural sea water. Chlorinity = 19‰; salinity = 34·33‰ (calculated from Barnes' data, 1954). Artificial sea water (Aquarium Systems) could also be used.

TECHNIQUE

Daughter sporocysts are removed from the haemocoel or the digestive gland of *L. saxatilis*, and washed in filtered sea water containing penicillin and streptomycin. Only mature sporocysts of about the same size and containing fully formed metacercariae should be selected. These are placed individually in small covered glass dishes with 5 ml of medium (either modified M199, artificial or natural sea water), and maintained at 3–4°C.

N.B.: Pascoe *et al.* (1970) maintained 20 sporocysts in 5 ml of medium at either 4, 10 or 20°C.

Regardless of the metabolic criteria used, the sporocysts began to degenerate almost immediately in the non-nutrient media (either artificial or natural sea water), but had a high rate of metabolic activity and remained apparently healthy for up to 36 d in modified Medium 199 (Richards et al., 1972). Pascoe et al. (1970) assessed the media strictly on the longevity of sporocysts at various temperatures and concluded that the modified M199 was most suitable for long term maintenance at 4 or 10°C. For instance at 10°C, maximal survival of sporocysts in modified M199 was 56 d compared with 16 d in natural sea water and 10 d in artificial sea water.

These studies may be generally applicable to other daughter sporocysts and possibly rediae obtained from marine gastropods.

XI. ASPIDOGASTREA

A. *ASPIDOGASTER CONCHICOLA*

Members of this subclass of trematodes usually live in unionid bivalves. However, Michelson (1970) found a representative species, *Aspidogaster conchicola*, in two species of fresh water gastropods, *Viviparous malleatus* and *V. japonicus*, in the U.S.A. He maintained *A. conchicola* axenically for up to 135 d in a modified Unionid Ringer's solution.

1. *Detailed Technique*

1° *Aspidogaster conchicola*
(Michelson, 1970)

MEDIUM

Unionid Ringer's solution (Ellis et al., 1931) in g litre^{-1}: NaCl, 1·53; KCl, 0·15; $CaCl_2$, 0·12; $MgCl_2$, 0·10; Na_2HPO_4, 0·09. To this is added clam blood (5% obtained aseptically from the unionid, *Anodonta implicata*), penicillin G (100 iu ml^{-1}), streptomycin sulphate (100 µg ml^{-1}) and sufficient $NaHCO_3$ to yield a pH of 7·9.

TECHNIQUE

A. conchicola are obtained aseptically from *Viviparous malleatus* or *V. japonicus* gastropods and maintained individually in 5 ml of the above medium in glass roller tubes (20 × 150 mm) sealed with non-reactive rubber stoppers and opened every 10 d.

Michelson maintained his cultures at 4, 20 and 25°C and average worm survival was 92, 65 and 49 d, respectively. Maximum worm survival at 4°C was 135 d. Many worms deposited eggs *in vitro* and most eggs hatched. Although the resulting larvae remained alive for up to 28 d, no adults developed from them.

XII. TROGLOTREMATIDAE

A. *PARAGONIMUS WESTERMANI*

Taylor and Baker (1968) reviewed early studies on the cultivation of *Paragonimus westermani* metacercariae. They provided details of the studies of Yokogawa *et al.* (1955, 1958; Taylor and Baker, 1968, pp. 187–188) who obtained postmetacercarial development and growth (ovigerous worms were not obtained) and worm survival up to 203 d. Recently, Kannangara (1974) cultivated excysted metacercariae of *P. westermani* in various media. Best success was obtained in his Medium No. 1 (40% human serum, 40% NCTC 135 with added 2% yeast extract, 10% human red blood cells and 100 iu ml^{-1} of penicillin and 100 μg ml^{-1} streptomycin. Cysts, transported by air from Japan to England, were excysted by the method of Wikerhauser (1960, section II, A, 2) for *Fasciola hepatica*. Excysted metacercariae were transferred to sterile plastic (11 ml) tubes containing 7·5 ml of Medium No. 1, agitated (Gallenkamp agitating incubator) at 40 rev. min^{-1} at 37°C. On d 29 worms were large and active, had adult proportions, their caeca were distended with red cells, and their testes were in the 16 cell stage. Worms in these cultures became contaminated by d 43 and died (according to Kannangara, contamination probably resulted from the frequent medium changes), but one worm showed imperfectly formed ova.

XIII. OPISTHORCHIIDAE

A. *CLONORCHIS SINENSIS*

As reviewed by Taylor and Baker (1968), Hoeppli and Chu (1937) maintained adult and immature *Clonorchis sinensis* for 5 months at 37°C in various sera diluted with an equal volume of Tyrode's solution. Because of certain disadvantages in using media containing sera, Sun (1969) explored the use of Medium 199 for the maintenance of adult *C. sinensis*. He maintained adults in M199 with or without the addition of 5% rabbit serum in test tube cultures at 37°C. The medium was renewed weekly. Flukes maintained in M199 with 5% serum showed maximal survival of 96 d and average survival of 21·4 d. In M199 alone, maximal survival was 28 d (average 15·1 d).

XIV. ISOPARORCHIIDAE

A. *ISOPARORCHIS HYPSELOBAGRI*

Isoparorchis hypselobagri lives as an adult in the swim bladder of the fresh water fish, *Wallago attu*. According to Nizami and Siddiqi (1975) the swim bladder habitat is relatively simple, is rich in oxygen and provides blood as the major nutrient for the trematode. These authors feel that *I. hypselobagri* provides a good physiological model and therefore a synthetic or chemically defined medium should be devised for its *in vitro* cultivation. They maintained flukes individually in sterile Ehrlenmeyer flasks containing 20 ml of a sterile balanced salt solution containing glucose (consult original paper for details). The medium was changed on alternate days and presumably cultures were maintained at room temperature. Under aerobic conditions (gas phase of air) maximal worm survival was 49 d compared with 30 d under "anaerobic" conditions (nitrogen gas bubbled into the medium). Worms lost much weight during cultivation, but this loss was reduced considerably when 5 ml of bovine blood (ACD, acid–citrate–dextrose as anticoagulant, Chapter 5: II, A, 1, 1°) was added to the glucose-saline medium. Mature worms continued to lay eggs during the first 10 d of culture, after which egg production deteriorated. Metabolites of the parasite lowered the pH from 7·0 to 5·5–4·5 (presumably within 1 or 2 d). The authors conclude that culture conditions are still suboptimal which results in the gradual deterioration of worms.

XV. MONOGENEA

Taylor and Baker (1968) cited Llewellyn's (1957) study on the maintenance of various adult and larval monogenes obtained from marine fishes. Llewellyn's technique involves maintaining these flukes in Petri dishes containing filtered sea water with no attempt to keep the cultures sterile.

Studies on axenic cultivation of monogenes appear not to be available.

Most recent *in vitro* studies have been concerned mainly with factors which influence the hatching of oncomiracidia; see Kearn (1975) and MacDonald (1975).

A. ENTOBDELLA HIPPOGLOSSI

Kearn (1974) described a procedure for maintaining the marine monogene, *Entobdella hippoglossi*. He simply detaches living specimens of *E. hippoglossus* from the halibut, *Hippoglossus hippoglossus*, and maintains them at 4–5°C in filtered Plymouth sea water. At this temperature some of the parasites remain alive for 2 weeks and lay large numbers of eggs.

REFERENCES

Barnes, H. (1954). Some tables for the ionic composition of sea water. *Journal of Experimental Biology* 31, 582–588.

Basch, P. F. (1969). *Cotylurus lutzi* sp.n. (Trematoda: Strigeidae) and its life cycle. *Journal of Parasitology* 55, 527–539.

Basch, P. F. and DiConza, J. J. (1973). Primary cultures of embryonic cells from the snail *Biomphalaria glabrata*. *American Journal of Tropical Medicine and Hygiene* 22, 805–813.

Basch, P. F. and DiConza, J. J. (1974). The miracidium-sporocyst transition in *Schistosoma mansoni*: surface changes *in vitro* with ultrastructural correlation. *Journal of Parasitology* 20, 935–941.

Basch, P. F. and DiConza, J. J. (1975). *Cotylurus lutzi* (Strigeidae): pre- and postadult stages cultured *in vitro*. *Journal of Invertebrate Pathology* 26, 263–264.

Basch, P. F., DiConza, J. J. and Johnson, B. E. (1973). Strigeid trematodes (*Cotylurus lutzi*) cultured *in vitro*: production of normal eggs with continuance of life cycle. *Journal of Parasitology* 59, 319–322.

Bayne, C. J. (1972). On the reported occurrence of an ecdysone-like steroid in the freshwater snail, *Biomphalaria glabrata* (Pulmonata; Basommatophora), intermediate host of *Schistosoma mansoni*. *Parasitology* 64, 501–509.

Bell, E. J. and Smyth, J. D. (1958). Cytological and histochemical data for evaluating development of trematodes and pseudophyllidean cestodes *in vivo* and *in vitro*. *Parasitology* 48, 131–148.

Berntzen, A. K. (1966). A controlled culture environment for axenic growth of parasites. *Annals of the New York Academy of Science* 139, 176–189.

Berntzen, A. K. and Macy, R. W. (1969). *In vitro* cultivation of the digenetic trematode *Sphaeridiotrema globulus* (Rudolphi) from the metacercarial stage to egg production. *Journal of Parasitology* 55, 136–139.

Buecher, E. J., Perez-Mendez, G., Hansen, E. L. and Yarwood, E. (1974). Sulfhydryl compounds under controlled gas in culture of *Schistosoma mansoni* sporocysts. *Proceedings of the Society for Experimental Biology and Medicine* 146, 1101–1105.

Bruce, J. I., Weiss, E., Stirewalt, M. A. and Lincicome, D. R. (1969). *Schistosoma mansoni*: glycogen content and utilization of glucose, pyruvate, glutamate, and citric acid cycle intermediates by cercariae and schistosomules. *Experimental Parasitology* 26, 29–40.

Chow, K. and Chiu, J. K. (1972). (Preliminary report of *Schistosoma japonicum in vitro* cultivation.) (5th Scientific Meeting, Chinese Soc. Microbiol., 17 Dec. 1972, Nat. Taiwan Univ., Taipei. Abstract.) *Chinese Journal of Microbiology* 5, 131–132. (In Chinese.) Quoted in *Helminthological Abstracts* 44, 880 (1975).

Clegg, J. A. (1965). *In vitro* cultivation of *Schistosoma mansoni. Experimental Parasitology* 16, 133–147.

Clegg, J. A. and Smithers, S. R. (1972). The effects of immune rhesus monkey serum on schistosomula of *Schistosoma mansoni* during cultivation *in vitro. International Journal for Parasitology* 2, 79–98.

Coles, G. C. (1972). Oxidative phosphorylation in adult *Schistosoma mansoni. Nature* 240, 488–489.

Colley, D. G. and Wikel, S. K. (1974). *Schistosoma mansoni*: simplified method for the production of schistosomules. *Experimental Parasitology* 35, 44–51.

Contos, N. and Fried, B. (1976). Histochemical observations on the body surface of *Leucochloridiomorpha constantiae* (Trematoda) cultivated *in vitro. Proceedings of the Helminthological Society of Washington* 43, 88–89.

Cowper, S. G., Fletcher, K. A. and Maegraith, B. G. (1972). An improved apparatus for the maintenance of *Schistosoma mansoni* and *Plasmodium knowlesi* or other blood protozoa in a continuous flow medium. *Annals of Tropical Medicine and Parasitology* 66, 67–73.

Dawes, B. (1954). Maintenance *in vitro* of *Fasciola hepatica. Nature, Lond.* 174, 654–655.

DiConza, J. J. and Basch, P. F. (1974). Axenic cultivation of *Schistosoma mansoni* daughter sporocysts. *Journal of Parasitology* 60, 757–763.

DiConza, J. J. and Hansen, E. L. (1973). Cultivation of *Schistosoma mansoni* daughter sporocysts in arthropod tissue cultures. *Journal of Parasitology* 59, 211–212.

Dixon, K. E. (1966). The physiology of excystment of the metacercaria of *Fasciola hepatica* L. *Parasitology* 56, 431–456.

Ellis, M. M., Merrick, A. D. and Ellis, M. D. (1931). The blood of North American freshwater mussels under normal and adverse conditions. *Bulletin of the United States Bureau of Fisheries* (Document 1097) 46, 509–542.

Eveland, L. K. (1972). *Schistosoma mansoni*: conversion of cercariae to schistosomula. *Experimental Parasitology* 32, 261–264.

Eveland, L. K. and Morse, S. I. (1975). *Schistosoma mansoni*: *in vitro* conversion of cercariae to schistosomula. *Parasitology* 71, 327–335.

Foster, G. R. (1970). A suggested medium for maintaining *Fasciola hepatica* prior to *in vitro* experimentation. *Zeitschrift für Parasitenkunde* 34, 177–178.

Fried, B. and Contos, N. (1973). *In vitro* cultivation of *Leucochloridiomorpha constantiae* (Trematoda) from the metacercaria to the ovigerous adult. *Journal of Parasitology* 59, 936–937.

Gazzinelli, G., Oliveira, C. C. De, Figueiredo, E. A., Pereira, L. H., Coelho, P. M. Z. and Pellegrino, J. (1973). *Schistosoma mansoni*: biochemical evidence for morphogenetic change from cercariae to schistosomule. *Experimental Parasitology* 34, 181–188.

Gilbert, B., Rosa, M. N. Da, Borojevic, R. and Pellegrino, J. (1972). *Schistosoma mansoni*: *in vitro* transformation of cercariae into schistosomula. *Parasitology* 64, 333–339.

Hansen, E. L. (1975). Secondary daughter sporocysts of *Schistosoma mansoni*: their occurrence and cultivation. *Annals of the New York Academy of Science* 266, 426–436.

Hansen, E. L., Perez-Mendez, G., Long, S. and Yarwood, E. (1973). *Schistosoma mansoni*: emergence of progeny-daughter sporocysts in monoxenic culture. *Experimental Parasitology* 33, 486–494.

Hansen, E. L., Perez-Mendez, G. and Yarwood, E. (1974a). *Schistosoma mansoni*: axenic culture of daughter sporocysts. *Experimental Parasitology* 36, 40–44.

Hansen, E. L., Perez-Mendez, G., Yarwood, E. and Buecher, E. J. (1947b). Second-generation daughter sporocysts of *Schistosoma mansoni* in axenic culture. *Journal of Parasitology* **60**, 371–372.

Hoeppli, R. and Chu, H. J. (1937). Studies on *Clonorchis sinensis in vitro*. In "Festschrift Bernhard Nocht", pp. 199–203. Hamberg: Institut für Schiffs ünd Tropenkrankheiten.

Howell, M. J. (1968). Excystment and *in vitro* cultivation of *Echinoparyphium serratum*. *Parasitology* **58**, 583–597.

Howell, M. J. (1970). Excystment of the metacercariae of *Echinoparyphium serratum* (Trematoda: Echinostomatidae). *Journal of Helminthology* **44**, 35–36.

Howell, M. J. and Bourns, T. K. R. (1974). *In vitro* culture of *Trichobilharzia ocellata*. *International Journal for Parasitology* **4**, 471–476.

Hsu Shih-Ê (1974). Culture of *Schistosoma japonicum in vitro*, with special reference to egg production and development. *Acta Zoologica Sinica* **20**, 231–242. (In Chinese.) Quoted in *Helminthological Abstracts* **45**, 463 (1976).

Ito, J., Yasuraoka, K. and Komiya, Y. (1955). Studies on the survival of *Schistosoma japonicum in vitro*. 1. Survival in blood or serum media. *Japanese Journal of Parasitology* **4**, 12–18.

Jacqueline, E. and Biguet, J. (1973). Méthode simple pour l'isolement d'oeufs de *Schistosoma mansoni* aseptiques et infectieux. *Annales de Parasitologie Humaine et Comparée* **48**, 623–626.

Jaw, C. Y. and Lo, C. T. (1974). *In vitro* cultivation of *Echinostoma malayanum* Leiper, 1911. *Chinese Journal of Microbiology* **7**, 157–164.

Kannangara, D. W. W. (1974). *In vitro* cultivation of the metacercariae of the human lung fluke *Paragonimus westermani*. *International Journal for Parasitology* **4**, 675–676.

Kannangara, D. W. W. and Smyth, J. D. (1974). *In vitro* cultivation of *Diplostomum spathaceum* and *Diplostomum phoxini* metacercariae. *International Journal for Parasitology* **4**, 667–673.

Kearn, G. C. (1974). Nocturnal hatching in the monogenean skin parasite *Entobdella hippoglossi* from the halibut, *Hippoglossus hippoglossus*. *Parasitology* **68**, 161–172.

Kearn, G. C. (1975). Hatching in the monogean parasite *Dictyocotyle coeliaca* from the body cavity of *Raja naevus*. *Parasitology* **70**, 87–93.

Lancastre, F. and Golvan, Y. (1973). L'utilisation des cultures *in vitro* dans l'étude de *Schistosoma mansoni*. 1. Le maintien en survie des adultes. *Annales des Parasitologie Humaine et Comparée* **48**, 307–313.

Lewert, R. M. and Para, J. B. (1966). The physiological incorporation of carbon-14 in *Schistosoma mansoni* cercariae. *Journal of Infectious Diseases* **116**, 171–182.

Llewellyn, J. (1957). The larvae of some monogenetic trematode parasites of Plymouth fishes. *Journal of the Marine Biological Association of the United Kingdom* **36**, 243–259.

Lo, C. T. and Cross, J. H. (1974). *In vitro* cultivation of *Fasciolopsis buski*. *Southeast Asian Journal of Tropical Medicine and Public Health* **5**, 252–257.

Locatelli, A. and Paoletti, C. (1970). Coltivazione *in vitro* di *Fasciola hepatica* adulte: produzione di uova. Nota preventiva. *Atti dela Societa Italiana delle Scienze Veterinarie*, 1969, **23**, 879–881. Quoted in *Helminthological Abstracts* **41**, 102 (1972).

McCowen, M. C., Galloway, R. B. and Young, R. A. L. (1968). *In vitro* cultivation of *Schistosoma mansoni* from the cercarial stage to young adults. (Abstract.) International Congress of Tropical Medicine and Malaria (8th), Teheran, Sept. 7–15, Abstracts and reviews, pp. 1096–1097.

MacDonald, S. (1975). Hatching rhythms in three species of *Diclidophora* (Monogenea) with observations on host behavior. *Parasitology* **71**, 211–228.

190 BERNARD FRIED

Macy, R. W., Berntzen, A. K. and Benz, M. (1968). *In vitro* excystation of *Sphaeridiotrema globulus* metacercariae, structure of cyst, and the relationship to host specificity. *Journal of Parasitology* **54**, 28–38.

Magzoub, M. (1973). Recovery and *in vitro* cultivation of adult *Schistosoma bovis* from experimentally infected mice. *Journal of Zoology* **170**, 139–142.

Mao, S. P. and Lyu, K. L. (1957). (Studies on the cultivation of *Schistosoma japonicum in vitro*.) *Medskaya Parasitology* **26**, 166–172. (In Russian: English summary pp. 171–172.) Quoted in Taylor and Baker (1968).

Michaels, R. M. and Prata, A. (1968). Evolution and characteristics of *Schistosoma mansoni* eggs laid *in vitro*. *Journal of Parasitology* **54**, 921–930.

Michelson, E. H. (1970). *Aspidogaster conchicola* from fresh-water gastropods in the United States. *Journal of Parasitology* **56**, 709–712.

Muftic, M. (1969). Metamorphosis of miracidia into cercariae of *Schistosoma mansoni in vitro*. *Parasitology* **59**, 365–371.

Nizami, W. A. and Siddiqi, A. H. (1975). Studies on *in vitro* survival of *Isoparorchis hypselobagri* (Digenea: Trematoda). *Zeitschrift für Parasitenkunde* **45**, 263–267.

Osuna-Carillo de A, A. and Guevara-Pozo, D. (1974). Cultivo de helmintos parásitos. I. Primeros resultados con un medio básico para el cultivo *in vitro* de *Fasciola hepatica* L. *Revista Ibérica de Parasitología* **34**, 137–140.

Pascoe, D., Richards, R. J. and James, B. L. (1968). Oxygen uptake, metabolic rate, reduced weight, length, and number of cercariae in starving sporocysts of *Cercaria dichotoma*. *Experimental Parasitology* **23**, 171–182.

Pascoe, D., Richards, R. J. and James, B. L. (1970). The survival of the daughter sporocysts of *Microphallus pygmaeus* (Levinsen, 1881) in a chemically defined medium. *The Veliger* **13**, 157–162.

Pullin, R. S. V. (1970). A study of the environment of larval *Fasciola hepatica* L., with a view to developing a defined medium for *in vitro* culture. PhD Thesis; University of York, England.

Pullin, R. S. V. (1971). Composition of the haemolymph of *Lymnaea truncatula*, the snail host of *Fasciola hepatica*. *Comparative Biochemistry and Physiology* **40A**, 617–626.

Pullin, R. S. V. (1973). Preliminary experiments with a defined culture medium for larval *Fasciola hepatica*. *Journal of Helminthology* **47**, 181–189.

Ractliffe, L. H., Guevara-Pozo, D. and Lopez-Roman, R. (1969). *In vitro* maintenance of *Fasciola hepatica*: a factorial approach based on egg production. *Experimental Parasitology* **26**, 41–51.

Ramalho-Pinto, F. J., Gazzinelli, G., Howells, R. E., Mota-Santos, T. A., Figueiredo, E. A. and Pellegrino, J. (1974). *Schistosoma mansoni*: defined system for stepwise transformation of cercaria to schistosomule *in vitro*. *Experimental Parasitology* **36**, 360–372.

Richards, R. J., Pascoe, D. and James, B. L. (1972). Variations in the metabolism of the daughter sporocysts of *Microphallus pygmaeus* in a chemically defined medium. *Journal of Helminthology* **46**, 107–116.

Rohrbacher, G. H. (1957). Observations on the survival *in vitro* of bacteria-free adult common liver flukes, *Fasciola hepatica*. *Journal of Parasitology* **43**, 9–18.

Schiller, E. L. (1965). A simplified method for the *in vitro* cultivation of the rat tapeworm, *Hymenolepis diminuta*. *Journal of Parasitology* **51**, 516–518.

Schiller, E. L., Bueding, E., Turner, V. M. and Fisher, J. (1975). Aerobic and anaerobic carbohydrate metabolism and egg production of *Schistosoma mansoni in vitro*. *Journal of Parasitology* **61**, 385–389.

Senft, A. W. and Senft, D. G. (1962). A chemically defined medium for maintenance of *Schistosoma mansoni*. *Journal of Parasitology* **48**, 551–554.

Sewell, M. M. H. and Purvis, G. M. (1969). *Fasciola hepatica*: the stimulation of excystation. *Parasitology* **59**, 4P.

Smith, M. and Webbe, G. (1974a). Culture of *Schistosoma haematobium* and the effects of immune sera *in vitro*. *Transactions of the Royal Society of Tropical Medicine and Hygiene* **67**, 9–10.

Smith, M. and Webbe, G. (1974b). Damage to schistosomula of *Schistosoma haematobium in vitro* by immune baboon sera and absence of cross-reaction with *Schistosoma mansoni*. *Transactions of the Royal Society of Tropical Medicine and Hygiene* **67**, 70–71.

Smith, M., Clegg, J. A. and Webbe, G. (1976). Culture of *Schistosoma haematobium in vivo* and *in vitro*. *Annals of Tropical Medicine and Parasitology* **70**, 101–108.

Stirewalt, M. A. and Uy, A. (1969). *Schistosoma mansoni*: cercarial penetration and schistosomule collection in an *in vitro* system. *Experimental Parasitology* **26**, 17–28.

Stirewalt, M. A., Minnick, D. R. and Fregeau, W. A. (1966). Definition and collection in quantity of schistosomules of *Schistosoma mansoni*. *Transactions of the Royal Society of Tropical Medicine and Hygiene* **60**, 352–360.

Sun, T. (1969). Maintenance of adult *Clonorchis sinensis in vitro*. *Annals of Tropical Medicine and Parasitology* **63**, 399–402.

Taylor, A. E. R. and Baker, J. R. (1968). "The Cultivation of Parasites *in Vitro*." Blackwell Scientific Publications, Oxford and Edinburgh, 377 pp.

Tiba, Y., Holanda, J. G., Ramalho-Pinto, F. J., Gazzinelli, G. and Pellegrino, J. (1974). Schistosomula (*Schistosoma mansoni*) obtained *in vitro*: viability in culture and infectivity for mice. *Transactions of the Royal Society of Tropical Medicine and Hygiene* **68**, 72.

Voge, M. and Jeong, K. (1971). Growth *in vitro* of *Cotylurus lutzi* Basch 1969 (Trematoda: Strigeida), from tetracotyle to patent adult. *International Journal for Parasitology* **1**, 139–143.

Voge, M. and Seidel, J. S. (1972). Transformation *in vitro* of miracidia of *Schistosoma mansoni* and *S. japonicum* into young sporocysts. *Journal of Parasitology* **58**, 699–704.

Weinstein, P. P. and Jones, M. F. (1956). The *in vitro* cultivation of *Nippostrongylus muris* to the adult stage. *Journal of Parasitology* **42**, 215–236.

Weinstein, P. P. and Jones, M. F. (1959). Development *in vitro* of some parasitic nematodes of vertebrates. *Annals of the New York Academy of Science* **77**, 137–162.

Wikerhauser, T. (1960). A rapid method for determining the viability of *Fasciola hepatica* metacercariae. *American Journal of Veterinary Research* **21**, 895–897.

Wikerhauser, T. and Cvetnić, S. (1967). Survival of young and sexually mature adult *Fasciola hepatica* in various cell-free media with and without mammalian cell cultures. *Experimental Parasitology* **20**, 200–204.

Wikerhauser, T., Cvetnić, S. and Brudnjak, Z. (1970). Further study of the survival of young *Fasciola hepatica* in cell cultures. *In* "H. D. Srivastave commemoration volume." (Singh, K. S. and Tandan, B. K. eds.) Izatnagar, U. P.: Indian Veterinary Research Institute, pp. 279–281. Quoted in *Helminthological Abstracts* **41**, 301 (1972).

Wilson, R. A., Pullin, R. and Denison, J. (1971). An investigation of the mechanism of infection by digenetic trematodes: the penetration of the miracidium of *Fasciola hepatica* into its snail host *Lymnaea truncatula*. *Parasitology* **63**, 491–506.

Yasuraoka, K. and Kojima, K. (1970). *In vitro* cultivation of the heterophid trematode, *Metagonimus yokogawai*, from the metacercaria to adult. *Japanese Journal of Medical Science and Biology* **23**, 199–210.

Yasuraoka, K., Kaiho, M., Hata, H. and Endo, T. (1974). Growth *in vitro* of *Parvatrema timondavidi* Bartoli (1963) (Trematoda: Gymnophallidae) from the metacercarial stage to egg production. *Parasitology* **68**, 293–302.

Yokogawa, M., Oshima, T. and Kihata, M. (1955). Studies to maintain excysted metacercariae of *Paragonimus westermani in vitro*. *Journal of Parasitology* **41**(6), Section 2 (Supplement): 28.

Yokogawa, M., Oshima, T. and Kihata, M. (1958). Studies to maintain excysted metacercariae of *Paragonimus westermani in vitro*. II. Development of the excysted metacercariae maintained *in vitro* at 37°C for 203 days. *Japanese Journal of Parasitology* **7**, 51–55.

Chapter 9

Cestoda

MARIETTA VOGE

*Department of Microbiology and Immunology, School of Medicine,
University of California, Los Angeles, U.S.A.*

During the past eight years considerable effort has been expended on growing tapeworms axenically *in vitro* and some progress has been made. One of the important features of this progress is simplification of the culture systems. Screw cap tubes or various flasks have been substituted successfully for more complex apparatus. Simplification reduces the total cost and the time spent by the researcher, and increases the ability to produce large numbers of organisms for various biochemical studies, as well as for the preparation of antigens.

Progress has also been achieved in the determination of some essential factors for growth, and other parameters of their biology.

The following represents an attempt to relate most of the accomplishments and some of the shortcomings.

I. TETRARHYNCHIDEA

The difficulty of obtaining live material from sharks, while having access to a well-equipped biology laboratory, may explain the dearth of experimental work with shark tapeworms! The only study pertinent to *in vitro* culture is that by Voge and Edmonds (1969) who were able to hatch oncospheres from coracidia of *Lacistorhynchus tenuis*. The technique is very simple and is based on lowering the osmolarity of the sea water to 600 mosmol by 50% dilution with distilled water. Thus, after the eggs have been allowed to become embryonated in sea water at room temperature and the coracidia have escaped from the eggs, coracidia are pipetted into small glass dishes and an equal volume of distilled water is added. The ciliary beat very quickly becomes disturbed and motility eventually ceases. The oncospheres contract and expand within the ciliary coat and in about 15 minutes emerge through a small opening.

II. PSEUDOPHYLLIDEA

A. *SPIROMETRA*

Berntzen and Mueller (1972) described development of *Spirometra mansonoides* from the plerocercoid to the gravid adult in a culture system and medium very similar to that already published (see Taylor and Baker, 1968, p. 208). Modifications included the addition of 0.1 ml cat bile litre^{-1} of culture medium and the omission of solution 6 from the medium after 120 h of culture. Plerocercoids developed to gravid adults at 18 d which was the earliest day when eggs were released into the culture medium. The Australian and Malayan strains of *Spirometra* behaved similarly *in vitro*, maturing and releasing eggs at about the same time.

Production of sparganum growth factor (SGF) was tested (Chang *et al.*, 1973) by growing spargana *in vitro* in medium 199 with foetal calf serum or with monolayers of various cell types. The growth-promoting effect of SGF released in the medium was tested on hypophysectomized rats. Spargana were maintained for several months on amnion cell monolayers without serum in the medium and axenically with varying amounts of serum. SGF apparently had no effect on the growth of monolayers, but when injected into rats it produced weight gains comparable to those conferred by injection of serum from worm-infected rats. It should be noted that spargana fared well in axenic cultures containing only 3–5% foetal calf serum but did not survive in axenic cultures without serum.

Tachovsky *et al.* (1973) reported on an additional modification of culture medium for the growth of spargana, beginning with scoleces cut from spargana grown in hamsters. This method differs from the original one developed by Mueller (see Taylor and Baker, 1968, p. 206) by the use of Minimal Essential Medium, Eagle (MEM), supplemented with 10% foetal calf serum (inactivated?) and 10% tryptose phosphate broth. Culture vessels were Falcon plastic flasks, each containing 50 ml of medium. Flasks were incubated in a 5% carbon dioxide atmosphere. Medium was completely exchanged twice weekly. Spargana of *S. mansonoides*, incubated in medium 199, were used by Tkachuck and Weinstein (1976) to study uptake of vitamin B_{12}.

III. CYCLOPHYLLIDEA

A. *MONIEZIA*

The development of an effective hatching technique for oncospheres of an anoplocephalid cestode (Caley, 1975) should make possible the *in vitro* cultivation of the larval stages usually occurring in oribatid mites. It is likely that the technique for hatching (1, 1°), perhaps with some modifications, might also be effective for the oncospheres of other anoplocephalid genera and species.

1. *Detailed Technique*

1° *Moniezia expansa*, hatching of oncospheres
(Caley, 1975)

Gravid proglottids are washed thoroughly and placed in 0·8% sodium chloride. Eggs are dissected from proglottids and placed in a test tube with small glass beads. The tube is shaken vigorously to remove the egg shell and underlying membrane. Removal of egg shells brings about activation of most oncospheres within the pyriform apparatus. They are transferred to a solution consisting of chymotrypsin (Sigma, α-chymotrypsin Type II, crystallized) 10 mg ml^{-1} in 0·8% sodium chloride (?) at 25°C, pH 7·0 (?). In less than two minutes the pyriform apparatus swells and begins to dissolve, and the oncospheres are freed.

Note: Diluent for chymotrypsin was not specified in Caley's paper but sodium chloride or other salines at appropriate pH should serve to prepare this enzyme solution.

B. *HYMENOLEPIS*

Taylor and Baker (1968) state that cultivation of beetle stages of

Hymenolepis has yet to be achieved. Fortunately, this has been accomplished for several species within the last ten years. Partial development of *H. citelli* from oncospheres hatched *in vitro* to the hollow ball stage was reported by Voge (1972). Eggs were dissected from gravid proglottids washed repeatedly in saline with antibiotics. Outer shells were broken mechanically and oncospheres were hatched in a sterile trypsin solution (Berntzen and Voge, 1965). However, until recently these results could not be repeated at will.

The addition of *freshly prepared* reducing agents to the culture medium at the time of inoculation elicited growth from the oncosphere to fully developed cysticercoids infective to mammalian hosts (Voge and Green, 1975; see Section III, B, 1, 1°). The method involves the use of a medium originally devised for the growth of cockroach cells with foetal calf serum, lactalbumin hydrolysate, and yeast extract as additives. Without reducing agent development was not initiated; the concentration of reducing agent was critical and some reducing agents, such as glutathione, were ineffective.

Monoxenic cultivation of *H. diminuta* cysticercoids with rat fibroblasts, a somewhat laborious procedure, was reported by Graham and Berntzen (1970; see Section III, B, 1, 2°). Oncospheres were hatched and freed from egg shell debris by a sugar gradient technique. The culture medium was NCTC 109 with 20% horse serum in a gas phase of 20% O_2, 10% CO_2, 70% N_2. Resulting cysticercoids were apparently infective to rats. The authors state that no growth was obtained in the presence of air and that whole eggs or egg shells in the cultures inhibited cysticercoid growth.

Axenic cultivation of *H. diminuta* cysticercoids (Voge, 1975; see Section III, B, 1, 3°) was achieved in the medium used for the growth of *H. citelli*. Again, reducing agents were essential for initiation and completion of development. Oncospheres were hatched in a trypsin and α-amylase solution as described by Berntzen and Voge (1965) and culture vessels were screw cap tubes. An incubation temperature of 28°C was optimal; 25°C was too low for continuous normal growth. As with *H. citelli*, air was used as gas phase. The presence of some whole eggs or egg shells in the culture tubes did not inhibit growth as reported by Graham and Berntzen (1970).

Voge *et al.* (1976; see Section III, B, 1, 4°) reported the effect of different gas phases on the growth of *H. diminuta* cysticercoids. When culture tubes were gassed with 100% N_2 or with 95% N_2 and 5% CO_2, development from oncosphere to cysticercoid was nearly twice as rapid as it was in air. Also, of the various reducing agents tested, only L-cysteine, coenzyme A, and homocysteine effected complete development while partial development occurred in the presence of ascorbic acid and dithiothreitol.

The adult phase of *H. diminuta* was cultured by Turton (1972) who transferred 6 d-old worms from rats into Schiller's (1965; see Section

III, B, 1, 5°) medium modified by the inclusion of defibrinated horse, instead of rabbit, blood. Growth was also obtained in Hanks's Balanced Saline, horse serum, yeast and liver extracts with or without nutrient agar. In both these media adult worms with oncospheres were obtained but knotting and fracturing of the worms was observed. Growth in monophasic media with or without liver extract showed that the presence of liver extract significantly promoted growth. Apparently, none of the worms in the different media contained fully developed, infective eggs and, disappointingly none of the media yielded better results than those of Schiller (1965). Turton did however suggest that 3 d-old worms should be used to assess the efficacy of growth media rather than 6 d-old worms.

Because the use of young worms grown in the rat presents "clean-up" problems, Roberts (1973; see Section III, B, 1, 7°) studied the effect of various surface-active agents on bacterial and other contaminants of the worm surface. Worms exposed briefly to low concentrations of benzalkonium chloride or hexachlorophene subsequently developed *in vitro* as did controls and tetracyclines were not deleterious. Roberts also determined that defibrinated sheep's blood can be substituted for rabbit blood in the medium. Earlier, Roberts and Mong (1969; see Section III, B, 1, 6°) showed that comparable growth of *H. diminuta in vitro* was observed with and without O_2; varying concentrations apparently did not affect growth.

DeRycke (1975) studied survival *in vitro* of fully developed cysticercoids of *Hymenolepis microstoma* dissected from beetles. Cysticercoids were kept in Eagle's Basal Medium or Ringer's solution for varying periods and their infectivity to mice was tested. It was concluded that maintenance *in vitro* adversely affected the organisms.

Seidel (1975; see Section III, B, 1, 8°) described the requirements of oncospheres growing to infective cysticercoids. Contrary to the situation with *H. citelli* and *H. diminuta*, *H. microstoma* develops without the addition of reducing agents, which produce structural abnormalities and inhibit completion of development. In all other respects the medium and culture methods were comparable with those used for *H. diminuta*. Cysticercoids grown *in vitro* were excysted according to the method of Rothman (1959) and placed in culture. Medium and culture procedures were as described by Seidel (1969, 1971; see Section III, B, 1, 8°). Adult worms which developed from cysticercoids grown *in vitro* contained gravid segments with normal eggs infective to beetles. Thus, the whole life cycle of *H. microstoma* was completed *in vitro*. The culture medium for growth of the adult was biphasic, containing nutrient agar with a fluid overlay of Triple Eagle's medium (TEM, Gibco) with 20% horse serum in screw cap tubes. Haemin was an essential additive to the medium as no strobilization occurred without it.

A more complex medium for the growth of the adult *H. microstoma* was described by Evans (1970; see Section III, B, 1, 10°). It contained Eagle's medium, horse serum, liver extract, and ox bile.

Several workers have studied excystation of cysticercoids of *H. microstoma*. Goodchild and Davis (1972) determined that prolonged exposure to artificial gastric juice shortened the time required in trypsin and bile salt solution. Lippens-Mertens and DeRycke (1973; see Section III, B, 1, 9°) also concluded that pepsin, trypsin, and bile salts were all necessary to achieve maximal speed of excystation, and Davis (1975) showed that excystation levels were related to age of the cysticercoids.

Sinha (1973; see Section III, B, 1, 12°) determined the effect of pepsin, trypsin, various bile salts, and pancreatin, singly or in various combinations, upon excystment of *Hymenolepis nana* cysticercoids. He concluded that rapid excystment occurred in a trypsin–bile salt solution after pretreatment with pepsin-HCl, and that trypsin or bile salts alone did not cause excystation.

Axenic development of *H. nana* cysticercoids from oncospheres hatched *in vitro* to cysticercoids infective to mice was reported by Seidel and Voge (1975; see Section III, B, 1, 11°). Culture medium and procedures were as described for *H. citelli* (see above) except that a gas phase of 95% N_2 and 5% CO_2 was used instead of air. Also, for complete development of cysticercoids, reducing agent had to be omitted from the culture medium.

1. *Detailed Techniques*

1° *Hymenolepis citelli* cysticercoids
(Voge and Green, 1975)

MEDIUM

(i) Base medium: A modification of that of Landureau (1966, 1968) for the growth of cockroach cells.

	g		g
L-Arginine HCl	2·0	L-Valine	0·152
L-Aspartic acid	0·200	$MgSO_4 . 7H_2O$	0·670
L-Glutamic acid	1·0	KCl	0·894
α-Alanine	0·120	$CaCl_2$	0·484
β-Alanine	0·0445	NaCl	7·42
L-Cysteine*	0·101	$NaHCO_3$	0·424
L-Glutamine	0·559	$NaH_2PO_4 . H_2O$	0·010
Glycine	0·200	Glucose	2·5
L-Histidine HCl	0·404	Trehalose	6·9
L-Leucine	0·249	α-Ketoglutaric acid	0·365

* Subsequently omitted.

	g		g
L-Lysine	0·0124	Citric acid :	0·0153
L-Methionine	0·492	Fumaric acid	0·0058
L-Proline	0·748	L-Malic acid	0·058
L-Serine	0·083	Succinic acid	0·0059
L-Threonine	0·020	Yeast extract	1·0
L-Tyrosine	0·362	Lactalbumin hydrolysate	3·5

The above are dissolved in 900 ml double-distilled water, filtered through a Millipore filter 0·22 μm pore size and stored at $-25°C$ until used.

Non-inactivated foetal calf serum, 100 ml (stored at $-25°C$) is added just before use. The pH is adjusted to 7·0 with 5% sodium bicarbonate (w/v) solution. Streptomycin, 300 μg ml^{-1}, and penicillin, 300 iu ml^{-1} are added.

(ii) L-cysteine stock solution (reducing agent), $7·9 \times 10^{-2}$ M, is prepared in distilled water, sterilized by filtration, and kept frozen ($-25°C$) in small aliquots until use.

TECHNIQUE

Gravid proglottids from worms grown in hamsters are washed in three changes of sterile 0·85% sodium chloride containing about 300 iu ml^{-1} penicillin and 300 μg ml^{-1} streptomycin. Proglottids are kept in each wash for 15 min. Segments are then transferred to depression slides. Earle's saline (see Chapter 1: IV, B, 1, f) with sodium bicarbonate (pH 7·0–7·2) is added to the wells and eggs are teased from segments with sterile fine forceps. Eggs are pipetted into sterile screw cap tubes containing glass beads, shaken manually for about 4 min and a small drop of the contents is then checked under the microscope. If most of the egg shells are broken, the saline containing the broken eggs is transferred to a small Petri dish with the hatching solution, 25,000 units Tryptar ml^{-1} (Armour Pharmaceutical) in Earle's saline, pH 7·0, and kept at room temperature. Contents are checked every 5 min for hatched oncospheres. When most oncospheres have hatched, the fluid in the dish is swirled gently. Most of the unbroken eggs and empty shells concentrate in the centre at the bottom of the dish. These are then discarded by pipette and the hatched oncospheres pipetted to a wash in a large Petri dish half filled with base medium. Oncospheres are then concentrated near the centre of the dish by swirling and thence pipetted (approximately 100–200 per tube) into the culture tubes (130 by 10 mm screw cap tubes) containing 6·0 ml of the culture medium without L-cysteine. Freshly thawed L-cysteine reducing agent is added to cultures with a 9 in (23 cm) Pasteur pipette; one drop of fluid equals 0·025 ml. Optimal final L-cysteine concentration in the medium is 1.9×10^{-3} M. Culture tubes are then sealed with Parafilm and kept in an inclined position at 26°C.

During the first week of culture, medium is not changed but thereafter half the medium is replaced with fresh medium three times a week. L-cysteine stock solution is added, 1 drop ml^{-1}, to the fresh medium. Culture tubes are then resealed. As growth proceeds, individual cultures appear to contain too many organisms and they may be "thinned" by transferring some organisms to fresh cultures or by discarding them.

At 30 d all cysticercoids have withdrawn the scolex but another 10 d are needed before they are infective to hamsters.

2° *Hymenolepis diminuta* cysticercoids
(Graham and Berntzen, 1970)

MEDIUM

NCTC 109 (Gibco), modified to contain 0·3% glucose (w/v), and supplemented with 20% horse serum (v/v) is used for establishment of primary monolayer rat fibroblast cultures. Antibiotics in the medium are 200 iu ml^{-1} of penicillin G and 250 μg ml^{-1} of streptomycin sulphate.

TECHNIQUE

Eggs, obtained from worms grown in Sprague–Dawley laboratory rats, are sterilized by immersion in a 1 : 5000 aqueous Zephiran chloride solution for 20 min (Hundley and Berntzen, 1969); eggs are then washed five times in sterile distilled water and stored (time not stated) at 21°C in flasks until needed or used immediately. Eggs are hatched by first breaking the shells in a 20 ml Kimax screw cap tube containing 0·5 ml of 3 mm glass beads and 3 ml of Earle's salt solution (Chapter 1: IV, B, 1, f); the tube is placed on a vortex mixer for 30 sec. The solution with broken eggs is then transferred to the hatching solution consisting of Earle's salt solution with 1% trypsin (w/v) and 1% bacterial amylase (w/v) buffered with 5% sodium bicarbonate (w/v) to pH 7·2. After 40 min oncospheres are washed four times by alternate centrifugation (speed not stated), siphoning, and resuspension in Earle's saline at room temperature. A sugar gradient is then prepared as follows: from a sterile sugar concentrate (method of sterilization not stated) composed of 500 g sucrose in 350 ml water, dilutions of sugar concentrate in water (presumably sterile, distilled) are made and layered with the hatched oncospheres in a 50 ml conical centrifuge tube so that 15 ml of a 3 : 1 sugar dilution is contained at the bottom of the tube, followed by 10 ml of 2 : 1 dilution, 15 ml of 1 : 1 dilution and finally 2 ml of the oncospheres in Earle's saline; the tube is centrifuged (1·5–2 min, 1000 rpm). Oncospheres, free of debris, are then recovered from the 1:1 sugar layer (i.e. beneath the top layer of the gradient which contains membranes and dead oncospheres) and washed six times in Earle's salt solution by alternate centrifugation, siphoning, and resuspension as before. The last rinsing solution is adjusted to contain approximately 250–300 oncospheres per drop. Fibroblast monolayer cultures are prepared from 18 d embryos of Sprague–Dawley rats according to Paul (1960) and maintained in screw cap Kimax serum culture bottles (160 × 50 × 50 mm) with 10 ml of medium at 37°C. Cultures are gassed with a mixture of 70% N_2, 20% O_2, and 10% CO_2. These are inoculated with 5000–6000 washed oncospheres, transferred to 30°C and kept undisturbed for 3–4 d. Two ml of old medium is then removed and replaced with fresh medium; this is repeated weekly thereafter: if it becomes acid, medium is replaced more frequently. Cultures are re-gassed after each medium change and returned to incubator.

Complete development of cysticercoids is achieved at the end of 25 d. No development occurs if fibroblasts are omitted.

3° *Hymenolepis diminuta* cysticercoids
(Voge, 1975)

MEDIUM

The culture medium, antibiotics, and preparation of L-cysteine stock solution are as described above (Section III, B, 1, 1°).

TECHNIQUE

Eggs of *H. diminuta* are dissected from surface-sterilized gravid proglottids as described (Section III, B, 1, 1°). After mechanical breakage of egg shells, oncospheres are hatched in a sterile solution of 1% trypsin or 25,000 units of Tryptar ml^{-1} (Armour Pharmaceutical), and 1% bacterial amylase in Earle's saline (see Chapter 1: IV, B, 1, f) with sodium bicarbonate at pH 7·0 at room temperature.

After 20 min, when most of the oncospheres should have hatched, oncospheres are transferred to a wash of base medium (see Section III, B, 1, 1°) and thence to screw cap tubes containing 6 ml of medium. L-cysteine is added to tubes from freshly thawed stock to give a final concentration of $8·93 × 10^{-4}$ M to $1·62 × 10^{-3}$ M. Tubes are sealed with Parafilm and incubated in an inclined position at 28°C. A lower incubation temperature retards growth unduly and may give rise to abnormalities. Medium change is as described for *H. citelli*. At 18 d most organisms have a withdrawn scolex but cysticercoids require an additional 6–7 d before they are infective.

4° *Hymenolepis diminuta* cysticercoids
(Voge, Jaffe, Bruckner, and Meymarian, 1976)

MEDIUM

The culture medium is the same as described in Section III, B, 1, 1°.

TECHNIQUE

This system differs from Voge (1975) only by the addition of a gas phase of $100\% N_2$ or $95\% N_2$ and $5\% CO_2$. After the addition of L-cysteine, tubes are gassed for 1 min and then sealed with Parafilm. Tubes are re-gassed after each medium change. The use of N_2 in the presence of suitable reducing agent causes organisms to develop nearly twice as rapidly as they do in air. Without appropriate reducing agent there is no development regardless of the gas phase used.

Again, no growth is obtained with reduced glutathione, D-cysteine, L-cystine, L-methionine and others. Complete development occurs in the presence of L-cysteine, homocysteine, and coenzyme A, all in 1 mM concentrations.

5° *Hymenolepis diminuta*
(Schiller, 1965)

MEDIUM

A diphasic medium, consisting of a blood-agar base (Novy and MacNeal, 1903; Nicolle, 1908) overlaid with Hanks's balanced salt solution (Chapter 1: IV, B, 1, g) is employed. The medium is prepared as follows: 16 g Difco nutrient agar and 3·5 g NaCl are dissolved in 700 ml distilled water. After autoclaving, this solution is mixed thoroughly with 300 ml sterile, defibrinated rabbit blood (inactivated: 30 min, 56°C). The blood-agar mixture is then dispensed to sterile, cotton-stoppered 50 ml Ehrlenmeyer flasks in quantities of 10 ml per flask. After gelation of the blood-agar, 10 ml of Hanks's solution adjusted to pH 7·5 with $NaHCO_3$, and containing penicillin (100 iu ml^{-1}) and streptomycin (100 μg ml^{-1}), are added to each flask. The medium is then preincubated (32°C, 24 h) to permit diffusion and to assure freedom from bacterial contamination. Before inoculation with tapeworm larvae, the fluid phase of the medium is readjusted to pH 7·5 with NaOH after saturation for 10 min with a gas mixture of 97% N_2 and 3% CO_2. The pH is determined with a glass electrode during continuous flow of the gas mixture through the solution.

TECHNIQUE

H. diminuta cysticercoids, dissected from experimentally infected flour beetles 16 d after infection, may be excysted artificially by incubating them in undiluted ox bile (30 min, 37°C). Before being inoculated into the culture medium the excysted larvae are washed 3 times in sterile normal saline containing penicillin (100 iu ml^{-1}) and streptomycin (100 µg ml^{-1}).

From 10 to 15 excysted tapeworm larvae are introduced into each of several flasks and the flasks placed in a metabolic shaking incubator. The cultures are incubated at 37°C, under a gas mixture of 97% N_2 and 3% CO_2, delivered at approximately 100 ml min^{-1}. The vessels are oscillated at about 30 cycles min^{-1}.

After 6 d of continuous incubation, the supernatant fluid from each flask is poured into a sterile Petri dish and the young worms are transferred aseptically to fresh media by means of a stainless-steel dissecting hook. The worms are again transferred on the eighth day. At this time powdered glucose is added to the medium, and subsequently throughout the period of incubation (1 mg ml^{-1} of fluid overlay). On d 10 the worms are separated and transferred to individual flasks without further change in the glucose content of the medium. Thereafter, transfers are made every 24 h. On d 20 the volume of Hanks's saline per flask is increased to 20 ml, with powdered glucose added again (1 mg ml^{-1}). During subsequent cultivation the procedures for preparation of the medium remain unchanged.

By this means Schiller grew *H. diminuta* from the cysticercoid to the adult stage within 24 d *in vitro*. Viable eggs were produced (demonstrated by infectivity to beetles). One worm survived *in vitro* for 62 d and continued to produce eggs from d 24.

6° *Hymenolepis diminuta* adults
(Roberts and Mong, 1969)

MEDIUM AND TECHNIQUE

Schiller's (1965; see Section III, B, 1, 5°) medium is used and his technique somewhat modified to allow for control of the gas phase and measurements of O_2 as the purpose of the study is to determine growth of the worms in the presence or absence of O_2. Before addition to the blood agar, Hanks's (see Chapter 1: IV, B, 1, g) solution is gassed for 10 min with the desired gas mixture (0, 1, 5, or 20% O_2) by means of a fritted glass gas dispersion tube. Then the container is sealed until addition of the solution to the culture flasks.

Hanks's solution is dispensed to culture flasks with a Cornwall automatic pipetter, and the Hanks's solution flask vented during that time. The flask is then resealed and the culture flasks are pre-incubated (24 h, 34°C) with the desired gas phase before the addition of the worms. The cultures are maintained (37°C) in a Dubnoff water bath with continuous flow and circulation of the gas mixture. A gas outlet in the gassing hood of the water bath makes possible the analysis of outflowing gas for presence of oxygen. For further details of these analyses, the original paper should be consulted.

7° *Hymenolepis diminuta* adults
(Roberts, 1973)

MEDIUM AND TECHNIQUE

Schiller's (1965; see Section III, B, 1, 5°) medium was modified to test the efficacy of horse or sheep blood instead of rabbit blood and of sheep serum. Defibrinated sheep blood is an adequate substitute but sheep serum is not favourable.

Worms removed from rats can be sterilized by 2–3 sec rinses in 0·1% Phisohex (Winthrop lab.) in Hanks's basal salt solution (Chapter 1 : IV, B, 1, g). This procedure does not interfere with subsequent *in vitro* growth of the worms.

8° *Hymenolepis microstoma*, life cycle
(Seidel, 1969, 1971, 1975)

(i) Cysticercoids

MEDIUM

Medium composition is as described above (Section III, B, 1, 1°) except that L-cysteine is omitted.

TECHNIQUE

Eggs are dissected from surface-sterilized gravid segments (see Section III, B, 1, 1°) in Earle's (see Chapter 1 : IV, B, 1, f) salt solution and shaken with glass beads in a test tube to remove the thin outer shell (Berntzen and Voge,

1965). They are then transferred to a 150 by 25 mm centrifuge tube containing 5% sodium bicarbonate in 0·85% sodium chloride and 50,000 units of Tryptar ml^{-1} (Armour Pharmaceutical) which had been gassed for 10 min with 95% N_2 and 5% CO_2. After the addition of oncospheres the gas is bubbled through the hatching solution for 20 min through Tygon tubing connected to a 22 μm Millipore gas filter with Teflon tubing on the filter side into which sterile Pasteur pipettes are inserted. After 20 min the tube is centrifuged (5000 rpm, 2 min); the hatched oncospheres are transferred from the sediment into a wash of culture medium and then to screw cap tubes, each containing 6 ml of medium, sealed with Parafilm and incubated at 28°C. After the first week of culture, half of the medium is replaced three times a week. After 16 d structural development appears to be complete; cysticercoids are infective to mice at 21 d and can also be excysted *in vitro*.

Addition of reducing agents to cultures gives rise to structural abnormalities.

(ii) Adults

MEDIUM

Twenty ml nutrient agar (Difco) slants, with or without 5% human blood, are prepared and overlaid with 20 ml of Triple Eagle's Medium (TEM, Gibco) with 30% inactivated (56°C, 30 min) horse serum. When used with nutrient agar only (without blood) haemin must be incorporated in the overlay, or segmentation and subsequent growth will be inhibited.

Haemin solution is prepared by dissolving 1·0 mg of equine haemin in 1– 2 ml triethanolamine, adding double-distilled water to make a 1·0 mg 100 ml^{-1} solution which is sterilized by filtration and added to cultures to give a final haemin concentration of 0·5–1·0 mg litre^{-1}. The pH of the fluid overlay is adjusted to 7·0–7·2 with sodium bicarbonate.

NCTC 135 (Gibco) with 20% horse serum may be used instead of the Triple Eagle's overlay but growth is then somewhat reduced.

TECHNIQUE

Cysticercoids from 21 d cultures (see above) are excysted in sterile solutions consisting of a 1% pepsin solution (adjusted to pH 1·5 with N HCl made up in 0·85% NaCl) and 0·3% taurocholate in a 0·5% trypsin solution also in 0·85% NaCl (Rothman, 1959). Exposure to pepsin was for about 10 min, and to the trypsin solution for 15–20 min at 37°C. Juvenile worms are then washed once in adult worm culture medium and transferred to 150 by

25 mm screw cap tubes containing agar with overlay, 5–10 organisms per tube. Before introduction of the organisms, culture tubes with media are pre-incubated at 37°C for 24 h. Tubes with organisms are sealed with Parafilm and incubated at 37°C in an inclined position. One half of the fluid medium is replaced every second day during the first 10 d of culture and daily thereafter. Organisms are transferred to fresh agar slants on d 10, 14, and 21. When worms have grown to about 10 mm, cultures are "thinned" to contain no more than three worms per tube. After 14 d *in vitro*, pre-oncospheres may be observed in distal segments. Fully developed eggs infective to beetles (*Tribolium*) are contained in some segments after 3 weeks of culture (Seidel, 1975).

Corkscrewing and fragmentation of the strobila occur frequently, especially when culture tubes are agitated.

9° *Hymenolepis microstoma*, cysticercoid excystment
(Lippens-Mertens and DeRycke, 1973)

Omission of pepsin, trypsin, or bile salts decreases the speed of excystation *in vitro*. Maximal and speediest excystation occurs when cysticercoids are placed for 6 min at 37°C in a solution of 100 ml Hanks's saline (Chapter 1: IV, B, 1, g), 20 ml 0.2 N hydrochloric acid and 1 g pepsin, pH 1.5. Cysticercoids are then rinsed three times in Hanks's saline and transferred to the excysting solution at 37°C containing 0.5% trypsin and 0.3% sodium glycotaurocholate in Hanks's pH 7.2. After 6 min almost all cysticercoids have excysted.

10° *Hymenolepsis microstoma* adults
(Evans, 1970)

MEDIUM

Basic medium is prepared by mixing 60 ml of Eagle's (Gibco) medium (containing 100 iu ml^{-1} of penicillin, 100 µg ml^{-1} of streptomycin, and 2 µg ml^{-1} of *n*-butyl-*p*-hydroxybenzoate), 10 ml of liver extract, and 30 ml of horse serum to make 100 ml. The pH is adjusted to 7.6 with 0.2 N sodium hydroxide. One ml of whole ox bile is added to 100 ml of basic medium. Fresh ox bile is harvested and stored at $-15°C$ until use. Liver extract from sheep or hamsters is prepared according to the method of Sinha and Hopkins for lamb liver extract (see Taylor and Baker, 1968, p. 236).

TECHNIQUE

Roller tubes (size not stated) containing 5 ml of freshly prepared medium are gassed (30 sec) with 95% N_2 and 5% CO_2, sealed, and placed in a roller drum at 37·5°C to rotate for 2·5 h at 9 rev h^{-1} before inoculation with worms.

Cysticercoids dissected from beetles are washed 3 times in sterile balanced salt solution (modified Hanks's BSS, details not given), and transferred to 2·5 ml of 1% pepsin (w/v) in BSS, pH 1·6; after 11 min they are washed three times in sterile BSS, placed in 2·5 ml BSS containing 0·5% trypsin (w/v) and 0·3% sodium glycotaurocholate (w/v) (pH 7·2), 3–3·5 min). This solution is now diluted by one half with BSS, and again by one half 1 min later. After another min excysted organisms are washed three times in BSS and 20–30 worms transferred to each culture. Tubes are then gassed with the above mixture for 30 sec, sealed with Esco RHW rubber bungs, and placed on the roller drum. After 7 d the largest worms are separated out and returned in groups of 1–3 worms to culture tubes containing fresh medium. Medium is also changed on d 9 and d 11, and daily thereafter. Tubes are gassed after each medium change and then resealed. After 16 d gravid proglottids with apparently normal eggs are seen in many of the worms. No gravid segments are formed when ox bile is omitted from the medium.

11° *Hymenolepis nana*, cysticercoids
(Seidel and Voge, 1975)

MEDIUM AND TECHNIQUE

Culture medium, preparation of L-cysteine stock solution and surface sterilization of gravid proglottids obtained from mice are as described for *H. citelli* (see Section III, B, 1, 1°). Eggs dissected from proglottids in Earle's saline (Chapter 1: IV, B, 1, f) are transferred to a test tube with 3 mm glass beads and the thin outer shell removed by shaking for several min. Saline with oncospheres is then transferred to a solution of 7500 units of Tryptar (Armour Pharmaceutical) in 25 ml of Earle's saline at an initial pH of 3·0. The pH is raised slightly every 5 min with sodium bicarbonate to reach 7·0 within 20 min. During this period the solution is continuously gassed with 95% N_2 and 5% CO_2. Hatched oncospheres are then transferred to a Petri dish with culture medium and inoculated into 130 by 10 mm screw cap tubes containing 6 ml of medium. Tubes are then gassed with 95% N_2 and 5% CO_2 for 1 min and sealed with Parafilm. L-cysteine is not an essential component of the medium but may be added at 1 mM final concentration for the first 10 d of culture (28°C). Half the medium is exchanged after 5 d and three times a week thereafter. Tubes are gassed after each medium change. Full morphological development is attained at d 15 of culture. Cysticercoids are infective to mice a few days later. The use of air as a gas phase does not promote complete development.

12° *Hymenolepis nana*, cysticercoid excystment
(Sinha, 1973)

TECHNIQUE

Prolonged stay in excysting solutions is deleterious to cysticercoids. As sodium deoxycholate is very toxic to the larvae its presence in commercial bile salts must be kept in mind. The technique for excystation *in vitro* is as follows: cysticercoids are dissected from beetles in Hanks's basal salt solution (see Chapter 1: IV, B, 1, g) and transferred to 1% pepsin in Hanks's saline, pH 1·7, at 37°C.

After 12–15 min cysticercoids are washed 3 times in Hanks's saline at 37°C and are transferred to 0·5% trypsin and 0·3% sodium glycotauro-cholate in Hanks's saline (pH 7·2, 37°C). The majority of cysticercoids will excyst within 8–10 min. They are then washed 3 times in Hanks's saline (at 37°C) before use.

2. *Studies with Organisms Grown* In Vitro

Relatively few studies have been published on the use of organisms grown *in vitro* for investigations on metabolism, physiology, or other parameters. In an elegant study by Lackie (1976) cysticercoids of *H. diminuta* were grown *in vitro* and then injected into various insects to study the haemocytic defence reaction elicited by the worms in insect tissues as well as by latex beads. Results obtained clearly indicated the existence of a basic similarity between the surface of the worms and the surface of host tissues so that the insect hosts do not recognize the parasites as "non-self".

Adult *H. microstoma* grown *in vitro* were used by DeRycke and Evans (1972) to determine the influence of osmotic pressure of the ambient medium on the development of young adults. Freezing point depression (FPD) values of the culture media ranged from 188–650 mosmol litre^{-1}. The worms grew equally well within the FPD values of 269–392 mosmol litre^{-1} while at 650 mosmol litre^{-1} no growth took place. The authors suggest that *H. microstoma* might have a regulatory system which can operate only between the stated favourable limits.

The vitamin B_6 requirement for growth of the strobilate phase of *H. diminuta* was investigated by Roberts and Mong (1973). The authors incorporated an antimetabolite of B_6 into the culture medium and noted that growth of the worms was completely inhibited. This inhibition was entirely reversed by the presence of equimolar pyridoxine. Thus, good evidence was presented that *H. diminuta* has a direct nutritional requirement for this vitamin.

Tofts and Meerovitch (1974) found that farnesyl methyl ether (FME), a mimic of insect juvenile hormone, produced a pronounced inhibition of weight gain by *H. diminuta in vitro*. Study of neurosecretory cells in the scoleces suggested that FME triggered premature release of neurosecretory substances which upset a control mechanism in the germinative tissue of the neck region.

C. *MESOCESTOIDES*

Studies on the larval development *in vitro* of *Mesocestoides corti* are unusual in that the morphology of post-oncospheral growth stages was unknown until described from forms reared *in vitro* (Voge, 1967). As yet, no similar *in vivo* sequence has been published. Organisms reared *in vitro* are viable and apparently normal as they give rise to normal tetrathyridia when transferred to a different medium at the appropriate time (Voge and Seidel, 1968). Indeed, observations from *in vitro* studies also support the hypothesis of the requirement for two different hosts in the development of the larval stages as the tetrathyridial stage does not develop unless blood (or haemoglobin) is added to the medium.

Hart (1968) showed that tetrathyridia can be used as an effective model for studies on regeneration under fairly controlled *in vitro* conditions. Tetrathyridia maintained *in vitro* were also used by Heath (1970) to study their ability to synthesize purines and pyrimidines from simple precursors, and Heath and Hart (1970) showed that labelled orotic acid is incorporated into nucleic acid pyrimidines when tetrathyridia are maintained in a medium lacking preformed purines and pyrimidines. Mueller (1972) was able to keep some tetrathyridia alive *in vitro* for 7·5 months at 3–5°C.

1. *Detailed Techniques*

1° *Mesocestoides corti*, postoncospheral stages
(Voge, 1967)

MEDIUM

Triple Eagle's Medium (or NCTC 135, Gibco) is used with 30% horse serum (v/v; inactivated; 56°C, 30 min) plus antibiotics (penicillin, 300 iu ml^{-1}; dihydrostreptomycin, 300 µg ml^{-1}; pH 6·8–7·0). Culture vessels are screw cap tubes, 10 by 100 mm, each containing 3 ml of medium with air as the gas phase.

TECHNIQUE

Gravid proglottids from the faeces of dogs or skunks are washed (30 min each) in Petri dishes with Earle's saline (see Chapter 1: IV, B, 1, f) containing penicillin 900 iu ml^{-1}, and dihydrostreptomycin 900 µg ml^{-1}. Sterile glassware and solutions are used in all procedures.

Proglottids are transferred to depression slides, the uterine capsule is freed from surrounding tissue, and squeezed gently with forceps to release the oncospheres. Two drops of hatching solution (Tryptar, 25,000 units ml^{-1} in Earle's saline with bicarbonate, pH 7·0) are added to each well of the depression slides containing oncospheres. The slides are covered with a sterile Petri dish lid and kept at room temperature. Oncospheres hatch within 30–40 min. A drop of Earle's saline is added to each well to concentrate the oncospheres at the well bottom, and they are then pipetted into the culture tubes. These are sealed with a double layer of Parafilm and kept at room temperature (24–28°C) in a slanted position. One ml of medium is exchanged once a week for the first two weeks of culture and three times a week thereafter. After each medium change tubes are resealed with Parafilm. At five weeks of culture some organisms measure about 300 µm and contain calcareous corpuscles as well as sucker primordia. They resemble small tetrathyridia and are ready for transfer to the mouse host or to the medium which elicits complete development to the tetrathyridial stage.

Note: In NCTC 135, development to the pre-tetrathyridial stage is usually delayed by two weeks.

2° *Mesocestoides corti*, tetrathyridia from postoncospheral stages
(Voge and Seidel, 1968)

MEDIUM

Blood agar slants made of 5% human citrated blood in nutrient agar (Difco) are prepared in 16 by 125 mm screw cap tubes and overlaid with 8 ml of Triple Eagle's Medium (Gibco) with horse serum and antibiotics as above, pH 7·0.

TECHNIQUE

Post-oncospheral stages 300–400 µm in length are transferred to culture tubes. Tubes are sealed with Parafilm and incubated in an inclined position at room temperature. Fluid medium is changed once a week. Within 3–4 weeks fully developed tetrathyridia may be seen. Asexual multiplication of these tetrathyridia occurs 10 d after transfer of the organisms to 0·5 ml whole human blood overlay with 6 ml Triple Eagle's Medium (as above).

2. *Studies with Organisms Grown* In Vitro

The biosynthetic potential of parasitic helminths is virtually unknown, especially when it is realized that an inappropriate environment may not reveal the various capabilities an organism might have. A good illustration of this latter point was presented by Heath (1970) and by Heath and Hart (1970) who showed that tetrathyridia had the genetic potential for the synthesis of purines and pyrimidines *de novo*, but that this ability was masked when the culture medium was deficient in purine and pyrimidine derivatives. Further, under conditions of stress imposed by cutting the head region of the tetrathyridia in order to study regeneration, Heath and Hart (1970) showed that in the deficient medium the tetrathyridial fragments regenerated the scolex and incorporated labelled orotic acid. Thus, some kind of metabolic control is operative but its precise nature requires further study.

Hart (1968) showed that very small fragments with no more than one sucker are able to regenerate a whole tetrathyridium *in vitro*. However, tail fragments or portions without sucker material do not regenerate. Tetrathyridia produced in this way can reproduce asexually and produce young adults if placed in the appropriate medium. Thus, tetrathyridia are most useful for the study of regeneration of flatworms, as environmental conditions can be controlled more carefully than with planarians for example, and individual features of the axenic environment may be altered to study growth requirements. With refinements of methodology and instrumentation it should be possible to punch out small fragments in various areas of the scolex to determine more precisely their regenerative potential.

Tolerance and longevity of tetrathyridia as studied by Mueller (1972) illustrate additional problems which heretofore have not been sufficiently investigated. The ability of tetrathyridia to survive for many months at refrigerator temperature not only makes these worms a convenient research tool but also raises the question of the existence of comparable tolerances in other flatworm parasites. We have noted in our laboratory that "cooling" of tetrathyridia for several weeks will induce a burst of growth if these organisms are then slowly "warmed" and transferred to fresh culture medium. This phenomenon requires further study.

The results observed by Kowalski and Thorson (1976) on the effect of various lipids on tetrathyridia *in vitro* are difficult to interpret because their culture conditions were not the most favourable. The use of rat blood in the culture medium may have inhibited growth considerably. Rats are usually poor hosts for tetrathyridia and often are totally refractory to this infection.

D. *TAENIA*

Heath and Smyth (1970) described a culture method for cysticerci of *Taenia hydatigena*, *T. ovis*, *T. pisiformis*, and *T. serialis*. The importance of using serum from the natural intermediate host in the culture medium was well demonstrated; serum of young animals was more growth-promoting than that from older animals. Greatest success was achieved with *T. pisiformis* which was grown to cysticerci with hook-bearing scoleces. Heath and Elsdon-Dew (1972) studied *T. saginata* and *T. taeniaeformis in vitro*.

Heath (1973) simplified the original culture method by successfully substituting foetal calf serum for the natural host sera. Packed blood cells of rabbits were also added to the culture medium. The scolex formed in all four species but hooks were incompletely developed. During early growth of *T. pisiformis* it was noted that larvae divided into two viable organisms, each of which then developed a scolex.

Featherston (1971) determined the factors necessary for high rates of evagination of cysticerci of *T. hydatigena in vitro* and noted that shedding of much of the larval tissue did not occur as it did *in vivo*. *In vitro* hatching of taeniid oncospheres was used by Laws (1967) as a means to determine the efficacy of ovacidal compounds. Laws (1968) also used *in vitro* hatching techniques to observe the sequence of events and the mechanisms of the hatching process and to assess the effect of the tonicity of the hatching fluid.

Beveridge and Rickard (1975) investigated the hatching mechanism *in vitro* of *Anoplotaenia dasyuri* from the Tasmanian devil and concluded that the pattern and requirements of this species were similar to those described for other taeniids.

Organisms grown *in vitro* were used by Schiller (1973) to study abnormal scolex development in larval *T. crassiceps*, and Esch and Smyth (1976) grew the young strobilate phase of this worm from cysticerci. Heath (1976) used cysticerci of *T. pisiformis* grown *in vitro* to immunize rabbits. Schiller (personal communication) improved the method for culturing large numbers of *T. crassiceps* larvae and compared *in vivo* and *in vitro* development.

1. *Detailed Techniques*

1° *Taenia hydatigena*, *T. ovis*, *T. pisiformis* and *T. serialis*, cysticerci from oncospheres
(Heath and Smyth, 1970)

MEDIUM

The basic culture medium, Medium 858 (Healy *et al.*, 1955), is supplemented to contain $5 \cdot 3$ mg ml^{-1} glucose and $6 \cdot 0$ mg ml^{-1} potassium ions. Inactivated (56°C, 30 min) serum from the natural host is added, 20% per volume, e.g. rabbit serum for the growth of *T. pisiformis* and *T. serialis* and lamb serum for *T. ovis* and *T. hydatigena*. Serum from animals which may have harboured infections with these cysticerci should not be used as it will prevent growth. The pH is adjusted to $7 \cdot 2$ with $2 \cdot 8\%$ sodium bicarbonate (w/v) and the final medium gassed with 10% O_2 and 5% CO_2 in N_2.

TECHNIQUE

Eggs are dissected from gravid proglottids and pretreated in a screw cap roller tube for 1 h at 37–39°C with a solution containing 1% pepsin (w/v) and 1% concentrated hydrochloric acid (v/v) in $0 \cdot 85\%$ sodium chloride (w/v) in distilled water. The tube is then centrifuged (3 min, 3000 rev min^{-1}) the supernatant drawn off and the hatching solution [1% pancreatin, 1% sodium bicarbonate and 5% whole bile, either rabbit or sheep depending on the host in which the oncospheres normally hatch (stored -10°C, thawed just before use), in distilled water] added. After 30 min incubation the roller tubes are centrifuged as above, the supernatant is discarded and the oncospheres washed and centrifuged three times in culture medium with penicillin 1000 iu ml^{-1}, streptomycin sulphate 1000 µg ml^{-1}, and nystatin 1000 iu ml^{-1}. Activation of oncospheres is estimated by adding a drop of $0 \cdot 1\%$ neutral red to a drop of fluid with oncospheres: this stains all oncospheres which have escaped from their membranes.

Leighton tubes containing 10 ml of culture medium are inoculated with about 5000 hatched oncospheres, gassed with the gas mixture (see medium above) and the capped tubes placed horizontally in a roller device (1 rev min^{-1}, 37–39°C). Medium is changed twice weekly; early growth stages are concentrated by gentle centrifugation (details not given). As larvae become bigger they require more space and should be transferred at 10–15 d *in vitro* to larger flasks (200 ml Kimax bottles) containing 100 ml of medium.

Note: Authors state that about 5000 oncospheres should be placed in culture to yield 10 mature larvae as a large proportion of oncospheres does not develop. The above medium and technique are especially suitable for the growth of *T. pisiformis* which attains scolex development. The other three species undergo early development only but scolex primordia are formed in *T. ovis* after 16–32 d *in vitro*.

2° *Taenia taeniaeformis* and *T. saginata*, cysticerci from oncospheres
(Heath and Elsdon-Dew, 1972)

MEDIUM AND TECHNIQUE

The basic medium is NCTC 135 (Gibco) supplemented with 20% serum (v/v) from the natural host, preferably from 6 week-old calves for the culture of *T. saginata*; rat serum is used for *T. taeniaeformis*. *T. saginata* eggs are collected and oncospheres hatched as described above (Section III, D, 1, 1°) except that *T. taeniaeformis* eggs are not pretreated with pepsin, and a mixture of 40% pig bile (v/v), 40% calf bile (v/v) and 20% rabbit bile (v/v) is used in the hatching solution.

Leighton tubes containing 10 ml of the culture medium are inoculated with oncospheres, their caps screwed on tightly, and incubated stationary at 37°C with air as a gas phase. Medium is changed twice weekly, gentle centrifugation being used to concentrate the larvae. Young larvae with a cavity are present in these cultures at 10 d.

Note: Calf serum macromolecules (α- and β-globulins) can be substituted for whole serum, although development is not then as advanced.

3° *Taenia ovis, T. serialis, T. taeniaeformis, T. hydatigena, T. pisiformis*
(Heath, 1973)

MEDIUM

NCTC 135 (Gibco) is supplemented with dextrose to 400 mg ml^{-1}. Penicillin G (100 iu ml^{-1}), streptomycin sulphate (100 µg ml^{-1}) and nystatin (100 iu ml^{-1}) are added. One percent (v/v) packed rabbit blood cells washed three times in phosphate-buffered saline (Chapter 1: IV, B, 1, b) is added with inactivated (56°C, 30 min) foetal calf serum at 50% concentration for the first 7 d, 20% for the next 14 d, and 10% thereafter.

TECHNIQUE

Eggs are collected from gravid proglottids in 0·85% sodium chloride and stored overnight at 4°C. The supernatant is removed and eggs are treated with 1:5000 Hibitane (chlorhexidine gluconate, I.C.I.) for 20 min. They are then rinsed three times in sterile distilled water and are hatched as described by Heath and Smyth (see Section III, D, 1, 1°), except that the pretreatment and hatching solutions are filtered with a 0·2 μm Millipore filter before use, and that a tissue culture roller is not used.

Culture vessels are 30 ml or 250 ml Falcon disposable flasks with air as a gas phase. After inoculation with oncospheres flasks are kept stationary at 37–39°C. Medium (volume not given) is changed twice weekly. As growth proceeds, the volume of fluid per larva is progressively doubled at each change by removing 50% of the larvae, until each larva ultimately receives 10 ml of culture fluid at each change.

All five species develop into immature cysticerci at a rate comparable to that *in vivo*; in *T. pisiformis* scolex differentiation is completed after 35 d *in vitro*.

4° *Taenia hydatigena*, cysticerci—evagination
(Featherston, 1971)

TECHNIQUE

Cysticerci, collected from sheep carcasses, are exposed to a solution containing 100 mg trypsin (or pancreatin) and 5 ml dog bile made up to 100 ml with Hanks's basal salt solution (Chapter 1: IV, B, 1, g), pH 6·9, prewarmed to 38°C.

The solution with cysticerci is kept at 38°C. Evagination of most of the cysticerci occurs almost instantaneously and the scoleces remain evaginated. Pretreatment with acid pepsin is not necessary.

5° *Taenia crassiceps*, larval multiplication
(E. L. Schiller, personal communication)

MEDIUM AND TECHNIQUE

A diphasic medium is prepared, consisting of 70% Difco nutrient agar and 30% fresh, defibrinated rabbit blood, overlaid with Hanks's balanced salt solution (see Chapter 1: IV, B, 1, g) containing penicillin (200 iu ml^{-1}) and streptomycin (200 μg ml^{-1}); the pH is adjusted to 7·8 with sodium bicarbonate. The blood agar is dispensed to sterile 50 ml Ehrlenmeyer flasks, 25 ml per flask. After gelation of the agar 5 ml of Hanks's solution are added to each flask. The flasks are stoppered with cotton and preincubated (34°C, 24 h). Glucose (1·0 mg ml^{-1}) is added just before each flask is inoculated with 5 immature buds measuring about 0·5 mm in diameter. Flasks are incubated in a Dubnoff shaking incubator (about 100 cycles min^{-1}, 37°C) with 95% N_2 and 5% CO_2. Worms are transferred to fresh medium every 72 h.

With this system buds reach a diameter of 3·0–5·0 mm in approximately two weeks. The scolex is completely developed and minute buds may be seen at the abscolex pole. During the longest period of continuous cultivation (162 d), each bud produced an average of 157·3 progeny.

6° *Taenia crassiceps*, young adults from cysticerci
(Esch and Smyth, 1976)

MEDIUM

The liquid phase consists of 100 parts medium 858 (Difco), 25 parts of inactivated foetal calf serum (Gibco), and 12·5 parts of 5% powdered yeast extract (Gibco or Oxoid). To one litre of this medium are added 22 ml of 30% glucose (w/v), 5 ml of 2% potassium chloride (w/v), 100 iu ml^{-1} of penicillin G and 100 µg ml^{-1} of streptomycin sulphate.

Culture vessels are 30 ml plastic flasks containing 5 ml of bovine serum, coagulated at 75°C for 60–90 min to form a soft substrate permitting migration of cysticerci. The overlay is 10 ml of the fluid medium above, and the gas phase is 10% O_2, 5% CO_2, or 10% CO_2 in N_2.

TECHNIQUE

Larvae are transferred from mice into Petri dishes with warm Hanks's balanced salt solution (see Chapter 1: IV, B, 1, g). Larvae are then passed through 2 washes of Hanks's solution. Fifty ml of Hanks's solution with 10 mg ml^{-1} of pepsin, pH 1·9, are added to the cysticerci in a water bath shaker at 38·9°C. After 20 min larvae are washed three times in warm Hanks's solution, transferred to the evaginating solution (Hanks's saline with 3 mg ml^{-1} pancreatin, 0·1 mg ml^{-1} sodium taurocholate and 0·15 mg ml^{-1} trypsin) and incubated in a shaking water bath (38·9°C, 3–20 h). They are then washed twice in warm Hanks's solution and transferred to a Petri dish with warm fluid culture medium. Larvae with evaginated scoleces bearing well-developed hooks are inoculated into culture flasks, 3–6 larvae per flask; the flasks are gassed and kept in a shaking water bath at 38·9°C. The liquid overlay is changed every 48 h. After 13 d of culture, young worms have proglottids with cirrus primordia and genital pores.

2. *Studies with Organisms Grown* In Vitro

Laws (1967) clearly demonstrated the usefulness of the *in vitro* hatching technique to test the efficacy of various chemicals in killing taeniid eggs. Effects such as inhibition of hatching, dehydration, and dissolution of the whole egg can be studied *in vitro* and the hatching technique, with appropriate untreated controls, provides a relatively inexpensive means of evaluating test substances.

As shown by Laws (1968) and others before him, hatching techniques may aid in the understanding of the essential processes involved in the transfer to and establishment of a parasite in the host. The structure of the coats surrounding the taeniid oncosphere, the differential susceptibility of the membranes to various chemicals, and the physical phenomena which accompany the release of the oncosphere are described by Laws in detail.

Much interesting information may be a by-product of attempts to culture an organism *in vitro*. The division into two organisms during the early development of *T. pisiformis*, described by Heath (1973), is most intriguing and deserves careful checking of the development *in vivo*. To find these small division stages would probably be difficult and might necessitate the examination of heavily infected animals.

Organisms grown *in vitro* could also be useful in genetic studies as shown by Schiller (1973) in preliminary studies with *T. crassiceps*.

The obvious usefulness of *in vitro* techniques in the study of immunity was realized by Heath (1976) who was able to induce resistance in rabbits to infection with *T. pisiformis* by injecting them subcutaneously with killed larvae reared from oncospheres after 6–9 d *in vitro*. Concentrated culture medium in which larvae had grown was also effective. The exogenous antigens produced by 10 d larvae *in vitro* were partially characterized.

E. *ECHINOCOCCUS*

Heath and Smyth (1970; Section III, E, 1, 1°) have hatched the oncospheres of *E. granulosus* and grown them to early cystic stages with cavity. The protoscoleces and brood capsules were placed into mammalian cell cultures or cell-free media by Brudnjak *et al.* (1970; Section III, E, 1, 2°) who obtained cystic development from protoscoleces and survival of organisms for several weeks.

The evagination of protoscoleces was investigated by Wikerhauser (1969; Section III, E, 1, 3°) using a variety of solutions and procedures.

Studies on the factors controlling strobilization and maturation were published by Smyth *et al.* (1967; Section III, E, 1, 4°) who obtained sexually mature adults without gravid segments. Similar studies by Smyth and Davies (1974a; Section III, E, 1, 5°) review the basic problems and methodologies in the development of the adult worm *in vitro*.

Smyth and Davies (1974b) studied physiological strains isolated from hydatids in sheep and horse. Protoscoleces from horse hydatids failed to produce segments *in vitro* while those from sheep hydatids did so.

1. Detailed Techniques

1° *Echinococcus granulosus*, cystic larvae from oncospheres
(Heath and Smyth, 1970)

MEDIUM AND TECHNIQUE

The hatching of oncospheres, culture medium and procedures are as outlined for the four species of *Taenia* (Section III, D, 1, 1°). Spherical organisms with a cavity are present by 10 d. Apparently, no additional development occurs with this medium.

2° *Echinococcus granulosus*, cystic development
(Brudnjak, Cvetnić and Wikerhauser, 1970)

MEDIUM

Monolayers of rhesus monkey kidney cells are grown in Parker 199 (Gibco) with 2·5% calf serum (v/v), in Bich–Demeter square bottles, each containing 15 ml of medium, with 100 iu penicillin ml^{-1} and 100 µg streptomycin ml^{-1}.

TECHNIQUE

Hydatid cysts from pigs livers are cut open, their contents transferred to sterile Petri dishes and checked for the presence of brood capsules and protoscoleces. These are then transferred to centrifuge tubes. Hydatid sand is allowed to settle. The supernatant is removed and the sediment is similarly rinsed twice in the liquid culture medium; 0·5 ml of sediment is inoculated into each culture bottle. Cultures are incubated at 37°C with air as a gas phase. Liquid medium is renewed when it becomes acid. At 40 d of culture vesicular protoscoleces with a laminated envelope may be seen. Brood capsules develop vesicles to which protoscoleces are attached.

Note: Several other media with or without cells were used. Best results, however, were obtained with the medium described above.

3° *E. granulosus*, evagination of protoscoleces
(Wikerhauser, 1969)

TECHNIQUE

Hydatid cysts from pigs are opened and protoscoleces removed to Petri dishes or centrifuge tubes. Scoleces are exposed for 10 min to 0·5% pepsin solution and then transferred to 0·4% trypsin solution with 20% whole ox bile in a Petri dish at 38°C. After 1–2 h in this solution, most scoleces have evaginated. Dog bile may be used instead of ox bile.

4° *E. granulosus*, adults from protoscoleces
(Smyth, Miller and Hawkins, 1967)

MEDIUM

A liquid medium composed of 20% hydatid fluid (v/v) in medium 858 (Gibco) is used over bovine serum, coagulated in 250 ml milk-dilution bottles (Kimax) in an oven at 100°C for 30 min or in a boiling waterbath for 10 min. The liquid medium contains 100 iu ml^{-1} penicillin and 100 μg ml^{-1} streptomycin, and the glucose level is raised to 2·5 mg ml^{-1}. The culture bottles are gassed with 5% CO_2 and 10% O_2 in N_2. At the time of segmentation, the medium is supplemented with 5% beef embryo extract.

TECHNIQUE

Protoscoleces are excysted by exposure for 15 min to 0·025% pepsin (w/v) in Hanks's salt solution (Chapter 1: IV, B, 1, g) pH 2·0 followed by transfer to a solution with 0·3% pancreatin (w/v), 0·1% trypsin (w/v) and 5% dog bile (v/v) in Hanks's saline, pH 7·6 for 1 h. Worms are then washed in several changes of Hanks's saline and finally in liquid medium before transfer to culture vessels. Medium is changed every 72 h. Worms with three segments develop in this system in 22 d. No eggs are present as fertilization has not occurred.

5° *Echinococcus granulosus*, adults
(Smyth and Davies, 1974a)

MEDIUM

The diphasic medium is similar to that described above (Section III, E, 1, 4°), modified as follows. The serum base is prepared by heating to 75–76° (instead of 100°C) in an oven for 30–60 min. Calf serum may be substituted for bovine serum. The liquid phase may be any one of the following: Parker 858, CMRL 1066, or NCTC 135 (Gibco), each supplemented with 3·75 ml of 30% glucose (w/v), 2·5 ml of 2% potassium chloride (w/v), 50 ml of foetal calf serum (inactivated; 56°C, 30 min) and 2·5 ml of 5% yeast extract (w/v) per 200 ml of base medium. Antibiotics are added (100 iu ml^{-1} penicillin, 100 µg ml^{-1} streptomycin). Milk dilution (Kimax) bottles or 300 ml Falcon flasks are used as culture vessels.

TECHNIQUE

The procedures described above (Section III, E, 1, 4°) have been modified and improved. Before removing protoscoleces from a cyst, the surface is painted with a 1% solution of iodine (w/v) in 95% ethanol (v/v); this procedure is repeated twice. Remove about 50% of the fluid with a syringe and 17–19 gauge needle, *wearing protective glasses as you do so*. Cut open cyst, fold back wall and remove protoscoleces, most of which are found on the bottom of the cyst. Protoscoleces may be stored at 4°C in 10 times their volume of hydatid fluid in sterile screw cap tubes, preferably for not more than 2 d.

Viability of protoscoleces should be checked. Dead protoscoleces are usually brownish in colour, while live ones are clear. At least 60% of protoscoleces should be viable for a successful culture.

Excystment of protoscoleces is carried out at 37–38°C; excystment solutions are sterilized by filtration before use. The pepsin solution should be 0·05%. Protoscoleces should be rinsed once in this solution, then transferred to fresh pepsin in a water bath (37–38°C), and gently shaken for 15–45 min. They should then be washed at least four times in Hanks's saline (see Chapter 1: IV, B, 1, g) with antibiotics, allowing 15 min (or more) for each wash. Scoleces are transferred to sodium taurocholate (0·2 mg ml^{-1}) in Hanks's BSS, pH 7·4, with 20% culture medium to be used in the experiment and kept in this solution for about 18 h (temperature not stated).

Worms are then washed in Hanks's saline or in culture medium. Remove solution, leaving 5–10 ml depending on the number of protoscoleces present. Culture vessels are inoculated with several thousand scoleces. After 34 d cells are found in the uterus and the worms are only 2 d behind those developing in dogs. No embryonated eggs are produced in culture.

IV. ACCOMPLISHMENTS, CRITIQUE AND SUGGESTIONS

Significant advances during the last seven years include the determination of several essential single environmental factors for the axenic cultivation of tapeworms. Among these are haemin for the growth of adult *Hymenolepis*, reducing agents for the development of some larval *Hymenolepis*, and the importance of osmolarity in hatching as well as culture for some tapeworm species. However, most of this information is still fragmentary and the precise functions of haemin and reducing agents in the *in vitro* systems remain to be elucidated.

Considerable simplification of culture apparatus has resulted from improvements of media as well as of techniques of handling the living material. Thus, whenever possible, screw cap tubes or various flasks are preferable to complex apparatus which in itself might discourage potential students of *in vitro* technique.

The importance of using homologous host sera and the detrimental effect of sera from animals infected with the parasite species to be grown *in vitro* have been established and constitute important aspects of technique. More importantly, the potential use of macromolecules in place of serum represents a very intriguing and useful idea which requires thorough testing with a variety of organisms.

The post-oncospheral growth of taeniid cestodes is a distinct advance but development comparable to that achieved for *T. pisiformis* has yet to be attained. One wonders why it should be necessary to inoculate culture vessels with several thousand oncospheres, whether the death of a large proportion of these contributes to the growth of the remaining ones or whether the culture system is not "large" enough to supply so many organisms with their needs.

While methods are usually described in sufficient detail, frequent omissions of certain points must be noted. First, there is often no mention of whether or not the serum used should be inactivated. This applies especially to foetal calf serum which we, in our laboratory, use without prior inactivation; other workers inactivate this serum when it is used for other purposes. Second, the volume of culture fluid used in each culture vessel is often not stated. Third, the time period for gassing of culture vessels is frequently omitted. Additional points which require attention are the diluents used to make up solutions such as hatching fluids. It is not always clear whether this is a saline and if so, which kind. The storage conditions for culture media is usually not specified. Prolonged exposure to refrigerator temperature can be deleterious. Freezing in small amounts and thawing just before use may be the best procedure unless fresh medium is made up each time. Whatever the means of storage, it should be described.

While it is obvious that considerable progress has been made, we must look to the future for essential studies in those areas which will enable us to use tapeworms grown *in vitro* more effectively in experiments involving physiology, biochemistry, and immunology. One of our greatest needs is the development of minimal media and of defined media. Most culture media in use now contain too many complex substances (liver extracts, yeast extract, sera, etc.). Further, many of the defined components present in commercial media such as certain amino acids or vitamins may not be needed by the organisms.

Unfortunately, the establishment of minimal media is a fairly costly, time-consuming, and perhaps boring activity so that little attention has been directed to it. In our laboratory we have been able to eliminate yeast extract from our medium for the growth of cysticerci of *H. diminuta* but lactalbumin hydrolysate was shown to be an essential component (M. Voge, unpublished observations). There is as yet no defined medium available for tapeworm culture, although Taylor managed to obtain juvenile worms from cysticerci of *T. crassiceps* in Eagle's medium and maintained them for 25 d (Taylor, 1963; Taylor *et al.* 1966). Some media have been simplified but serum is still an essential component of all. Perhaps the use of macromolecules might at least eliminate the need for whole serum.

Several workers have mentioned the importance of purity of the substances used in culture media. This is a real concern because of the potentially toxic or inhibitory effects of trace contaminants. The possible presence of impurities should be kept in mind at all times.

The behaviour of organisms growing *in vitro* has been noted by several workers. In our experience, behaviour can furnish a useful clue to the adequacy of the *in vitro* environment. Rapid contractions and twisting usually mean that the *in vitro* situation is either harmful or irritating. Lack of motility of live organisms may also denote inadequate surroundings. More observations on behaviour in different *in vitro* environments would be of interest.

Very little information is available on analysis of media, in which worms have been growing, to determine what substances are utilized and are therefore present in smaller amounts than initially, or to see what metabolic by-products have been released into the medium.

Most importantly, only a few workers have used worms grown *in vitro* in studies of biochemistry and physiology. There is still a tendency to remove worms from the host and place them for a few minutes into some saline solution before beginning the experiment. Others may remove worms from the host and transfer them for a few hours to a culture medium. Usually the shock to the organism induced by this drastic change in environment is not considered. Experimental conclusions reached on the basis of these techniques are of dubious validity and very often may yield misinformation rather than meaningful facts.

REFERENCES

Berntzen, A. K. and Mueller, J. F. (1972). *In vitro* cultivation of *Spirometra* spp. (Cestoda) from the plerocercoid to the gravid adult. *Journal of Parasitology* **58**, 750–752.

Berntzen, A. K. and Voge, M. (1965). *In vitro* hatching of oncospheres of four hymenolepidid cestodes. *Journal of Parasitology* **51**, 235–242.

Beveridge, I. and Rickard, M. D. (1975). Studies on *Anoplotaenia dasyuri* Beddard, 1911 (Cestoda: Taeniidae), a parasite from the Tasmanian devil: observations on the egg and metacestode. *International Journal of Parasitology* **5**, 257–267.

Brudnjak, Z., Cvetnić, S. and Wikerhauser, T. (1970). Cystic development of the protoscoleces and brood capsules of *Echinococcus granulosus* in cell cultures and cell-free media. *Veterinarski Arhiv, Zagreb.* **40**, 292–296.

Caley, J. (1975). *In vitro* hatching of the tapeworm *Moniezia expansa* (Cestoda: Anoplocephalidae) and some properties of the egg membrane. *Zeitschrift für Parasitenkunde* **45**, 335–346.

Chang, T. W., Raben, M. S., Mueller, J. F. and Weinstein, L. (1973). Cultivation of the sparganum of *Spirometra mansonoides in vitro* with prolonged production of sparganum growth factor. *Proceedings of the Society of Experimental Biology and Medicine* **143**, 457–459.

Davis, B. O. Jr. (1975). *Hymenolepis microstoma* (Cestoda): Effects of cysticercoid age on morphology, excystation and establishment. *Acta Parasitologica Polonica* **23**, 229–236.

DeRycke, P. H. (1975). Maintenance of *Hymenolepis microstoma* (Cestoda) cysticercoids *in vitro*. *Acta Parasitologica Polonica* **23**, 291–297.

DeRycke, P. H. and Evans, W. S. (1972). Osmoregulation of *Hymenolepis microstoma II*. Influence of the osmotic pressure on the *in vitro* development of the young adult. *Zeitschrift für Parasitenkunde* **38**, 147–151.

Esch, G. W. and Smyth, J. D. (1976). Studies on the *in vitro* culture of *Taenia crassiceps*. *International Journal of Parasitology* **6**, 143–149.

Evans, W. S. (1970). The *in vitro* cultivation of *Hymenolepis microstoma* from cysticercoid to egg producing adult. *Canadian Journal of Zoology* **48**, 1135–1137.

Featherston, D. W. (1971). *Taenia hydatigena*. II. Evagination of cysticerci and establishment in dogs. *Experimental Parasitology* **29**, 242–249.

Goodchild, C. G. and Davis, B. O. Jr. (1972). *Hymenolepis microstoma* cysticercoid activation and excystation *in vitro* (Cestoda). *Journal of Parasitology* **58**, 735–741.

Graham, J. J. and Berntzen, A. K. (1970). The monoxenic cultivation of *Hymenolepis diminuta* with rat fibroblasts. *Journal of Parasitology* **56**, 1184–1188.

Hart, J. L. (1968). Regeneration of tetrathyridia of *Mesocestoides* (Cestoda: Cyclophyllidea) *in vivo* and *in vitro*. *Journal of Parasitology* **54**, 950–956.

Healy, G. M., Fisher, D. C. and Parker, R. C. (1955). Nutrition of animal cells in tissue culture. X. Synthetic medium no. 858. *Proceedings of the Society for Experimental Biology and Medicine* **89**, 71–77.

Heath, D. D. (1973). An improved technique for the *in vitro* culture of taeniid larvae. *International Journal of Parasitology* **3**, 481–484.

Heath, D. D. (1976). Resistance to *Taenia pisiformis* larvae in rabbits: immunization against infection using non-living antigens from *in vitro* culture. *International Journal of Parasitology* **6**, 19–24.

Heath, D. D. and Elsdon-Dew, R. (1972). The *in vivo* culture of *Taenia saginata* and *Taenia taeniaeformis* larvae from the oncosphere with observations on the role of serum for *in vitro* culture of larval cestodes. *International Journal of Parasitology* **2**, 119–130.

Heath, D. D. and Smyth, J. D. (1970). *In vitro* cultivation of *Echinococcus granulosus, Taenia hydatigena, T. ovis, T. pisiformis* and *T. serialis* from oncosphere to cystic larva. *Parasitology* **61**, 329–343.

Heath, R. L. (1970). Biosynthesis *de novo* of purines and pyrimidines in *Mesocestoides* (Cestoda) I. *Journal of Parasitology* **56**, 98–102.

Heath, R. L. and Hart, J. L. (1970). Biosynthesis *de novo* of purines and pyrimidines in *Mesocestoides* (Cestoda) II. *Journal of Parasitology* 56, 340–345.

Hundley, D. F. and Berntzen, A. K. (1969). Collection, sterilization, and storage of *Hymenolepis diminuta* eggs. *Journal of Parasitology* 55, 1095–1096.

Kowalski, J. C. and Thorson, R. E. (1976). Effects of certain lipid compounds on growth and asexual multiplication of *Mesocestoides corti* (Cestoda) tetrathyridia. *International Journal of Parasitology* 6, 327–331.

Lackie, A. M. (1976). Evasion of the haemocytic defense reaction of certain insects by larvae of *Hymenolepis diminuta* (Cestoda). *Parasitology* 73, 97–107.

Landureau, J. C. (1966). Cultures *in vitro* de cellules embryonnaires de blattes (insectes dictyoptères). *Experimental Cell Research* 41, 545–556.

Landureau, J. C. (1968). Cultures *in vitro* de cellules embryonnaires de blattes (insectes dictyoptères). *Experimental Cell Research* 50, 323–337.

Laws, G. F. (1967). Chemical ovacidal measures as applied to *Taenia hydatigena*, *T. ovis*, *T. pisiformis*, and *Echinococcus granulosus*. *Experimental Parasitology* 20, 27–37.

Laws, G. I. (1968). The hatching of taeniid eggs. *Experimental Parasitology* 23, 1–10.

Lippens-Mertens, F. and DeRycke, P. (1973). Excystation of *Hymenolepis microstoma*. II. Relative influences of proteolytic enzymes and bile salts. *Zeitschrift für Parasitenkunde* 42, 61–67.

Mueller, J. F. (1972). Survival and longevity of *Mesocestoides* tetrathyridia under adverse conditions. *Journal of Parasitology* 58, 228.

Nicolle, C. (1908). Culture du parasite du bouton d'orient. *Compte Rendu hbd. Séanc. Acad. Sci.* 146, 842–843.

Novy, F. G. and MacNeal, W. J. (1903). The cultivation of *Trypanosoma brucei*. A preliminary note. *Journal of the American Medical Association* 41, 1266–1268.

Paul, J. (1960). "Cell and Tissue Culture." Livingstone, Edinburgh & London, pp. 312.

Roberts, L. A. and Mong, F. N. (1969). Developmental physiology of cestodes. IV. *In vitro* development of *Hymenolepis diminuta* in presence and absence of oxygen. *Experimental Parasitology* 26, 166–174.

Roberts, L. A. and Mong, F. N. (1973). Developmental physiology of cestodes. XIII. Vitamin B_6 requirement of *Hymenolepis diminuta* during *in vitro* cultivation. *Journal of Parasitology* 59, 101–104.

Roberts, L. S. (1973). Modifications in media and surface sterilization methods for *in vitro* cultivation of *Hymenolepis diminuta*. *Journal of Parasitology* 59, 474–479.

Rothman, A. (1959). Studies on the excystment of tapeworms. *Experimental Parasitology* 8, 336–364.

Schiller, E. L. (1965). A simplified method for the *in vitro* cultivation of the rat tapeworm *Hymenolepis diminuta*. *Journal of Parasitology* 51, 516–518.

Schiller, E. L. (1973). Morphologic anomalies in scoleces of larval *Taenia crassiceps*. *Journal of Parasitology* 59, 122–129.

Seidel, J. S. (1969). Development *in vitro* of *Hymenolepsis microstoma* from cysticercoid to adult. 44th Annual Meeting, American Society of Parasitologists, Abstract No. 150.

Seidel, J. S. (1971). Hemin as a requirement in the development *in vitro* of *Hymenolepis microstoma* (Cestoda: Cyclophyllidea). *Journal of Parasitology* 57, 566–570.

Seidel, J. S. (1975). The life cycle *in vitro* of *Hymenolepis microstoma* (Cestoda). *Journal of Parasitology* 61, 677–681.

Seidel, J. S. and Voge, M. (1975). Axenic development of cysticercoids of *Hymenolepis nana*. *Journal of Parasitology* 61, 861–864.

Sinha, D. P. (1973). *In vitro* excystation of cysticercoids of *Hymenolepis nana*. *The Annals of Zoology* **9**, 41–51.

Smyth, J. D. and Davies, Z. (1974a). *In vitro* culture of the strobilar stage of *Echinococcus granulosus* (sheep strain): a review of basic problems and results. *International Journal of Parasitology* **5**, 631–644.

Smyth, J. D. and Davies, Z. (1974b). Occurrence of physiological strains of *Echinococcus granulosus* demonstrated by *in vitro* culture of protoscoleces from sheep and horse hydatid cysts. *International Journal of Parasitology* **4**, 443–445.

Smyth, J. D., Miller, H. J. and Hawkins, A. B. (1967). Further analysis of the factors controlling strobilization, differentiation, and maturation of *Echinococcus granulosus in vitro*. *Experimental Parasitology* **21**, 31–41.

Tachovsky, T. G., Hare, J. D. and Ritterson, A. L. (1973). A culture method permitting *in vitro* growth of *Spirometra mansonoides* spargana. *Journal of Parasitology* **59**, 937–938.

Taylor, A. E. R. (1963). Maintenance of larval *Taenia crassiceps* (Cestoda : Cyclophyllidea) in a chemically defined medium. *Experimental Parasitology* **14**, 304–310.

Taylor, A. E. R. and Baker, J. R. (1968). "The Cultivation of Parasites *in vitro*." Blackwell Scientific Publications, Oxford & Edinburgh, pp. 377.

Taylor, A. E. R., McCabe, M. and Longmuir, I. S. (1966). Studies on the metabolism of larval tapeworms (Cyclophyllidea : *Taenia crassiceps*). II. Respiration, glycogen utilization and lactic acid production during culture in a chemically defined medium. *Experimental Parasitology* **19**, 269–275.

Tkachuck, R. D. and Weinstein, P. P. (1976). Comparison of the uptake of vitamin B_{12} by *Spirometra mansonoides* and *Hymenolepis diminuta* and the functional groups of B_{12} analogs affecting uptake. *Journal of Parasitology* **62**, 94–101.

Tofts, J. and Meerovitch, E. (1974). The effect of Farnesyl Methyl Ether, a mimic of insect juvenile hormone, on *Hymenolepis diminuta in vitro*. *International Journal of Parasitology* **4**, 211–218.

Turton, J. A. (1972). The *in vitro* cultivation of *Hymenolepis diminuta* : the culture of six day old worms removed from the rat. *Zeitschrift für Parasitenkunde* **40**, 333–346.

Voge, M. (1967). Development *in vitro* of *Mesocestoides* (Cestoda) from oncosphere to young tetrathyridium. *Journal of Parasitology* **53**, 78–82.

Voge, M. (1972). Axenic development of postembryonic stages of *Hymenolepis citelli* (Cestoda). American Society of Parasitologists, 47th Annual Meeting, Abstract No. 203.

Voge, M. (1975). Axenic development of cysticercoids of *Hymenolepis diminuta* (Cestoda). *Journal of Parasitology* **61**, 563–564.

Voge, M. and Edmonds, H. (1969). Hatching *in vitro* of oncospheres from coracidia of *Lacistorhynchus tenuis* (Cestoda : Tetrarhynchidea). *Journal of Parasitology* **55**, 571–573.

Voge, M. and Green, J. (1975). Axenic growth of oncospheres of *Hymenolepis citelli* (Cestoda) to fully developed cysticercoids. *Journal of Parasitology* **61**, 291–297.

Voge, M. and Seidel, J. S. (1968). Continuous growth *in vitro* of *Mesocestoides* (Cestoda) from oncosphere to fully developed tetrathyridium. *Journal of Parasitology* **54**, 269–271.

Voge, M., Jaffe, J., Bruckner, D. and Meymarian, E. (1976). Synergistic growth promoting action of L-cysteine and nitrogen upon *Hymenolepis diminuta* cysticercoids *in vitro*. *Journal of Parasitology* **62**, 951–954.

Wikerhauser, T. (1969). Istraživanja devaginacije protoskoleksa *Echinococcus granulosus*. *Veterinarski Arhiv, Zagreb* **39**, 268–271.

Chapter 10

Nematoda Parasitic in Animals and Plants*

EDER L. HANSEN AND JAMES W. HANSEN

561 Santa Barbara Rd., Berkeley, California, U.S.A.

I. SCOPE AND APPLICATION OF CULTURE

The recent work reviewed in this chapter covers 54 species of nematodes. Most of these were included in the compilation by Taylor and Baker (1968). Some new groups have been added, specifically Anisakidae, Mermithidae and plant parasites. Essentially, only work using axenic culture methods is included, short term *in vitro* maintenance for toxicity studies or biochemical preparations being excluded.

During the past eight years the main emphasis has been on culturing specific stages in the nematode life cycle for production of antigens or for study of developmental controls. Nematodes release antigenic materials

* In this chapter the details of cultivation methods are dispersed throughout Section III; the composition of many of the solutions and media used are given in Chapter 1 (Eds).

during their development, which can be recovered from the culture medium. Specific experiments on production and characterization of immunogenic metabolic antigens are indicated in Section III.

In using cultures for study of developmental controls the nematodes need to be maintained only through the portion of the life cycle during which moulting occurs. Control appears to be by hormones analogous to the moulting and juvenile hormones of insects. Related molecular events are investigated by incubating cultures in the presence of actinomycin-D, mitomycin-C and puromycin which interfere with DNA, RNA and protein synthesis. Particular findings with *Aphlenchus*, *Neoaplectana*, *Nippostrongylus*, *Phocanema* and *Trichinella* are indicated in Section III.

II. CULTURE PROCEDURES AND MEDIA

The culture procedures are still largely those described by Taylor and Baker (1968); they are reviewed by Leland (1970b) and by Silverman and Hansen (1971). In this section we discuss some aspects of culture procedures that are of general application.

A. SOURCE OF NEMATODES

Axenic cultures can be successful only when started with well-grown individuals. When free living stages are used they should be obtained from cultures maintained under standardized conditions. Procedures for such xenic cultures are not described herein except in so far as they relate directly to axenic cultivation. When parasitic stages are used they should be removed promptly from freshly excised host organs. Nematodes must not be exposed to inhibiting conditions during preparation of the inoculum, otherwise they may be irreversibly committed to arrested development. Nematodes collected from source material are thoroughly washed in a Baermann-type apparatus (see Chapter 1, V) or individually isolated by micro-manipulation and then axenized.

B. AXENIZATION

Procedures for axenization differ depending on whether the culture is to be started with infective eggs or larvae, with nematodes from the intestine, or with a developing stage from host tissue. Host tissue is relatively sterile and the parasite can be recovered free of contaminating organisms by using aseptic techniques. Nematodes from non-sterile environments must be axenized by immersion in antiseptics and antibiotics. Treatment should be kept to the minimum that will accomplish complete removal of bacteria and fungal spores.

Antiseptic chemicals, even though used for brief periods, may be inhibitory to nematode development. However, when the organism is taken from a highly contaminated situation, antiseptics are usually a necessary first step. Their use must be followed by thorough washing with several changes of water. The following treatments have been reported by various authors:

Benzylalkonium chloride (zephiran), 3·4% for 15 min;
Formalin, 0·5% for 15 min;
Hyamine 1622 (Rohm and Haas), 0·4% for 10 sec 0·001% for 15 min;
Hydrochloric acid, 0·1 N for 10 min;
Iodine, 0·2% for 10 sec;
Mercuric chloride, 0·2% for 2 min;
Merthiolate, 0·1% for 15 min;
Peracetic acid, 0·5% for 15 min;
Peroxide, 3% for 2 min;
Sodium alkylaryl sulphonate, 0·1% for 15 min;
Sodium hypochlorite, 0·05%–0·5% for 10 min.

The final axenization is done by treatment with antibiotics, using high concentrations for short periods followed by longer periods at lower concentrations. Developing stages or egg laying adults should not be exposed to high concentrations. The following concentrations (ml^{-1}) have been reported by various authors:

Penicillin 300–10,000 iu and Streptomycin 250 μg–10 mg together
  with one of the following:
Actidione ®, 100 μg;
Amphotericin-B (Fungizone ®), 2·5–10 μg;
Chlorotetracycline, 50 μg;
Kanamycin, 1000 μg;
Neomycin, 5 mg;
Nystatin (Mycostatin ®), 100–5000 iu.

Following the antibiotic treatment the eggs or nematodes are again thoroughly rinsed in water or physiological saline. Foreign particles must be completely removed as these carry contaminating spores. Use of broad spectrum antibiotics can avoid the appearance of antibiotic resistant organisms. To overcome persistant organisms it may be necessary cautiously to use chemical antiseptics. It is not possible to recommend one treatment over another due to the wide variations in source contamination and in manipulation techniques of various workers. The only suggestion is the use of frequent sterility checks and careful observation of the nematodes in culture to detect any signs of inhibition.

C. INOCULUM

The success of a culture depends upon characteristics of the inoculum. While there are particular considerations for each species, we can deal here only with some general aspects that can influence the outcome.

1. *Exsheathment of Infective Larvae*

Infective larvae of some species start exsheathing as soon as they are introduced into a favourable medium at a suitable temperature. Others require exsheathment treatments (Lackie, 1975). Sodium hypochlorite, 0.5%, is used, or agents that mimic the effect of the natural environment. Except for nematodes, such as *Dictyocaulus*, for which proteolytic enzymes (Parker and Croll, 1976), hydrochloric acid or duodenal fluids (Mapes, 1972) are used, high levels of CO_2 at 37°C–40°C provide the effective stimulus. Optimal conditions of gassing and pH, and the catalytic effect of reducing agents, tetraborate (Slocombe and Whitlock, 1969) and other ions differ for each species (Cypress *et al.*, 1973). Exsheathment rate is also affected by conditions of culture and storage of infective larvae (Meza-Ruiz and Alger, 1968).

The response to an exsheathing stimulus has been examined in detail for *Haemonchus contortus* (Rogers and Sommerville, 1968). The neurohormonal response appears to be mediated through noradrenaline (Rogers and Head, 1972). An enzyme involved in exsheathment has been identified as leucine aminopeptidase (Rogers and Brooks, 1976), although others using somewhat different methods of stimulus and assay have not found this enzyme in exsheathing fluids (Ozerol and Silverman, 1972; Slocombe, 1974). Aminopeptidase was produced also during the fourth moult of *Phocanema*. The enzyme did not act externally and mammalian leucine aminopeptidase was not an adjunct for artificial exsheathment.

2. *Size of Inoculum*

It is the practice to inoculate cultures heavily, usually with more than 1,000 organisms ml^{-1}. There is great variation in the rate of maturation and relatively few nematodes reach an advanced stage. However, if lower numbers are used there is greater uniformity in development. Furthermore there seems to be an inhibiting interaction between different stages and development proceeds further if the nematodes are separated in culture (Das, 1968; Zimmerman and Leland, 1971; Rose, 1973).

3. Metabolic Products

Among waste products ammonia is probably the most inhibitory. In a highly buffered medium it may not be made evident by a change in pH. The periodic replenishment of medium may not be adequate to keep the ammonia content always below inhibitory concentrations and in some situations gas exchange systems may be required.

D. MEDIUM COMPONENTS

Media for *in vitro* cultivation must replace the diverse nutrients that occur in soluble and particulate form in the *in vivo* environment. A medium should be concentrated and contain particles of about 0·5 μm size to provide for an adequate rate of intake of nutrients. This high intake rate is required particularly for the extensive growth after the fourth moult and the high rate of egg production. Media for parasitic nematodes usually consist of various mixtures of the following: an embryo extract (usually chick) at 10%–90%, serum at 10%–50%, and fresh or powdered liver extract, all added to a base of physiological salt solution, peptone solution or chemically defined medium. Killed bacteria can be used for free living stages of gastro-intestinal nematodes. A substrate of sterile excised organ fragments is suitable for facultative parasites. A substrate of tissue culture cells has been used in cultivation of Strongyloidea and filarial larvae.

1. Serum

Human serum or homologous serum from helminth-free hosts has been widely used. Recent reports indicate that commercially available serum can be used, including agammaglobulinaemic, foetal calf or horse serum.

Serum has both stimulating and inhibiting effects. Inhibition is due in part to immunoglobulins and can be decreased by "inactivation" (56°C for 30 min). Inactivation eliminated a specific inhibitory effect of serum in culture of free living larvae of *Cooperia punctata*. Because of the great variation in serum, a new batch should be introduced into culture media with adequate controls and compared with a previously found "good" batch. Whenever satisfactory batches of serum are discovered it is good practice to stockpile enough for several months' work.

2. Embryo Extracts

Preparation and use of chick and other extracts is reviewed by Leland (1970b). Chick embryo extract (CEE, see Chapter 1: IV, A, 1) has wide general usefulness. It is aseptically prepared by homogenizing 11–13 day old embryos in physiological saline or chemically defined medium. The best results are obtained with freshly prepared CEE. However this is rarely practical for an extensive investigation and the procedure is adopted of preparing frozen aliquots of the final medium with fresh CEE, which are then thawed individually just before use. The serum in mixed media aids in keeping the CEE in a fine suspension suitable for ingestion by nematodes.

CEE stocks slowly form a precipitate on storage and lose activity. Various species of nematodes differ in their response to this loss. Thus, stored commercial CEE was suitable for culture of Aphlenchoidea but not *Strongyloides*.

3. Liver Extracts

Commercially available dehydrated liver extracts (e.g. Sigma liver concentrate) are convenient to use. However cultures appear to be better if extracts are prepared freshly (see Chapter 1: IV, A, 1).

Partially denatured liver proteins can be effective supplements. They were designed as stable extracts to provide starting material for fractionation in order to identify nematode nutritional requirements. Raw liver extract (RLE), used in media for *Neoaplectana*, is a mild acid extract (protein 15 mg ml^{-1}). Heated liver extract (HLE) is extracted at pH 7 and heated. It is used in media for Rhabditoidea, Aphlenchoidea and *Neoaplectana* and prepared as follows (Sayre *et al.*, 1963). Freshly excised liver (horse preferred because of large size, cheapness and freedom from fat) is homogenized with equal parts of water, w/v. Aliquots of 150 ml are heated with stirring and maintained at 53°C for 6 min, cooled and centrifuged at 39,900 g for 30 min. The supernatant is sterilized by filtration through a membrane of 0·22–0·3 μm pore size and then stored frozen in 10–20 ml aliquots. The protein is 45–55 mg ml^{-1}.

4. Yeast Extract

Extract of autolysed yeast, included in media as a source of vitamins, is available commercially as a powder. An extract of fresh yeast cells (Lower and Buecher, 1970) proved to be a stable supplement for culture of several Rhabditoidea, *Neoaplectana* spp. and *Trichinella spiralis* (Berntzen, personal communication), thus extending the range of natural products that have possible value as components of media. The washed yeast cells were suspended in phosphate buffer, fractured, and centrifuged. The extract was filtered through a 0·3 μm membrane and stored in frozen aliquots (protein 38 mg ml^{-1}).

5. *Peptones*

Peptone preparations, commercially available as powders and granules, provide a means of introducing high levels of amino acids into media. There appear also to be additional stimulating factors and the effectiveness of different peptones differs. Casein hydrolysate is widely used. Lactalbumin hydrolysate, the preferred peptone in tissue culture, has been used by Rose (1973). We found soy peptone (Humko-Sheffield) to be best of a range of commercial products and used this in a basal peptone yeast medium containing 3% peptone and 3% Difco yeast extract. A 2-fold stock is made in water, adjusted to pH 7 if necessary, and sterilized by autoclaving (121°C, 10 min) in aliquots of 20 ml. The stock is diluted with water and supplements of serum, chick embryo extract or liver extract added as required.

6. *Defined Media*

Several defined media have been used in cultivating parasitic nematodes. These include the commercially available tissue culture media such as NCTC 109, M199 and Eagle's Minimum Essential Medium (Taylor and Baker, 1968, pp. 351–356). Sometimes extra vitamins and amino acids are added from commercially available stocks (see catalogue of Grand Island Biological). Special media, such as Berntzen's M-102 (Taylor and Baker, 1968, p. 251) and Denham's W-20 (Denham, 1967), have been designed with a broader range of components to satisfy possible nutritional requirements. These include metabolic intermediates, cofactors, and nucleic acid substituents. The commercially available *C. briggsae* maintenance medium, CbMM (Grand Island Biological), contains a broad range of components at high levels. It was designed for cultivation of microphagous nematodes. It has been used for *Ascaris suum* and *Haemonchus contortus* but for prolonged culture of vertebrate parasites the ionic balance of the medium should be adjusted to that of physiological saline (as described for culture of *Hymenolepis*, Hansen and Berntzen, 1969).

CbMM is a suitable basal medium for culture of Rhabditoidea and Aphlenchoidea. Other defined media have been designed for *Aphlenchoides* (Myers and Balasubramanian, 1973) and *Neoaplectana* (Jackson, 1973b). The former medium differs from CbMM in balance of amino acids and in containing organic acids and additional cofactors. The salts are at a lower level. The latter medium contains higher concentrations of amino acids and glucose, and lower concentrations of vitamins. The formulae of the three media have been tabulated by Platzer (1978).

Defined media for insect tissue culture systems have higher concentrations of amino acids, a higher K^+/Na^+ ratio, and a higher osmolality than vertebrate media. The several formulations have been tabulated by Hink (1976). Two of these, Schneider's and Grace's, both commercially available, have been used in culture of mermithid nematodes. Their application may possibly facilitate cultivation of larval filariae.

7. *Antibiotics*

While antibiotics are not used in media for nematodes that are in continuous culture, they are included in media used for isolation from an *in vivo* source. The concentrations used have been reported as penicillin, 100–400 iu ml^{-1}, plus streptomycin, 100–400 μg ml^{-1}, and one of the following:

Amphotericin-B (Fungizone ®), 225 μg ml^{-1}.
Cycloheximide (Actidione ®), 100 μg ml^{-1}.
Kanamycin, 50 μg ml^{-1}.
Neomycin, 225 μg ml^{-1}.
Nystatin (Mycostatin ®), 50–100 iu ml^{-1}.

These levels are at least twice those considered possibly inhibitory in tissue culture. Stock solutions are available at $100 \times$ tissue culture levels.

Inclusion of antibiotics in media should be avoided. Instead the procedures should be devised to achieve complete axenization of the original inoculum. Low level bacterial or fungal growth can occur in the presence of antibiotics and can be undetected yet producing inhibitory effects on the nematodes. Furthermore, in long term cultures there is a high possibility that cryptic organisms will be merely suppressed by the antibiotics and will later suddenly become evident. Tolerance of nematodes to antibiotics during growth and reproduction has not been established. There may also be differences between species. If antibiotics are to be included it is preferable to avoid cycloheximide and streptomycin and use broad spectrum, less toxic, antibiotics such as kanamycin at 100 μg ml^{-1} or Gentamicin (Schering) at 50 μg ml^{-1}.

E. PHYSICAL CONDITIONS

1. *Temperature*

Cultures of nematodes parasitic in vertebrates are usually incubated at temperatures within the range 37–40·5°C. Limiting or optimum temperatures have been established for only a few species. Development of *Haemonchus contortus* to fourth stage does not proceed below 37°C (Rogers and Sommerville, 1968). Phillipson (1973) noted that temperature limits for mating of *Nippostrongylus brasiliensis* were 34·8–38·4°C with an optimum at about 37°C. For *Anisakis* the temperature must be less than 38°C.

The free living stages of parasitic nematodes live at normal air temperature, and development may be inhibited if the eggs deposited in culture are held at 37°C (Yasuraoka and Weinstein, 1969). In contrast, the free living phase of *Strongyloides fülleborni* will develop at 37°C (Hansen *et al.*, 1975). Luminal autoinfection with this intestinal parasite can occur.

Nematode parasites of invertebrates and of plants are cultured at air temperatures and usually do not tolerate prolonged temperatures above about 32°C.

2. *Osmolality, pH, Redox, Gas Phase*

The culture environment must provide conditions that would be encountered by the parasite in its host, keeping in mind the changes in nematode physiology that are implicated in the successive sites selected by the parasite as it matures from larva to adult. Specific conditions of pH, redox, gas phase and osmolality are noted in Section III.

Control and measurement of pH of media is routine. A known gas phase can be provided with properly regulated gas flow. Sensors for determining dissolved oxygen tension are now available in a convenient form (Fatt, 1976). Improvements in development by cultivation under decreased oxygen may well be enhanced by adding sulphydryl compounds such as cysteine, reduced glutathione or the more stable dithiothreitol. Instruments are available for monitoring redox potential in the medium. Osmometers (based on freezing point depression) give rapid and reproducible determinations of the osmolality of the medium.

F. CULTURE IMPROVEMENTS

Improvements are most likely to come from a greater use of defined media, from increased use of a controlled gas phase, and from improved handling of the parasites.

Chemically defined media provide nutrients in a stable form and can replace a large part of the unstable and variable natural products. Based on metabolic studies the composition can be adjusted to provide proper levels of necessary nutrients. Commercially available media may be used as a starting point. Tissue culture media are usually too dilute and need to be augmented with additional components to provide for the greater requirements of the nematode, since the media must provide for a greater cell mass and for tissues undergoing integrated development.

Assay methods are needed to monitor labile natural components. To the extent that this lability is due to oxidation it should be prevented by the addition of reducing agents and by the use of a low oxygen content in the gas phase. There is also evidence that a gas phase with low oxygen and high carbon dioxide is essential for parasite development.

The need for proper handling of the parasites is of great importance but is rarely emphasized. It is probably responsible for most of the differences in results reported by different investigors. Anything which causes a setback, even temporary, is likely to be translated into a permanent decrease in growth and maturation.

With the improved instrumentation now available, it becomes eminently practical continually to monitor the physical properties of the medium and re-evaluate these with reference to the host environment. Only with this information will it be possible to provide adequately for the needs of the maturing nematode.

Finally we should look to broadening the range of species investigated in culture. New species *in vitro* introduce new characteristics and possibilities that in turn can give new insight into old problems.

III. RECENT CULTURE STUDIES

The species are listed in Table I. The descriptions of their cultivation are grouped by host: vertebrate, invertebrate or plant. The vertebrate parasites are arranged alphabetically. Taxonomic relationship is shown by super-family designation. Ten superfamilies are represented, half of the species being Strongyloidea.

A. NEMATODE PARASITES OF VERTEBRATES

1. *Ancyclostoma* spp.

(a) *A. caninum* (Strongyloidea). Cultivation of free living larvae and

TABLE 1. *Parasitic Nematoda: culture studies since Taylor and Baker (1968)*

Species	Initial stage	Final stage (application)	Reference
Ancylostoma caninum	Infective	L4	Leland (1970b); Banerjee (1972)
Ancylostoma caninum	Adult	(Ingestion)	Roche et al. (1971)
Ancylostoma caninum	Adult	(^{14}C uptake)	Wong and Fernando (1970)
Ancylostoma tubaeforme	Infective	Late L4	Slonka and Leland (1970)
Ancylostoma tubaeforme	Adult	Egg laying	Slonka and Leland (1970)
Angiostrongylus cantonensis	Infective	(Anthelminthics)	Moreau and Lagraulet (1972)
Anisakis sp.	Infective	(M-3)	Sommerville and Davey (1976)
Anisakis sp.	Infective	L4	Schulz (1974); Khalil (1969)
Anisakis marina	Infective	Adult and L2	Van Banning (1971); Grabda (1976)
Aphelenchus avenae	Egg	Continuous	Hansen et al. (1970)
Aphelenchus avenae	Adult	(Feeding stimulus)	Fisher (1975)
Aphelenchoides rutgersi	Larvae	Adults and larvae	Myers (1967)
Aphelenchoides rutgersi	Larvae	(Defined medium)	Myers and Balasubramanian (1973)
Aphelenchoides rutgersi	Larvae	(CEE requirement)	Thirugnanam (1976)
Ascaridia galli	Egg	Hatching	Chatterjee and Singh (1968)
Ascaridia galli	Egg	Early L3	Dick et al. (1973)
Ascaridia galli	Adult	(Protein uptake)	Shishova et al. (1973)
Ascaridia galli	Adult	(Amino acid uptake)	Dryushenko and Berdyeva (1974)
Ascaris suum	Egg	L3	Levine and Silverman (1969)
Ascaris suum	Egg	L4	Douvres and Tromba (1970)
Ascaris suum	Egg	(L3, M-3 antigens)	Guerrero and Silverman (1971)
Ascaris suum	Egg	(L2 antigen)	Guerrero et al. (1974)
Ascaris suum	L3	L4	Sylk et al. (1974)
Ascaris suum	Adult	(Metabolism)	Shishova et al. (1973); Dryushenko and Berdyeva (1974)
Ascaris suum	Adult	(Organ culture)	Hirumi et al. (1969)
Brugia pahangi	Microfilaria	M-1	Garrigues et al. (1975)

TABLE 1—*continued*

Species	Initial stage	Final stage (application)	Reference
Capillaria hepatica	Egg	L1	Solomon and Soulsby (1973)
Chambertia ovina	Infective	L5	Schulz and Dalchow (1967)
Contracaecum osculatum	Egg	Early adult	McClelland and Ronald (1974a)
Cooperia oncophora	Infective	Adult and Eggs	Leland (1969)
Cooperia oncophora	Infective	L4	Rose (1973)
Cooperia punctata	Egg	Adult and Eggs	Leland (1967)
Cooperia punctata	Infective	Adults and Fl-infectives	Zimmerman and Leland (1971)
Cooperia punctata	Infective	(Serum effect)	Ridley and Leland (1973)
Cooperia punctata	Infective	(pH effect)	Dick and Leland (1973)
Cooperia punctata	L4	(^{14}C uptake)	Slonka et al. (1973)
Cooperia punctata	L3–L5	(Exoantigens)	Leland (1975)
Cooperia punctata	L3–L5	(Anthelminthics)	Leland et al. (1975)
Dictyocaulus viviparus	Egg	Infectives	Croll (1973); Taira (1975)
Dirofilaria corynodes	Microfilaria	Late L1	Cupp (1973)
Dirofilaria immitis	Microfilaria	Late L1	Klein and Bradley (1974)
Dirofilaria immitis	2-day L1	M-2	Weinstein (1970)
Dirofilaria immitis	1-day L1	L3	Weinstein (1976)
Dirofilaria repens	Microfilaria	Late L1	Dhar et al. (1967)
Haemonchus contortus	Egg	Infective	Wang (1971)
Haemonchus contortus	Infective	Early adult	Schulz (1967); Gevrey and Gevrey (1974)
Haemonchus contortus	L3	(pH, pCO$_2$, SH, K$^+$)	Mapes (1969); Sommerville (1976)
Haemonchus contortus	Infective	(Immune serum)	Jakstys and Silverman (1969)
Haemonchus contortus	Infective	(Antigens L3, 4, 5)	Alger (1968); Neilson (1969, 1975)
Haemonchus placei	Infective	L4	Leland (1970b)
Hydromermis conopophaga	L2	Early L3	Poinar (1975)
Hyostrongylus rubidus	Egg	L2	Leland (1969)
Hyostrongylus rubidus	Infective	Adults and Eggs	Leland (1969); Rose (1973)
Meloidogyne incognita	Adult	Eggs	Shepperson and Jordan (1974)

Species	Stage/Process	Reference
Mesodiplogaster lheritieri	Egg	Weiser (1966)
Metastrongylus elongatus	Infective	Tarakanov (1973)
Necator americanus	L3, L4, Adult	Burt and Ogilvie (1975)
Nematospiroides dubius	Egg	Yasuraoka and Weinstein (1969); Muria (1975)
Nematospiroides dubius	L4, Adult	Dennis (1976)
Nematodirus helvetianus	Infective	Rose (1973)
Neoaplectana carpocapsae	Egg (Insect hormones)	Hansen *et al.* (1968)
Neoaplectana glaseri	L3 Continuous (Amino acids)	Jackson (1973b)
Neoaplectana glaseri	L3 (Folate metabolism)	Jackson and Platzer (1974)
Neomesomermis flumenalis	Infective L2 L3	Finney (1976a)
Nippostrongylus brasiliensis	Adult (Exoantigens)	Wilson (1967); Denham (1969)
Nippostrongylus brasiliensis	L3 (^{14}C-proline in M-4)	Bonner *et al.* (1971)
Nippostrongylus brasiliensis	Larvae (Gene control)	Bonner and Buratt (1976); Bonner *et al.* (1976)
Nippostrongylus brasiliensis	L1 (Sterol, Porphyrin)	Bolla *et al.* (1972a, b, 1974)
Nippostrongylus brasiliensis	Adult (Fertilization)	Phillipson (1973)
Nippostrongylus brasiliensis	Adult (Acetylcholine-esterase)	Burt and Ogilvie (1975)
Oesophagostomum columbianum	Infective M-4	Das (1968)
Oesophagostomum columbianum	Infective (L4 antigens)	Neilson (1972)
Oesophagostomum dentatum	Infective L4	Rose (1973)
Oesophagostomum dentatum	Infective M-4	Tarakanov and Kaarma (1972)
Oesophagostomum quadrispinulatum	Infective Adult and Eggs	Schulz and Dalchow (1969); Leland (1970a)
Oesophagostomum radiatum	Infective Late L4	Douvres (1970)
Oesophagostomum radiatum	Infective (L3L4 live vaccine)	Herlich *et al.* (1973)
Ostertagia circumcincta	Infective L5	Denham (1970)
Ostertagia circumcincta	Infective Adult and Eggs	Rose (1973)
Ostertagia circumcincta	Infective (Exoantigens)	Rose (1976)
Ostertagia ostertagi	Infective Adult and Eggs	Rose (1973); Douvres and Malakatis (1977)

TABLE 1—*continued*

Species	Initial stage	Final stage (application)	Reference
Paraspidodera uncinata	Egg	Infective, L3, Egg	Bondy (1967)
Phocanema decipiens	L4	M-4 (Neurosecretion)	Davey and Sommerville (1974)
Phocanema depressum	L4	M-4 (Hormones)	Rajulu *et al.* (1972)
Pelodera strongyloides	Egg	Continuous	Yarwood and Hansen (1968)
Pristionchus uniformis	Egg	Adult	Fedorko and Stanuszek (1971)
Reesimermis nielseni	Infective-L2	L3	Sanders *et al.* (1973)
Rhabditis maupasi	Dauer L3	Continuous	Brockelman and Jackson (1974)
Romanomermis culicivorax	Infective-L2	Early female	Finney (1976b)
Steinernema kraussei	—	Continuous	Weiser (1976)
Stephanurus dentatus	L4; Adult	Survival	Tromba and Douvres (1969)
Strongyloides fülleborni	Infective	M-3	Buecher *et al.* (1969)
Strongyloides fülleborni	Egg	Rhabditoid Adult and eggs	Hansen *et al.* (1975)
Syphacia muris	Egg	Hatching	Gulden and van Erp (1976)
Terranova decipiens	L4	Adult	Townsley *et al.* (1963)
Terranova decipiens	Egg	L4	McClelland and Ronald (1974b)
Thelazia spp.	Infective and adult	Survival	Čorba and Leštan (1969)
Toxocara canis	L2	Survival	de Savigny (1975)
Trichostrongylus spp.	Infective	L3	Rose (1973); Douvres and Malakatis (1977)
Trichostrongylus colubriformis	L4	(Exoantigen)	Rothwell and Love (1974)
Trichinella spiralis	Encysted larva	Moult (Hormones)	Shanta and Meerovitch (1970); Hitcho and Thorson (1971)
Trichinella spiralis	Adult	L1	Dennis *et al.* (1970); Despommier *et al.* (1975)
Trichuris trichura	Egg	L1	Ortiz-Valqui and Lumbreras-Cruz (1970)

Note: L1, L2, L3 and L4 = 1st, 2nd, 3rd and 4th stage larvae; M1, M2, M3 and M4 = 1st, 2nd, 3rd and 4th moult.

maintenance of adults is described by Taylor and Baker (1968, pp. 272–276). Recent studies with adults include those on ingestion (Roche *et al.*, 1971) and on uptake and incorporation of radioactive nutrients (Wong and Fernando, 1970).

Third stage larvae were cultured by Leland (1970b) to early fourth stage in Ae medium (see Section III, A, 10). Banerjee (1972) maintained third stage larvae in a modified Ae medium incubated at 25°C. Larvae survived 25 d and there was development of a primitive oral cavity. At 37°C most died within 48 h. Gonzalez and Guerra (1973) also report culture of third stage larvae. As inoculum they used clean infective larvae that had migrated into the condensed water on the lid above the faecal cultures.

(b) *A. duodenale* (Strongyloidea) (Taylor and Baker, 1968, p. 273). Infective larvae will survive for 100 d at 25°C in physiological saline (Kim, 1969). Data on survival of eggs and larvae in faeces are also presented.

(c) *A. tubaeforme* (Strongyloidea). This species has been cultured from egg to fourth stage larvae in Ae medium (Slonka and Leland, 1970). Cultures of eggs to third stage larvae were incubated at 25°C. At 11 d the temperature was raised to 38·5°C. Only 0·5% reached the fourth stage and died during the fourth moult. Serum was included in all media and was not inhibitory as had been noted with *Cooperia*. Substitution of feline serum did not improve the response. Other adults grown *in vivo* were maintained with egg laying for 41 d in Ae medium.

2. *Angiostrongylus cantonensis* (Strongyloidea)

Maintenance of adults is described by Taylor and Baker (1968, pp. 276–277). Moreau and Lagroulet (1972) used infective larvae collected from *Biomphalaria glabrata*. On exposing them for 6 h in NCTC 109 (Taylor and Baker, 1968, pp. 351–354) with 30% inactivated human serum an inhibitory effect of Tetramisole was demonstrated.

3. *Anisakis* spp. (Ascaroidea: Anisakidae)

Cultivation studies of anisakine nematodes have been stimulated by recognition of human anisakiasis. Third stage infective larvae occur in fish. Aquatic invertebrates serve as the first intermediate host; the seal is the definitive host.

All stages of the life cycle have been used to initiate cultures (Bier, 1976). Schulz (1974) reported culture of third stage *Anisakis* sp. to fourth stage in an Ae type medium (see Section III, A, 10) incubated under 10% CO_2 in air. Khalil (1969) used M199 (Taylor and Baker, 1968, pp. 355–356) supplemented with horse serum and liver extract (proportions not stated) and observed moulting in 4–5 d at 36°C. Worms increased in size, surviving for 40 d but there was no egg laying.

Van Banning (1971) reported development of *A. marina*. He was unsuccessful with the medium of Townsley *et al.* (1963) and used instead a pepsin digest of fresh liver enriched with citrated beef blood added at about 5% initially, increasing to 50% in older cultures. Not more than three larvae were placed in 2 ml of medium which was changed daily. Temperatures of 34–37°C were suitable but at 38°C there was a high mortality (gas phase is not reported). The first moult occurred at d 4. The fourth stage, during which there was considerable gonad formation, extended over 26–98 d. Maturation then took place rapidly over 7 d. Males measured 3·5–7·0 cm and females 4·5–15 cm. Many eggs were produced. Insemination was ensured by placing mature males with new females, 5–8 specimens together, in 15 ml of medium. Eggs hatched in 4–8 d as ensheathed larvae. These retained motility for 3–4 weeks in sea water at 13–18°C. Grabda (1976) designates the group as *Anisakis simplex*. She used a similar culture method and obtained egg laying adults. The development and morphology of the third and fourth stages and adults was described.

Sommerville and Davey (1976) reported that the third stage of *Anisakis* sp. does not feed and that ecdysis can be stimulated in Krebs-Ringer solution by provision of 5% CO_2 during the first 40 h of incubation.

4. *Ascaridia galli* (Ascaroidea)

Studies on hatching, larval culture, and uptake of radioactive nutrients are reviewed by Taylor and Baker (1968, pp. 271–272).

A rapid method of hatching eggs was described by Chatterjee and Singh (1968). Eggs were treated for 3 min in 1% NaOH and hatched at 41°C in 40 ml of normal saline containing 400 mg glucose, 250 mg bicarbonate, 10 mg trypsin and HCl to adjust the pH to 7·1–7·5. Fifty percent hatched in 30 min.

Culture from egg to third stage was carried out by Dick *et al.* (1973) in Ae type medium (see Section III, A, 10) with chick serum substituted for beef serum and the pH adjusted to 6·75 (range tested, 6·3–7·3). Cultures were initiated with larvae that had been hatched artificially as follows. Eggs were deshelled by treatment for 24 h at 31°C in a mixture of equal parts of 0·2% sodium hypochlorite and 4% NaOH. Early second stage larvae were released from the remaining vitelline membrane by treatment with 0·2% Tween-80 for 1 h at 38·5°C. Cultures were inoculated with 2000 larvae per ml, and incubated at 38·5°C with rotation. The medium was replaced once or twice weekly. Additional supplements of trypsinized chick embryo cells or extract of chick embryo gut tissue slightly improved the response. The best medium contained autoclaved bacteria. Second stage larvae survived for 112 d and reached a length of 1034 μm. Ecdysis of second stage larvae (L2) was induced after culture for 21 d by treatment for 2·5 h in salt solution saturated with CO_2. The exsheathment rate was low. A few larvae developed to early third stage (L3), surviving up to 14 d.

Studies on uptake of amino acids were reported by Dryushenko and Berdyeva (1974). Shishova *et al.* (1973) showed differences in ingestion and utilization of different proteins. The cuticle was found to be impermeable to protein.

5. *Ascaris suum* (Ascaroidea)

Artificial hatching, culture of larvae to third stage, metabolic studies with adults and attempted organ culture are described by Taylor and Baker (1968, pp. 261–271).

Douvres and Tromba (1970) have carried development of newly hatched larvae to fourth stage using an associated tissue culture. The medium consisted of 50% serum (calf preferred to swine) with 50% NCTC 109 containing 112·5 mg% yeast extract, 140·6 mg% Bacto-peptone and 140·6 mg% glucose (Douvres *et al.*, 1966; Taylor and Baker, 1968, p. 284). Hydrochloric acid was added to lower the pH from 7·6 to 6·9. The medium was overlaid in 20–40 ml amounts on swine kidney cell cultures. Larvae were added to 24 h cell cultures, 5000–10,000 ml^{-1}, and were transferred to new cell cultures at 8 and 11 d. By d 20 they had reached early fourth stage (L4). When terminated on d 30 the maximum length was 2·1 mm. In medium without the cell culture, larvae grew to less than half this size, reaching the early third stage. Leland (1970b) reported development to the third moult in Ae medium (see Section III, A, 10).

For the production of antigens *in vitro* a more defined medium is desirable. In Eagle's MEM (Taylor and Baker, 1968, p. 355) second stage larvae survive only 8 d. In MEM supplemented with 16–22% pig serum or 7% Difco liver powder infusion (protein 30 mg%) there was development to early third stage (Levine and Silverman, 1969). These cultures had been inoculated with 150 larvae ml^{-1}, gassed with CO_2, sealed and incubated with rotation at 38°C. Protective metabolic second and third stage larval antigens were harvested from 12 d cultures (Guerrero and Silverman, 1971). Guerrero *et al.* (1974) noted that egg albumen plus haemin is a satisfactory supplement in MEM for cultures to be used for isolation of functional antigens. *C. briggsae* maintenance medium, CbMM (see Section II, D, 6), with 5 × vitamins and supplemented with serum also supported development to the third stage.

Antigens from older larvae were obtained from cultures initiated with larvae recovered from lung tissue. Third stage larvae were cultured at 38°C with rotation in Eagle's MEM with supplement of 2% pig serum and 1% bovine serum albumin. Third and fourth stage larvae metabolic antigens were harvested in the culture fluid at 5 d (Guerrero and Silverman, 1971).

Fourth stage larvae were obtained by Sylk *et al.* (1974) in cultures initiated with third stage larvae removed from the lung of rabbits or guinea pigs 10 or 7 d, respectively, after infection. They were cultured in medium 199 (Taylor and Baker, 1968, pp. 355–356) with additional glucose (2·25 mg% w/v), and guinea pig or porcine serum (4% v/v). Medium (20 ml) was placed in 200 ml bottles, inoculated with 1000–2000 larvae, gassed with $N_2:O_2:CO_2$ (90:5:5) and incubated at 37°C on a roller drum. Transfers were made to fresh medium at 10 d intervals. Larvae moulted to fourth stage after 1–4 d. The reproductive tract was differentiated at 12 d, and by 52 d there was a 3-fold increase in length. Ecdysis did not take place under air or in the absence of serum. Larvae removed earlier than 7 d after infection survived 10 d but did not moult.

In several studies of ingestion and uptake by adults, Shishova *et al.* (1973) noted a greater rate of intake of denatured protein compared with serum. Dryushenko and Berdyeva (1974) report on uptake of amino acids.

The large gonads of *Ascaris* provide a reasonable quantity of starting tissue for organ and tissue culture. However, such cultures are handicapped by lack of data on the cell milieu in a nematode. Hirumi *et al.* (1969) tried Schneider's *Drosophila* medium with 10% foetal calf serum as well as *Mycoplasma* broth with 20% serum. In both media cells migrated to form a monolayer but there was only very limited cell division.

6. *Brugia pahangi*, *B. malayi* (Filaroidea)

Filarial worms present special problems for cultivation. The larvae start developing in an arthropod (usually insect), part of the time intracellularly, and continue on to maturity in a vertebrate host. The infective stages, i.e. early first stage larvae (microfilariae) and third stage larvae, are never free, passing directly between vector and vertebrate host. At least 85 filarial life cycles have been recorded. Cultivation has been attempted with some 16 species. In 1970 Weinstein summarized these studies (see also Taylor and Baker, 1968, pp. 301–314).

New work since 1968 has been directed to culture of the microfilariae with emphasis on the relation of the developing larva to its vector. Morphological changes of the nematode gut during development are associated with changes in intake of nutrients. Initially the young larva draws on soluble nutrients taken in through the outer cuticular surface. Late in the second stage the alimentary canal becomes functional and particles of host origin are ingested (Beckett and Boothroyd, 1970).

To initiate cultures the microfilariae must be separated unharmed from blood cells. Obeck (1973) found it convenient to allow blood to clot in shallow layers in large Petri dishes, the microfilariae moving with the expressed serum. Ah et al. (1974) prepared cell free microfilariae of Brugia pahangi and B. malayi after intraperitoneal growth in jirds. The jirds were opened aseptically 150–255 d after intraperitoneal inoculation with 100 infective L3 larvae obtained from crushed Aedes aegypti. The exposed viscera were washed in cold Hanks's solution (see Chapter 1: IV, B, 1, g) to collect microfilariae and adult worms. Adult worms were removed and microfilariae were sedimented gently in cold Hanks's solution to separate them from leucocytes.

When microfilariae are inoculated into a medium containing serum or blood and the temperature lowered they show the start of changes characteristic of the arthropod phase and reach the "sausage" form. Aoki (1971), using thick blood smears, noted that exsheathment was greatest at 15–20°C, occurring for Brugia pahangi within the first hour. Exsheathment declined at higher temperatures and was inhibited at 37°C. Dhar et al. (1967) cultured Dirofilaria repens to sausage stage in diluted heparinized infected blood.

For continued larval development the medium should be of insect tissue culture type. In Grace's medium supplemented with 20% heat inactivated pony serum and with a gas phase of increased carbon dioxide, Dirofilaria immitis developed to late first stage (Klein and Bradley, 1974; see also Timofeev, 1969). However development was highly variable and no complete moults were observed.

In culture with insect tissue cells (see Wood and Suitor cited by Taylor and Baker, 1968, p. 312) there is improved development but still without complete moults (Weinstein, 1970). Cupp (1973) briefly described cultures of Dirofilaria immitis and D. corynodes in cell cultures of Aedes vexans and Culex inornata. Development was enhanced by the cell cultures but still did not pass late first stage. Cultures were improved by the addition of insect haemolymph. Development of Brugia pahangi to late first stage in cultures of Aedes cells was recently reported by Garrigues et al. (1975). The microfilariae had first been treated with alkaline phosphatase.

Cultures have been more successful when initiated with nematodes that have already entered into the arthropod phase. Weinstein (1970) reported development of Dirofilaria immitis to the second moult and a length of 782 μm in cultures of Anopheles quadrimaculatus Malphighian tubules containing 2-d or 4-d old infections. The medium was NCTC 109 (Taylor and Baker, 1968, pp. 351–354) with serum, a gas phase of 5% CO_2 in air, at

28°C. He later (Weinstein, 1976) observed development to the third larval stage in cultures from 1-d old infections from *Aedes aegypti*. The infected Malphighian tubules were prepared for culture with pieces of terminal gut attached, its contractility serving as a marker for viability of the organ culture. A serum level of 40% was necessary.

For effective synxenic cultures, the associated cells or organ must be sustained in an active condition, recognizing the fact that cells in explanted organs gradually decline and degenerate. Even in a growing tissue culture there is a fall in metabolic activity after the cells have reached log phase of growth, and both cells and medium must be replenished at frequent intervals. Support of the cells or organ is a good indication that the medium could be adequate for development of the nematode. Flourishing associated cells affect the medium and the gas phase of the culture and, in certain systems, can promote growth and differentiation which does not take place in the medium alone (*cf* cercarial growth in snail tissue culture, Hansen, 1976).

Vertebrate serum is usually included in the medium. However for invertebrate cells serum can be inhibitory (Hansen, 1974). This inhibiting effect may be contributing to the difficulty of sustaining filariae through differentiation of their second larval stage.

The metabolism of microfilariae has been studied in vertebrate blood (Jaffe and Doremus, 1970). Exogenous amino acids were incorporated into proteins, adenosine and adenine were converted to phosphorylated derivatives and incorporated into RNA, and purine nucleotides were not synthesized from glycine or formate. If this metabolism is characteristic also of developing filarial larvae, purines and pyrimidines should be included in the medium if not already adequately supplied by the serum.

In view of the difficulty in achieving larval development *in vitro* the question arises of dependence on stimulus from host hormones. However, Gwadz and Spielman (1974) have shown that filarial development is independent of host endocrines. Larvae matured in allatectomized or decapitated mosquitos as well as in male mosquitos into which they had been injected.

The current interest in immunology of filarial infections emphasizes the need for effective cultivation of third stage infective larvae (Taylor and Baker, 1968, p. 303) as a source of third and fourth stage metabolic antigens.

7. *Capillaria hepatica* (Trichuroidea)

Incubation of eggs is mentioned by Taylor and Baker (1968, p. 225). Solomon and Soulsby (1973) described a 2-step process for stimulating hatching. Hypochlorite treated eggs were incubated for 2 h at 37°C in 0·1 M sodium bicarbonate containing 0·03 M sodium dithionite and 30 mg%

magnesium chloride under a gas phase of 10% CO_2 in nitrogen. They were transferred to MEM (Taylor and Baker, 1968, p. 355) and 2 h later 68% of the eggs hatched.

8. *Chambertia ovina* (Strongyloidea)

This was cultured by Schulz and Dalchow (1967) from infective larvae to fifth stage. The third moult occurred on d 6 and the fourth by d 31. The medium contained 30% CEE, 15% sheep liver extract, 10% pig serum and 45% nutrient broth. Cultures were incubated under 5% CO_2 in air, with rotation.

9. *Contracaecum osculatum* (Ascaroidea: Anisakidae)

Methods for cultivation from egg to early adult were similar to those used for *Terranova decipiens* (McClelland and Ronald, 1974a). Eggs were dissected from adults taken from the stomach of seals and incubated in sterile sea water. Second stage larvae hatched in 19–25 d. These were exsheathed in 0.05% sodium hypochlorite in sea water and cultured at 15°C in Eagle's MEM (Taylor and Baker, 1968, p. 355) with 20% foetal calf serum. Five ml of medium was used in tubes or in 35 mm Petri dishes, under air or 5% CO_2 in air. The tubes were placed in rollers at 1 rev min^{-1}. Initially there were 200–300 larvae ml^{-1} medium, the numbers being gradually reduced to 20–30 by 22 weeks. The medium was changed weekly. At 32 weeks the larvae had reached the infective stage and were 6–9 mm long.

The temperature was now raised to 35°C and the medium volume increased to 20 ml either in flasks or in 100×15 mm Petri dishes. By 6 weeks the culture contained fourth stage larvae. There were also a few early adults, 13–17 mm long, showing dimorphic genital primordia.

10. *Cooperia* spp. (Strongyloidea)

Cultivation of these trichostrongyles has been the most successful among nematode parasites. Methods leading up to development *in vitro* of adults from infective larvae are described by Taylor and Baker (1968, pp. 293–299). The medium, designated "Ae" by Leland (1963), consists of 50% fresh chick embryo extract (made in Earle's solution, see Chapter 1: IV, B, 1, f), 15% cysteine-casein solution, 5% Eagle's vitamin solution $100 \times$, 14% Earle's solution and 1% antibiotic solution. Liver extract (Sigma) is prepared at 2% in part of the Earle's and added to give a final level of 100 mg%. The casein solution is 2% casein and 25 mg% cysteine dissolved in N-NaOH and back titrated to pH 7·2. Antibiotics are also added to all of the Earle's solution. The medium is described by Taylor and Baker (1968, pp. 293–294).

Three *Cooperia* species have been cultured. They differ in the extent of segmentation of eggs produced *in vitro*. This may reflect a differential sensitivity to inhibitory effects of prolonged exposure of the eggs to 37–38°C (see *Nematospiroides dubius*). The greatest degree of segmentation, 16–32 cells, was observed with *C. punctata* and most of the recent work has been done with this species.

(a) *C. punctata* (Strongyloidea). Culture through the complete sequence from egg to egg-laying adult (Leland, 1967) was aided by the fact that an exsheathing stimulus was not needed. Eggs from fresh faeces were axenized in antibiotics and hatched in water or placed directly in the medium. They were cultured for 7–14 d at 22–26°C with rotation in Ae medium modified by omission of serum. Second stage larvae were noted at 3 d and the third stage at d 5. The cultures were then changed to complete Ae medium and the incubation temperature raised to 38·5°C. The fourth stage was observed at 14 d, fifth at 24 d, and later there were a few mature adults. Eggs were seen at 39 d. Egg laying continued for 100 d. In cultures started in complete Ae medium the larvae did not develop beyond the second stage. However in a medium compounded with heat inactivated serum, development proceeded through all stages but not as well as with the 2-step change in medium.

Embryonation of eggs produced *in vitro* was studied by Zimmerman and Leland (1971). In cultures initiated in Ae medium with infective larvae from faeces-vermiculite cultures, adults and embryonated eggs developed by 33 d and larvae were found 30 d later. These larvae were removed to Ae medium lacking serum and incubated at 27°C. Four reached the ensheathed third stage by 15 d and died 8 d later.

Thus all stages of the life cycle have been obtained *in vitro*, but not yet consecutively.

Leland (1970b) reviewed the preparation and role of the components of Ae medium. The original formulation called for homologous host serum from helminth free animals. However Ridley and Leland (1973) found that good cultures of *C. punctata* could be obtained with commercial agammaglobulinaemic serum. With foetal calf serum large adults developed if precipitation in the medium was avoided by using it without prefreezing. Commercial bovine serum was the least effective. None of these sera was heat inactivated.

The effect of pH on the medium was investigated by Dick and Leland (1973). Within the range 6·4 to 8·1 a pH of 7·6–7·8 was optimal. This is a little higher than the original formulation of 7·2–7·3 and allows for the production of acid metabolites.

Unless media were replaced at least weekly, eggs were not produced.

Exoantigens were collected in physiological saline (Zimmerman and

Leland, 1974) by harvesting 21-d cultures containing third and fourth stage larvae and early adults. The nematodes were suspended in saline and incubated at 38·5°C with rocking for 3 d. Antigens were separated from the saline by ultrafiltration and demonstrated by disc electrophoresis and immunodiffusion. Administration of the exoantigens to calves produced some protection (Leland, 1975).

Metabolism of fourth stage *C. punctata* was examined by Slonka *et al.* (1973). Labelled glucose was added to 11-d cultures and incubation continued for 2 d. The ^{14}C was incorporated into glycogen, lipid and non-lipid fractions, and nucleic acids, and, in the protein, was distributed in at least 12 amino acids.

Anthelmintic activity was evaluated in 28 d cultures (Leland *et al.*, 1975). The materials tested were introduced into the cultures and the effect evaluated after 7 d.

(b) *C. oncophora* (Strongyloidea) was more difficult to culture (Leland, 1968). In Ae medium most of the nematodes remained as fourth stage. Eggs were seen in 3 adults with only a few showing early cleavage. Rose (1973), using a modified Ae medium (see Section III, A, 21) found development only to the fourth stage.

11. *Dictyocaulus viviparus* (Strongyloidea)

Culture to early fourth stage in Earle's solution with peptone or liver extract is described by Taylor and Baker (1968, pp. 277–279). Croll (1973) confirmed the non-feeding of the larvae during development to the infective stage. Taira (1975) found that growth in tap water was stimulated by addition of 40 mM KCl.

12. *Dirofilaria repens, D. immitis, D. corynedes* (Filaroidea)

See Section III, A, 6.

13. *Haemonchus* spp.

(a) *H. contortus* (Strongyloidea). Development from egg to infective larvae, exsheathment, and development of third stage to fourth moult is described by Taylor and Baker (1968, pp. 290–296). Wang (1971) re-examined cultivation of the early larval stages and found that killed *E. coli* would support development to the infective stage. Ingestion of the bacteria was detected with suspended washed ^{3}H-adenosine *E. coli*.

The exsheathment of infective larvae has been intensively studied with regard to stimulatory conditions and the nature of the neuroendocrine response (see Section II, C). There have been several reports (e.g. Campbell *et al.*, 1973) that infective larvae of *H. contortus* and other species survived better in frozen storage if first exsheathed.

The exsheathed third stage larvae, though non-feeding, were influenced in their rate of development to fourth stage by ions and dissolved gasses in the suspending fluid. Mapes (1969), using half strength Ringer's solution, studied the effect of pH and pCO_2, and later (Mapes, 1970) noted that the rate of development was increased by the addition of sulphydryl compounds. Sommerville (1976) deleted individual salts and found that each, particularly potassium (ion), affected development to the fourth stage. The osmolality could range between 200 and 350 mosmol.

Development of mature adults has not been accomplished *in vitro*. Gevrey and Gevrey (1974) reported a low number of fifth stage in an Ae-type medium (see Section III, A, 10) containing 1% blood. Schulz (1967) obtained development of a few fifth stage males and females reaching 6·5 mm and 6·2 mm respectively. They were first seen in 26-d cultures and died without further development 5–10 d later. The medium consisted of 30% rabbit embryo extract, 15% pig liver extract, 10% calf serum and 15% nutrient broth. Cultures were incubated at 38·5°C under 13·5% CO_2 in air (pH 6·6).

Development to early fifth stage was used by Jakstys and Silverman (1969) as a measure of the effect of different types of serum, added at 10%, in chemically defined medium CbMM with 2× vitamins (see Section II, D, 6). Immune rabbit serum, prepared against axenically cultured free living nematodes, retarded development of fourth stage *H. contortus*. The effect was removed by heating the serum to 56°C for 30 min. Serum with a high titre of non-specific γ-globulin was stimulatory.

Metabolic antigens of third, fourth and fifth stage larvae cultured in CbMM with 5× vitamins were demonstrated by immunoelectrophoresis (Alger, 1968). Neilson (1969) used a balanced salt solution for collection of third stage antigens. After incubation under a gas phase of $CO_2:O_2:N_2$ (50:15:35) for 3 d, 70% reached third moult. He later reported (Neilson, 1975) that the metabolic antigens were not immunogenic when administered to 3-month old lambs.

(b) *H. placei* (Strongyloidea). Leland (1970b) obtained development of infective larvae to fourth stage in Ae medium (see Section III, A, 10).

14. *Hyostrongylus rubidus* (Strongyloidea)

Culture of infective larvae to fourth stage is described by Taylor and Baker (1968, pp. 298–300). Cultivation has now been carried to adults and egg laying. In an Ae-type medium (see Section III, A, 10) containing pig serum,

Leland (1969) found that adults developed by 26 d, and by 60 d eggs had been deposited, some showing early segmentation. Females and males reached a length of 6 mm and 4 mm respectively. Rose (1973), in modified Ae medium (see Section III, A, 21), obtained somewhat faster development, with adults by 22 d and egg laying at 27 d.

Most of the eggs deposited *in vitro* are infertile. Leland (1969) noted that females deposited eggs even when there were no males among the cultured adults. Eggs deposited by adults grown *in vivo* hatched and developed only to second stage larvae, possibly inhibited by the polymyxin B added to media containing the adults. Other eggs trapped *in utero* developed to third stage while being incubated at 38·5°C.

15. *Metastrongylus elongatus* (*M. apri*) (Strongyloidea)

See Taylor and Baker (1968, p. 279). Cultivation of this nematode would allow further investigation of its possible role in transmission of viruses. Culture of infective larvae to fifth stage was reported by Tarakanov (1973).

16. *Necator americanus* (Strongyloidea)

Culture of eggs to filariform larvae is described by Taylor and Baker (1968, pp. 272–274). Weinstein and Jones (1959) further described culture of filariform larvae to fourth stage in CEE-rat serum medium.

As a guide to the synthesizing ability of parasites in culture, Burt and Ogilvie (1975) studied the synthesis and release of acetylcholinesterase. Nematodes grown *in vitro* were maintained in Hanks's solution (see Chapter 1: IV, B, 1, g) with 0·5% lactalbumin hydrolysate. Enzyme production by adults was high for the first 2 d in culture and continued at a low level for 18 d. Production by fourth stage larvae was lower and none was detected in cultures of *in vivo* grown third stage or infective larvae. Worms died in unsupplemented tissue culture medium.

17. *Nematodirus* spp. (Strongyloidea)

This nematode was among several species cultured to fourth stage in Ae medium (see Section III, A, 10) (Leland, 1970b). In Rose's (1973) modification of Ae medium (see Section III, A, 21), cultures of *N. helvetianus* were improved but still with only partial development of the fourth stage. He used Hanks's solution (see Chapter 1: IV, B, 1, g) and the pH was lowered to 6·8.

18. *Nematospiroides dubius* (Strongyloidea)

Cultivation from egg to early adult is mentioned by Taylor and Baker (1968, pp. 300–301). Reproductive behaviour of adults grown *in vivo* and exsheathment stimulation by HCl are described.

Eggs deposited in cultures of trichostrongyles frequently fail to complete embryonation. While it is possible that eggs that fail in early cleavage are not fertile, a direct inhibitory effect of temperature of the culture was shown by Yasuraoka and Weinstein (1969). Eggs derived from *N. dubius* grown *in vivo* completed embryonation in Krebs-Ringer-Tris solution (see Chapter 8: VIII, A, 1, 1°, i) only if the temperature was lowered to 25°C so that the eggs were at 37°C for no longer than 20 h. Muria (1975) investigating a range of temperatures confirmed that temperatures above 30°C prevented hatching.

Filariform larvae reared in axenic culture at 26°C in a medium of 50% CEE with a solution of liver concentrate (2·6 mg ml^{-1}) are infective for germ-free mice (Weinstein *et al.*, 1969). The cycle could not be completed under germ free conditions because such faeces were unsuitable for filariform development.

The effects of exogenous hormones on moulting and egg laying were shown by Dennis (1976). Fourth stage larvae and adults removed from intestinal mucosa and lumen respectively, were exposed to hormones in Tyrode's solution (see Chapter 1: IV, B, 1, c) or M199 (Taylor and Baker, 1968, pp. 355–356) for 48 h. Synthetic juvenile hormone stimulated egg laying and inhibited moulting: α-ecdysone stimulated moulting; testosterone stimulated egg laying.

19. *Nippostrongylus brasiliensis* (*N. muris*) (Strongyloidea)

This was the first nematode parasite of vertebrates to be cultured from egg to adult (Taylor and Baker, 1968, pp. 286–289). Media designed for its culture have been useful in culture of other nematodes. Eggs are cultured to infective larvae at 26°C in CEE. For the parasitic stages, serum is added to the CEE and the temperature raised to 37·5°C. The medium is improved by the addition of vitamins and peptones. Adults develop to maturity but the eggs are infertile and there is no evidence of copulation. During the reproductive phase *in vivo* copulation apparently occurs frequently, but Phillipson (1973), after providing various surfaces *in vitro*, found that fertilization occurred only when the adults were in contact with intact mucosa.

The effect of specific nutrients on development of the free living stage was investigated by Bolla *et al.* (1972a, 1974). The materials were added to killed *Escherichia coli* cells in water. Using bacteria killed by formalin, addition of 50 µg ml^{-1} 7Δ-dehydrocholesterol increased development of third stage. Of the precursors, squalene, but not farnesol, had some effect. Addition of coproporphyrin to the medium resulted in the infective larvae reaching a size comparable to those grown in faecal culture. Labelled coproporphyrin was recovered from third stage larvae grown in the presence of ^{14}C-rat blood haematin (Bolla et al., 1974).

Heat killed bacteria (62–65°C, 30 min) with added sterol and haemoglobin supported development only to second stage larvae and were not as good as similarly supplemented live or formalin killed bacteria, which supported development to L3 (Bolla et al., 1972a). In contrast, certain free living rhabditid species can be cultured through the entire life cycle on briefly autoclaved bacteria with added sterol and Fe-porphyrins (reviewed by Nicholas, 1975). It is possible that nematode utilization of killed E. coli supplied in these media was influenced by the conditions of bacterial culture or by the different heating method employed, since both are critical factors.

Using a 2-step culture procedure, Bonner et al. (1971) investigated synthesis of cuticular proteins. Third stage larvae, which in this species are a feeding stage, were axenized and cultured for 48 or 72 h at 37–38°C in serum-CEE media containing ^{14}C-proline. They were then inoculated subcutaneously into rats and 88 h later were recovered from the intestine as fourth stage larvae. These were maintained in Krebs-Ringer-glucose or NCTC 109 (Taylor and Baker, 1968, pp. 351–354) with 20% human serum for 48 h to complete the fourth moult. Analysis of the cuticular proteins showed ^{14}C-proline and ^{14}C-hydroxyproline. The in vivo step had been introduced to achieve synchrony in the fourth stage development.

Genetic control of development to the third stage was shown by the effect of Actinomycin-D on larvae cultured in the presence of ^{3}H-uridine, ^{3}H-thymidine or ^{14}C-amino acids (Bolla et al., 1972b). Cultures were inoculated with first stage larvae that had been made synchronous by hatching into Krebs-Ringer-Tris-malate solution (see Chapter 8: VIII, A, 1, 1°, i). In a medium of formalin killed E. coli with 100 μg ml^{-1} cholesterol, control cultures developed to second stage rhabditiform larvae in 24 h and by d 6 filariform third stage larvae. In cultures containing 20 μg ml^{-1} actinomycin-D most larvae remained as second stage, and synthesis of DNA, RNA and proteins was inhibited. By adding inhibitors at intervals to the larval cultures, Bonner et al. (1976) then showed inhibition of specific mRNA synthesis in the second moult. Bonner and Buratt (1976) also showed that development and infectivity of third stage larvae were inhibited. Actinomycin, 1–10 μg ml^{-1}, was added to larvae cultured in NCTC 135 (see Grand Island Biological catalogue) with 20% inactivated calf serum incubated at 37°C under 5% CO_2 in air. Morphology and subcutaneous infectivity were examined after 1–3 d cultivation.

Wilson (1967), using immunoelectrophoresis, showed that antigens were produced by adults during 24 h incubation in Eagle's MEM (Taylor and Baker, 1968, p. 355) with amino acids and vitamins increased 4-fold.

Supplementation with 17% foetal calf serum prolonged production of antigens for 7–10 d but there was a marked decline after 5 d. Egg laying also continued for only 5 d in these cultures. Denham (1969) obtained metabolic antigens over a period of 6 d in a complex defined medium (Denham, 1967) supplemented with 10% horse serum.

Sommerville and Weinstein (1967) had shown that fourth stage larvae grown *in vivo* matured to adults in a minimal medium lacking serum but containing hydrolysates (Krebs-Ringer-Tris solution containing 1·5% sodium caseinate, 0·5% yeast extract and 0·25% glucose). Burt and Ogilvie (1975) used this medium and measured production of an "exoenzyme", acetylcholinesterase, as an indication of the synthesizing ability of worms for production of metabolic antigens. By comparison, in a simpler medium consisting of Hanks's solution-HEPES (see Grand Island Biological catalogue) with 0·5% lactalbumin hydrolysate, enzyme production was lower. It was still lower in completely defined tissue culture media. Enzyme production declined after 3 d. Examination at 3 d showed ultrastructural damage to intestinal cells and the excretory gland (Love *et al.*, 1975). These observations indicate changes in the worms during cultivation which may interfere with production of antigens.

20. *Oesophagostomum* spp. (Stongyloidea)

Culture of infective larvae to fourth stage larvae is described by Taylor and Baker (1968, pp. 279–300). The difficulty of obtaining development *in vitro* beyond mid-fourth stage may reflect a physiological change in the larvae which *in vivo* migrate at this time from cysts to the lumen of the caecum and colon.

(a) *O. columbianum.* Das (1968) cultured this nematode in an Ae type medium (see Section III, A, 10). Yeast extract and liver extract (Sigma) were found to be essential. Vitamin B_{12} was added. Larvae were exsheathed in hypochlorite solution and cultured at 37·5°C under 5% CO_2 in air. The tubes were rolled twice daily. Medium was replaced daily for 12 d, then at intervals of 2–3 d. Fourth stage larvae were seen on d 8. By d 34, 4% had developed to the fourth moult. The largest female measured 2·3 mm and the male 2·2 mm.

Culture in a simpler medium was undertaken by Neilson (1972) with a view to identifying antigens. The protein content of the medium was decreased by chromatographic fractionation of the supplements to exclude materials of molecular weight above 6000. The medium was composed of Earle's solution (see Chapter 1: IV, B, 1, f) with 40% CEE, 20% sheep serum and 2% Eagle's 100 × vitamin mixture (see Grand Island Biological catalogue). Serum and CEE were fractionated before mixing. The CEE had also been incubated at 4°C overnight with hyaluronidase and centrifuged

(100,000 g, 85 min) before being applied to a column of Biogel-P6 (Bio-Rad Laboratories). Larvae were axenized in antibiotics. They were exsheathed in Earle's saline with added 0·046 M sodium bicarbonate and 0·02 M sodium dithionate with a gas phase of $CO_2:O_2:N_2$ (50:10:40) and agitated for 20 h at 39°C. The inoculum was 50,000 to 50 ml medium. Cultures were then gassed with $CO_2:O_2:N_2$ (10:20:70) and incubated at 39°C with shaking. The medium was replaced every 4 d for 14 d. The fourth stage was reached by d 7 (24%). This stage survived 5–10 d and reached a length of 1·7 mm. Proteins increased in the harvested medium and reacted immunoelectrophoretically with antisera.

(b) *O. dentatum* (Strongyloidea). Tarakanov and Kaarma (1972) describe culture through the fourth moult. In a modified Ae medium (see Section III, A, 21), Rose (1973) obtained development into the fourth stage.

(c) *O. quadrispinulatum* (Strongyloidea) was cultured by Leland (1970a) in Ae medium (see Section III, A, 10) to the fourth stage. By replacing beef serum with pig serum development improved, 4% reaching the adult stage. Infertile eggs were found in one adult. The longest female and male measured 9·2 mm and 7·8 mm respectively. Cultures of 2 ml medium had been inoculated with 500–1000 larvae. They were incubated with rotation at 38°C under air, the medium having been pregassed with CO_2. Medium was replaced at weekly intervals. The fifth stage was observed at 31 d and cultures were continued for 138 d.

Schulz and Dalchow (1969) obtained a higher proportion of adults (5–25%) and found infertile eggs deposited at 90 d. The fifth stage was reached by 22–26 d and mature adults by d 74. Their medium contained 30% CEE, 10% swine serum, 15% fresh pig liver extract and 45% nutrient broth. The gas phase was 5% CO_2 in air. The inoculum, 200 per ml, was smaller than that used by Leland.

(d) *O. radiatum* (Strongyloidea). Douvres (1970) improved the yield of late fourth stage larvae to 90% by culturing with bovine embryonic kidney cells. The method used is as described for *Stephanurus dentatus* by Taylor and Baker (1968, p. 284). The medium was compounded with 50% bovine serum and NCTC 135 (see Grand Island Biological catalogue) replaced NCTC 109. Herlich *et al.* (1973) found that L3L4 larvae grown *in vitro* given by intraperitoneal injection to calves provided up to 90% protection.

21. *Ostertagia* spp. (Strongyloidea)

Cultivation by Leland to mature adults in Ae medium is described by Taylor and Baker (1968, pp. 293–300).

(a) *O. circumcincta*. Recent work has produced an improved rate of development. Denham (1970) used CEE mixed with 15% inactivated calf serum and obtained development to fifth stage in 21 d. The CEE was prepared at 50% using an equal volume of 2 × concentrated M199 ((Taylor and Baker, 1968, pp. 355–369) prepared from a 10× stock) and was extracted for 18 h at 4°C before centrifugation (2000 g, 1 h). Larvae (300–500), exsheathed in hypochlorite solution, were inoculated into 4 ml medium and incubated under air at 38°C with rotation. The medium was replenished at 2–3 d intervals. Rose (1973) modified Ae medium by replacing sodium caseinate with 700 mg% lactalbumin hydrolysate and using horse serum. By d 27 12% had developed to adults. Eggs were deposited and showed early segmentation. Antigens in harvested culture media gave a degree of protection to lambs (Rose, 1976).

(b) *O. ostertagi* (Strongyloidea) responded less well in Rose's medium. By 35 d 5% had matured but the females did not contain fully developed eggs. Recently Douvres and Malakatis (1977) reported culture to egg laying adults using a stepwise culture procedure and two new complex media developed in an attempt to culture *Trichostrongylus axei*. Infective larvae were exsheathed in an artificial rumen fluid with a sequence of gas applications and shaking; viz: $CO_2:N_2$ (40:60) for 10 min, shake 30 min, compressed air 5 min, shake 25 min. The larvae were cultured in 100 ml medium in roller bottles for 3 d at 37°C in the artificial rumen fluid under 5% CO_2 in air. Cultivation was continued under air in a complex medium consisting of Eagle's MEM (Taylor and Baker, 1968, p. 355) with additional defined components and supplements of yeast, peptone, 7·5% rabbit embryo extract, 7% calf liver extract and 25% inactivated calf serum. Pepsin, 2 mg ml^{-1}, was added and the medium acidified with HCl to pH 4·5. After 6 d the larvae were transferred to the same medium acidified to pH 6·0. This medium was replaced at intervals of 6–7 d and cultures were maintained for 81 d.

By d 2 all larvae had exsheathed. The third moult was noted on d 7, the fourth at 15–19 d, and egg laying at 34–41 d. With an inoculum of 105 in 100 ml 35% reached mature adults, the largest male and female measuring 6·6 mm and 7·25 mm respectively. Spermatozoa were packed in the seminal vesicles but were not found in the uterus. Numerous eggs were deposited and developed to morula stage, possibly inhibited by the culture temperature. Pandey (1972) had noted that development of infective larvae in faecal cultures was inhibited at 35°C.

22. *Paraspidodera uncinata* (Oxyuroidea)

This is a parasite of the lumen, caecum and colon of guinea pigs. Development of third stage ensheathed larvae within the eggs in physiological saline is described by Bondy (1967).

23. Pelodera strongyloides (Rhabditis strongyloides) (Rhabditoidea)

This is free-living on decaying organic matter, but can invade the orbit of the eye, and on skin can cause dermatitis. It can be cultured monoxenically on agar with *Escherichia coli*. Axenic cultures (Yarwood and Hansen, 1968) can be maintained through successive generations on sterile kidney fragments placed on nutrient agar, or in liquid medium consisting of peptone-yeast supplemented with heated liver extract (see Section II, D). Because a surface to support copulation must be provided, the liquid medium is placed as a film on nutrient agar or glass wool. Defined medium CbMM (Section II, D, 6) can be used if supplemented with 20% heated liver extract. Scott and Whittaker (1970), in using CbMM for their cultures, modified the inorganic salts. Potassium phosphates were omitted and sodium chloride was added at $1\cdot225$ mg litre^{-1}. The optimum temperature for axenic culture is 23–25°C. Reproduction continues to 30°C, but not at 33°C. However dauer larvae survived exposure to 45°C (Barriga, 1971). The favourable pH limits were found to be $4\cdot5$–$7\cdot0$ (Soroczan and Krauze, 1973).

24. Phocanema spp. (Ascaroidea: Anisakidea)

(a) *Phocanema decipiens* (see also *Terranova decipiens*). Moulting of fourth stage larvae to adults has been studied extensively. In the complex medium of Townsley et al. (1963) (see Section III, A, 28) ecdysis occurs at 3–6 d (Davey, 1965). In $0\cdot9\%$ saline the new cuticle is formed but not shed (Davey, 1969). Ecdysis can be stimulated by addition of juvenile hormone or by 10^{-6} M farnesyl methyl ether (Davey and Sommerville, 1974).

(b) *P. depressum*. This nematode occurs as larvae in the muscles of the shrew and as adults in the digestive tract of vultures. The larvae were cultured by the method of Townsley et al. (1963) (see Section III, A, 28) and used at 2 d for a study of moulting hormones (Rajulu et al., 1972).

25. Stephanurus dentatus (Strongyloidea)

Cultivation of infective larvae to late fourth stage in swine kidney cell cultures overlaid with NCTC 109 (Taylor and Baker, 1968, pp. 351–354) containing yeast extract, Bacto-peptone and glucose, with 50% swine serum, is described by Taylor and Baker (1968, pp. 280–286). Tromba and Douvres (1969) continued their work with this system and examined the survival of late stage larvae and adults grown *in vivo*. Larvae survived to 198 d, but without development. Survival was longer in the presence of the kidney cell cultures than in overlay medium alone. Adults survived 29 d and eggs were laid for 2–3 d. Some eggs were removed and cultured to infective larvae.

26. *Strongyloides* spp. (Rhabdiasoidea)

Cultivation of the free living phase of *S. fülleborni* and *S. ratti*, and the parasitic phase of *S. papillosus*, to fourth stage is described by Taylor and Baker (1968, pp. 260–261).

S. fülleborni in the free living phase larvae developed in faecal cultures to infective filariform larvae or alternatively to rhabditoid adults which produced a second generation of rhabditoid and filariform individuals (Beg, 1968). The proportion developing as rhabditoids increased under certain environmental conditions, including decreased O_2-increased CO_2 (Hansen et al., 1969). Rhabditoids developed at temperatures of 23–37°C, but were rare at 20°C. Gorkhall and Basir (1968) found that development was inhibited at 38–40°C. Barrett (1968) reported 20°C to be optimum for filariform larvae. Culture of the free living generation has been re-examined by Arizono (1974).

In axenic culture (Hansen et al., 1975) a mixed population developed, one-third filariform and two-thirds rhabditoid, comprising 45% female and 55% male. Segmented eggs were produced; a few were deposited but no insemination was observed. These cultures had been initiated with eggs axenized from freshly deposited faeces. The medium consisted of CbMM (see Section II, D, 6) with 25% CEE, 10% human serum and the sulphydryl compounds cysteine 0·66 mM plus reduced glutathione 0·33 mM, or dithiothreitol 0·33 mM. Cultures were incubated at 27–33°C under a gas phase of increased CO_2-decreased O_2 (Hansen et al., 1975). The CEE was freshly prepared and could not be substituted by stored CEE or CEE from a commercial source. In CbMM alone larvae remained as L2 or developed as filariforms.

Early parasitic larvae were cultured from infective larvae that had been exsheathed by penetrating a stretched skin of mouse embryo under aseptic conditions. In a defined basal medium, Berntzen's M115 (Taylor and Baker, 1968, p. 208), supplemented with serum and CEE and incubated at 37°C under a gas phase of $CO_2 : O_2 : N_2$ (10 : 5 : 85), there was an increase in length from 524 μm to 948 μm, the gonad increasing from 15 μm to 48 μm. Incomplete moults were observed at 5 d (Buecher et al., 1969).

27. *Syphacia muris* (Oxyuroidea)

Infective eggs of this rodent pinworm were found by Gulden and van Erp

(1976) to hatch in 0·1 M phosphate buffer containing 1% trypsin, 3% bile and 0·2% cysteine. Hatching was stimulated by 5–10% CO_2.

28. *Terranova decipiens* (Ascaroidea: Anisakidae)

See also *Phocanema decipiens* (Myers, 1975). The definitive host is the seal. Eggs pass out in the faeces and hatch to ensheathed larvae which are eaten by fish, possibly being carried by marine invertebrates. They encyst as fourth stage in fish muscle.

The first cultures were made using fourth stage larvae (Townsley *et al.*, 1963). Nematodes were removed from fresh cod muscle that had been surface sterilized with detergent and 70% alcohol. If hypochlorite was used at this stage it prevented subsequent maturation. This sensitivity to hypochlorite was also found with eggs (Bier, 1976). Cysts were placed in physiological saline at 35°C for 1 h to permit worms to free themselves. Larvae (5–10) were placed in 2 ml medium in 60 × 15 mm Petri dishes. The medium consisted of M199 (Taylor and Baker, 1968, pp. 355–356) with 10% beef embryo extract, 10% fresh beef liver extract and 5% glucose (pH 6·2, 35°C, gas phase air). The medium was replaced daily. Larvae moulted in 3–6 d. About 80% became mature adults in 16–20 d with eggs *in utero*.

McClelland and Ronald (1974b) cultured this nematode from egg to fourth stage. Cultures were initiated with eggs dissected from gravid adults from seals. The eggs were axenized with antibiotics in 0·9% saline and incubated at 15°C in 5 ml sterile sea water with antibiotics. They hatched in 10–14 d. The freshly hatched ensheathed second stage larvae were washed as above. They were cultured in Eagle's MEM (Taylor and Baker, 1968, p. 355) with 20% foetal calf serum at 15°C under air or 5% CO_2 in air. Five ml of medium were used in 35 × 15 mm Petri dishes, in 20 ml bottles, or in 150 × 16 mm tubes gassed before sealing. The tubes were rotated at 1 rev min^{-1}. Apparently all methods were equally effective. Medium was changed weekly. Cultures were started with 1000–5000 larvae; later the inoculum was decreased to 25. Larvae exsheathed in the medium over a period of a few hours to several weeks. The larvae grew to 31 mm and were similar morphologically to infective larvae in the cod. At 35 weeks the temperature was raised to 36°C. Six weeks later 80% had reached pre-adult stage. The genital primordia were differentiated but the vulva and the caudal alae and papillae of the male were lacking.

29. *Thelazia* spp. (Spiruroidea)

Survival of adults and infective larvae in phosphate buffered physiological saline, pH 7·2, was increased up to 100 d by addition of about 30% fresh egg white (Čorba and Leštan, 1969).

30. *Toxocara canis* (Ascaroidea)

Larvae maintained *in vitro* provided a source of metabolic antigens for serodiagnosis (de Savigny, 1975). Eggs were hatched by Fairbairn's method (Taylor and Baker, 1968, p. 262). Larvae incubated with rotation at 37°C survived as second stage for 18 months in Eagle's MEM (Taylor and Baker, 1968, p. 355). The harvested medium showed electrophoretically at least 2 antigens. In contrast *Ascaris suum* larvae survived only 21 d in the Eagle's medium.

Olson and Jones (1974) collected aseptic larvae by hatching the eggs with pressure and allowing the larvae to migrate.

31. *Trichinella spiralis* (Trichuroidea)

Cultivation of adults from decapsulated muscle larvae is described by Taylor and Baker (1968, pp. 249–255). Recent reviews of culture include those by Meerovitch (1970) and Tarakanov (1971).

The final moult starts after about 2 d in culture and multiple sheaths are shed. Their appearance differs with different culture conditions but in electron microscope studies there appeared to be four cuticles (Kozek, 1971). In investigating the hormonal control of this final moult Shanta and Meerovitch (1970) found that farnesyl methyl ester inhibited development of male genitalia. Hitcho and Thorson (1971) extracted compounds from the larvae that had an effect on moulting similar to that of insect hormones and were structurally related to them. The medium used consisted of 1 part inactivated rabbit serum and 8 parts CEE, with added B12 and mixed vitamins, and gas phase $CO_2 : O_2 : N_2$ (29 : 4 : 67).

Savel *et al.* (1969) cultured muscle larvae in Hanks's solution (see Chapter 1: IV, B, 1, g) with ^{14}C-amino acids for 66 h under 5% CO_2 in air and found the radioactive protein formed was antigenic and specific.

Dennis *et al.* (1970) collected young larvae aseptically by culturing adults for 16–20 h in M199 (Taylor and Baker, 1968, pp. 355–356) with 29% dialysed calf serum. The shed larvae were separated by passage through a 25 μm screen and used for intravenous infections. Similar larvae were injected by Despommier *et al.* (1975) into the thigh muscles of rats and mice to obtain synchronous *in vivo* populations for detailed study of growth rate. This intracellular phase has not been cultured *in vitro*.

32. *Trichostrongylus* spp. (Strongyloidea)

Cultivation of these nematodes has not been successful (Taylor and Baker, 1968, pp. 296–297). Douvres and Malakatis (1977) re-examined *T. axei* in culture with kidney cells and found that worms exsheathed but did not develop. Rose (1973) obtained no development of *T. colubriformis* in his modified Ae medium (see Section III, A, 21). Fourth stage 8 d-old *T. colubriformis* grown *in vivo* held in 0·85% sodium chloride for 6 h released antigens that were protective for guinea pigs (Rothwell and Love, 1974).

33. *Trichuris trichura* (Trichuroidea)

Ortiz-Valqui and Lumbreras-Cruz (1970) reported hatching in 0·43% saline after 20 d at 27°C, and larval development in Earles' solution (see Chapter 1: IV, B, 1, f) at 37°C.

B. NEMATODE PARASITES OF INVERTEBRATES

1. *Host: Mollusca*

Rhabditis maupasi (Rhabditoidea) occur as dauer larvae in the mantle cavity of the snail *Helix aspera*, having entered through the pneumostome. When the snail dies the nematodes rapidly produce a large population of dioecious adults and dauer larvae.

Axenic cultures have been maintained at 20°C on kidney-agar slants or in an aqueous peptone medium with dextrose and raw liver extract as for *Neoaplectana*. Dauer larvae were axenized in antibiotics followed by 6 h in 3% dextrose to allow time for expelling bacteria from the gut. This was followed by a 5 min wash in 0·1 M HCl (Brockelman and Jackson, 1974). The *in vivo* history suggested an inhibitory effect by the snail and led to the finding of a protein in snails that was inhibitory *in vitro* (Brockelman, 1975).

2. *Host: Insecta*

Nematodes of insects are facultative or obligate parasites. Many can be maintained in the laboratory in xenic cultures (see Poinar, 1975). Axenic cultivation had been attempted with a few species, primarily as an adjunct to their application in biological control. This has been most easily accomplished with the facultative parasitic Rhabditoidea which reproduce also in decaying matter. Under axenic conditions they can be cultured through successive life cycles by procedures similar to those used for free living nematodes. Mermithids, which are obligate parasites, can survive *in vitro* only under conditions that approach the host environment. They have not yet been cultured to maturity.

(a) Mermithoidea. Cultivation of mermithid nematodes requires the provision of nutrients that can be absorbed through the cuticle since the oesophagus is a narrow blind tube (Fig. 1). Furthermore the medium should approach the composition of host fluids because uptake by active transport

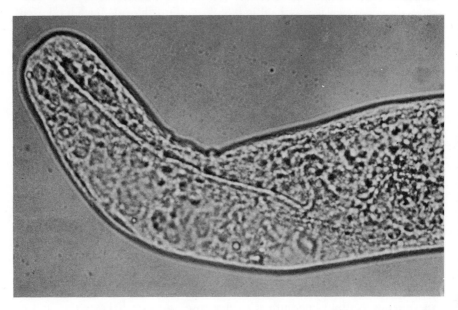

FIG. 1. *Romanomermis culicivorax* showing oesophageal tube; × 320. Parasitic juvenile cultured at 26°C for 6 weeks in diluted Grace's medium with 10% inactivated foetal calf serum. (Photograph courtesy of Jean Finney)

is likely to be influenced by the relative concentrations of components. This route of intake, unusual for nematodes, has been examined in detail. Gordon and Webster (1972) using *Mermis nigresens* showed incorporation of ^{14}C amino acids and glucose. The intake of ^{3}H labelled proteins from grasshopper haemolymph was of marginal significance. In experiments on carbohydrate uptake (Rutherford and Webster, 1974), juveniles were maintained aseptically in a solution containing 130 mM NaCl, 6 mM KCl, 1·1 mM glucose and 10·6 mM trehalose, with phosphate buffer to pH 8·0. Individuals were placed so that the anterior and posterior portions were immersed in different aliquots of solution, thus demonstrating cuticular intake.

Electron micrographs by Poinar and Hess (1976) of the body wall of early juvenile *Romanomermis culicivorax* from mosquitoes showed morphology of an absorptive surface (Fig. 2). The outer surface was reduced to membranes with pores large enough for the entrance of ferritin particles; the underlying hypodermal cells were bordered by microvilli, and contained ferritin (Poinar and Hess, 1977).

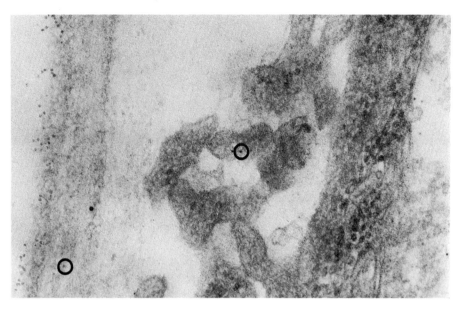

FIG. 2. Body wall of *Romanomermis culicivorax* showing ferritin particles (circled) in outer membrane and hypodermal cell; × 105,000. 4-d parasitic juvenile incubated 4–6 h at 22°C in Grace's medium with ferritin. (Photograph courtesy of G. O. Poinar Jr. and Roberta Hess)

(i) *Hydromermis conopophaga*. Infective larvae collected from a wild population of larval midges were cultured at 20°C in a medium consisting of pond water, bovine serum and CEE (3 : 1 : 1). The nematodes showed morphological change from infective larvae to parasitic forms (Poinar, 1975).

(ii) *Neomesomermis flumenalis*. This mermithid collected from wild populations of blackfly larvae, *Simulium* spp., was cultured using the same procedure as for *Romanomermis culicivorax* (see below) except that cultures were incubated at a lower temperature, 10°C. The nematodes reached a length of 8–9 mm (Finney, 1976a).

(iii) *Romanomermis culicivorax* (*Reesimermis nielseni*). This mermithid is available in large quantities from controlled breeding in laboratory colonies of *Culex pipiens* (Petersen, 1973). Axenic cultures were initiated with eggs or preparasitic L2 larvae, axenized in antibiotics. In culture the larva lost its active motility and assumed a characteristic parasite morphology. In

Schneider's *Drosophila* medium (Gibco 172) with foetal calf serum at 25°C, juveniles reached a length of 5–7 mm by d 25. The medium was replenished at intervals of 5–10 d (Sanders *et al.*, 1973). Culture studies have been continued by Finney (1976a, b) who reported that early female differentiation had been observed in 6-week old cultures incubated at 26°C in diluted Grace's medium (Gibco 159) with 10% inactivated foetal calf serum (5 ml Grace's, 2·2 ml water, 0·8 ml FCS). Stored material, characteristically present in cells of the trophosome, was however lacking in cultured mermithids. Other preliminary reports indicated that mermithid development was not promoted in synxenic culture with mosquito tissue.

(b) Rhabditoidea. (i) *Neoaplectana glaseri* (Steinernematidae). This facultative parasite was first established in axenic culture in 1944 using kidney slices placed on enriched nutrient agar slants (Taylor and Baker, 1968, pp. 255–261; Stoll, 1973). This method is still used. The cultures serve as a convenient source of dauer larvae (ensheathed L3; "survivalarvae") which migrate in large numbers on to the walls of the culture tube and can be stored for up to 3 years at 5°C in shallow layers of 5% glucose (Jackson, 1973a). They do not survive incubation at 37°C. Cultures are maintained by repeatedly moistening the agar with water or liquid medium or by loop inoculation to a fresh kidney slant. Kidney or liver slices can be used even after storage for a year if kept frozen in a sealed container.

Cultures can also be maintained in liquid medium by serial subculture. For successive inoculations young larvae must be selected in order to avoid ensheathed larvae which may fail to exsheath. Several liquid media have been used successfully. Jackson (1973a) used fresh beef heart infusion broth supplemented with 1% Pfanstiehl peptone (Pfanstiehl Chemical), 1%–3% glucose, and approximately 8% raw liver extract (RLE) (see Section II, D, 3), incubated at 20°C in 4 ml aliquots with slow shaking. In our laboratory *N. glaseri* and other *Neoaplectana* have been maintained in a liquid medium consisting of a base of 3% soy peptone with 3% yeast extract supplemented with RLE or 10% heated liver extract (see Section II, D, 3) or 20% fresh yeast extract (Lower and Buecher, 1970). The medium was distributed over glass wool lining either screw capped test tubes or separating funnels (Hansen and Cryan, 1966). The vessel was incubated horizontally and placed vertically for draining. A portion of the medium was replaced at intervals of 1–2 months. The harvest in the drained portion consisted of adults and larvae and a high proportion of the facultative dauer stage. We have also used the defined basal medium CbMM (see Section II, D, 6) supplemented with HLE or with CEE and serum (see Section III, C, 1).

Jackson (1973b) designed a defined medium (see Section II, D, 6) in which dauer larvae developed to adults producing limited numbers of first generation larvae. He showed a requirement for the nine mammalian amino acids plus arginine, and noted that maturation was delayed if non-essential amino acids were omitted. Later, Jackson and Platzer (1974) examined the requirement for folic acid and biotin, and the biosynthesis of purines, and demonstrated two enzymes involved in folate metabolism. Nucleic acid substituents were not required in defined medium for short term culture; however their requirement over successive generations had not been tested.

Actinomycin-D, 38 μg ml^{-1}, inhibited growth and development but not exsheathment (Despommier and Jackson, 1972).

(ii) *N. carpocapsae*. The Czechoslovakian strain was cultured axenically using the same procedures as for *N. glaseri* (Hansen *et al.*, 1968). It grew particularly well in deep aerated cultures (Buecher and Hansen, 1971). This species differs from *N. glaseri* in having smaller dauer larvae which exsheathed less readily in culture media.

In attempting to design a medium of defined components, cultures were maintained in CbMM with a defined supplement of 25 mg ml^{-1} γ-globulin plus 8 μg ml^{-1} haemin (Buecher *et al.*, 1970b). In other experiments using a low level supplement, 250 μg ml^{-1} of partially purified liver protein, 25 μg of protein was needed to grow an individual to maturity (Hansen *et al.*, 1968).

Similar culture procedures were suitable for the DD-136 strain of *N. carpocapsae* (Poinar and Hansen, 1972).

Juvenile hormone type compounds are toxic at 100 μg ml^{-1} to new larvae of *N. carpocapsae* in 24 h cultures (Hansen and Buecher, 1970).

(iii) *Neoplectana* spp. Tarakanov and Andreeva (1974) reported that liver slices were preferred to kidney for maintaining cultures on peptone agar. Kurashvili and Kakulia (1974) reported cultivation of two other species, *N. georgica* and *N. agriotos*.

(iv) *Steinernema kraussei* (Steinermatidae). Weiser (1976) reported cultivation on kidney or liver slices on agar slants (see also Mráček, 1977).

(v) *Mesodiplogaster lheritieri* (Diplogasteridae) was cultured axenically by Weiser (1966) on kidney-agar slants.

(vi) *Pristionchus uniformis* (Diplogasteridae) was cultured by Fedorko and Stanuszek (1971) on liver-agar slants. Eggs developed to non-reproducing adults.

C. PLANT PARASITIC NEMATODES

Plant parasitic nematodes are cultured monoxenically on aseptic plant roots or callus tissue or on fungal mycelia. Zuckerman (1971) has listed species in laboratory culture. Two migratory ectoparasites, *Aphlenchus avenae* and

Aphelenchoides rutgersi, have been established in continuous axenic culture, and females of one sedentary endoparasite, *Meloidogyne incognita*, have been maintained.

Plant parasitic nematodes ingest food material by means of a stylet which can penetrate plant or fungal cell walls. Penetration is, however, not a prerequisite of feeding. Stylet probing and the accompanying oesophageal pumping can be stimulated by components of the medium (Fisher, 1975).

1. *Aphelenchus avenae* (Aphelenchoidea)

This is similar to free living rhabditids in many respects. Adults are small, less than 1 mm long. Eggs are laid at a daily rate of about 5 for 30–40 d. The complete life cycle takes 8–9 d at 27°C. The more enriched of the mixtures used for culture of free living nematodes provide suitable media for *A. avenae*.

Axenic cultures are initiated with eggs or a gravid female axenized in antibiotics. Axenic nematodes can also be collected from a monoxenic culture on non-sporulating *Rhizoctonia solani* on potato-dextrose-agar. The larvae migrate from blocks of agar-mycelium and accumulate in droplets of water placed on the agar.

A suitable medium is composed of the defined basal medium CbMM (see Section II, D, 6) supplemented with 25% chick embryo extract (CEE) and 10% human or foetal calf serum (Hansen *et al.*, 1970). CEE from a commercial source can be used. At a level of 25% the CEE remains as a fine suspension. Higher levels form an unfavourable precipitate, and with lower levels there is a lower rate of reproduction.

Cultures can be maintained for several years at 20–25°C by distributing the medium over glass wool lining a screw capped test tube (Hansen and Cryan, 1966). At intervals of 1–2 months the medium and nematodes are drained off and the medium replenished. The residual nematodes on the glass wool serve as the inoculum for the new culture. Cultures can also be maintained by serial subculture of test tube cultures. The depth of medium should not exceed 3 mm unless there is vigorous aeration (Buecher and Hansen, 1971). The population of a culture comprises adults, larvae and unhatched eggs. Many larvae retain a loose third or fourth cuticle. The moult is completed and eggs hatch in the replenished medium.

Newly hatched larvae can be collected by allowing them to migrate through a 15 μm screen (Buecher *et al.*, 1973). When used as the inoculum in bioassay, comparisons can be made on the basis of the rate of growth to maturity and number of new larvae produced. In one such series attempting to replace crude supplements with more defined protein, CEE-serum was replaced by 18 mg ml^{-1} β-lipoprotein and resulted in a medium that supported low level reproduction (Buecher et al., 1970b).

There are several strains of A. avenae, including dioeceous as well as parthenogenetic forms. The strain first axenized consisted of females reproducing by meiotic parthenogenesis. Under the influence of environmental factors, including CO_2 or increased temperature (to 30°C), nematodes developed as males (Hansen et al., 1973). These males could inseminate females of an amphimictic population (Fisher and Triantaphyllou, 1976). Sex reversal was prevented by addition of 10 µg ml^{-1} mitomycin-C to the medium (Buecher et al., 1974).

2. *Aphelenchoides rutgersi* (*A. sacchari*) (Aphelenchoidea)

This was the first stylet feeding nematode to be cultured in a liquid medium (made from fresh liver extracts, Myers, 1967). The life cycle and cultivation of *Ap. rutgersi* is similar to that of *A. avenae*, and continuous cultures were maintained in CbMM (see Section II, D, 6) supplemented with CEE-serum (Buecher et al., 1970a). Myers et al. (1971) used a maintenance medium of 10% CEE in an autoclaved base containing 3% Bacto Soytone and 2% yeast extract.

A mixed population of nematodes from fungal cultures introduced into a completely defined medium (see Section II, D, 6) grew and produced a limited number of first generation larvae (Myers, 1968). With this response as an assay the requirements for components of the medium were examined. Synthesis of certain amino acids was shown by ^{14}C studies, but if these amino acids were omitted from the medium, reproduction was inhibited (Balasubramanian and Myers, 1971).

The nutritional role of CEE supplement was investigated by Thirugnanam (1976). Its activity was stable to trypsin but was lost after acid treatment or adsorption on charcoal; it could be restored by addition of nucleic acid substituents and 10 µg haemin.

Myers (1972) reported the use of cultures for evaluating nematicidal agents.

3. *Meloidogyne incognita* (Heteroderoidea)

Females were dissected from plant tissue and maintained in hanging drops in a medium of steamed soil infusion. They survived 30 d at 27°C. New egg masses were formed during the first week. Eggs hatched to infective second stage larvae (Shepperson and Jordan, 1974).

Excised gonads were cultured by Hirumi et al. (1968) in M 199 (Taylor and Baker, 1968, pp. 355–356) with foetal calf serum. Ova developed and hatched after 16 d in culture.

REFERENCES
Ah, H-S., McCall, J. W. and Thompson, P. E. (1974). A simple method for isolation of *Brugia pahangi* and *B. malayi* microfilariae. *International Journal of Parasitology* **4**, 677–678.

Alger, N. E. (1968). *Haemonchus contortus:* Somatic and metabolic antigens of third and fourth larval and adult stages. *Experimental Parasitology* **23**, 187–197.

Aoki, Y. (1971). Exsheathing phenomenon of microfilaria *in vitro*. *Tropical Medicine, Nagasaki* **13**, 170–179 (Japanese, English summary).

Arizono, N. (1974). Studies on the free-living generations of *Strongyloides*. I. Comparison of the test tube culture method and fecal culture method. *Japanese Journal of Parasitology* (*Kiseichugaku Zasshi*) **23**, 49 (abstract, Japanese).

Balasubramanian, M. and Myers, R. F. (1971). Nutrient media for plant-parasitic nematodes. II. Amino acid requirements of *Aphelenchoides* sp. *Experimental Parasitology* **29**, 330–336.

Banerjee, D. (1972). *In vitro* cultivation of third stage larvae of *Ancylostoma caninum* (Ercolani, 1859). *The Journal of Communicable Diseases* (*India*) **4**, 175–183.

Barrett, J. (1968). The effect of temperature on the development and survival of the infective larvae of *Strongyloides ratti* Sandground 1925. *Parasitology* **58**, 641–650.

Barriga, O. O. (1971). Dinamica de la poblacion de un nematodo protoparasito en cultivo de laboratorio. *Boletin Chileno de Parasitologia* **26**, 14–16.

Beckett, E. B. and Boothroyd, B. (1970). Mode of nutrition of the filarial nematode *Brugia pahangi*. *Parasitology* **60**, 21–26.

Beg, M. K. (1968). Studies on life cycle of *Strongyloides fülleborni* von Linstow 1905. *Annals of Tropical Medicine and Parasitology* (*Liverpool*) **62**, 502–505.

Bier, J. W. (1976). Experimental Anisakiasis: Cultivation and temperature tolerance determinations. *Journal of Milk, Food Technology* **39**, 132–137.

Bolla, R. I., Weinstein, P. P. and Lou, C. (1972a). *In vitro* nutritional requirements of *Nippostrongylus brasiliensis*—I. Effects of sterols, sterol derivatives and heme compounds on the free living stages. *Comparative Biochemistry and Physiology* **43B**, 487–501.

Bolla, R., Bonner, T. and Weinstein, P. (1972b). Genic control of the postembryonic development of *Nippostrongylus brasiliensis*. *Comparative Biochemistry and Physiology* **41B**, 801–811.

Bolla, R. I., Weinstein, P. P. and Lou, C. (1974). *In vitro* nutritional requirements of *Nippostrongylus brasiliensis*—II. Effects of heme compounds, porphyrins and bile pigments on the free-living stages. *Comparative Biochemistry and Physiology* **48B**, 147–157.

Bondy, R. (1967). Contribution to the knowledge of the exogenic phase of the nematode *Paraspidodera uncinata* (Rudolphi, 1819). *Helminthologia* (Prague) **7**, 21–27.

Bonner, T. P. and Buratt, M. (1976). The effect of actinomycin-D on development and infectivity of the nematode *Nippostrongylus brasiliensis*. *International Journal for Parasitology* **6**, 289–294.

Bonner, T. P., Weinstein, P. P. and Saz, H. J. (1971). Synthesis of cuticular protein during the third moult in the nematode *Nippostrongylus brasiliensis*. *Comparative Biochemistry and Physiology* **40B**, 121–127.

Bonner, T. P., Evans, K. and Kline, L. (1976). Cuticle formation in parasitic nematodes; RNA biosynthesis and control of moulting. *International Journal for Parasitology* **6**, 473–477.

Brockelman, C. R. (1975). Inhibition of *Rhabditis maupausi* (Rhabditidae: Nematoda) maturation and reproduction by factors from the snail host, *Helix aspera. Journal of Invertebrate Pathology* **25**, 229–237.

ILS

Brockelman, C. R. and Jackson, G. J. (1974). *Rhabditis maupasi*: Occurrence in food snails and cultivation. *Experimental Parasitology* 36, 114–122.

Buecher, E. J. and Hansen, E. L. (1971). Mass culture of axenic nematodes using continuous aeration. *Journal of Nematology* 3, 199–200.

Buecher, E. J., Hansen, E. L. and Berntzen, A. K. (1969). Environmental effects on *Strongyloides fülleborni* in culture. *American Society of Parasitologists, Annual Meeting*, Abstract, p. 64.

Buecher, E. J., Hansen, E. L. and Myers, R. F. (1970a). Continuous Axenic Culture of *Aphelenchoides* sp. *Journal of Nematology* 2, 188–190.

Buecher, E. J., Hansen, E. L. and Yarwood, E. A. (1970b). Growth of nematodes in defined medium containing hemin and supplemented with commercially available proteins. *Nematologica* 16, 403–409.

Buecher, E. J., Yarwood, E. and Hansen, E. L. (1973). A screen to separate young larvae of *Aphelenchus avenae* from adults and eggs. *Nematologica* 19, 565–566.

Buecher, E. J., Yarwood, E. and Hansen, E. L. (1974). Effects of mitomycin-C on sex of *Aphelenchus avenae* (Nematoda) in axenic culture. *Proceedings of the Society for Experimental Biology and Medicine* 146, 299–301.

Burt, J. S. and Ogilvie, B. M. (1975). *In vitro* maintenance of nematode parasites assayed by acetylcholinesterase and allergen secretions. *Experimental Parasitology* 38, 75–82.

Campbell, W. C., Blair, L. S. and Egerton, J. R. (1973). Unimpaired infectivity of the nematode *Haemonchus contortus* after freezing for 44 weeks in the presence of liquid nitrogen. *Journal of Parasitology* 59, 425–427.

Chatterjee, R. K. and Singh, K. S. (1968). *In vitro* culture and hatching of *Ascaridia galli* eggs. *Indian Journal of Veterinary Science and Animal Husbandry* 38, 517–523.

Čorba, J. and Leštan, P. (1969). Beitrag zur Züchtung von Augenhelminthin *in vitro*. *Biologia, Bratislava*, Seria B24, 634–638.

Croll, N. A. (1973). Feeding shown to be unnecessary in preinfective larvae of *Dictyocaulus viviparus*. *International Journal for Parasitology* 3, 571–572.

Cupp, E. W. (1973). Development of filariae in mosquito cell cultures. *WHO/Fil/73*, 111.

Cypress, R. H., Pratt, E. A. and Van Zandt, P. (1973). Rapid exsheathment of *Nematospiroides dubius* infective larvae. *Journal of Parasitology* 59, 247–250.

Das, D. N. (1968). Cultivation of parasitic stages of *Oesophagostomum columbianum* (Curtice, 1890) Stossich, 1899, *in vitro*. *Indian Veterinary Journal* 44, 1035–1044.

Davey, K. G. (1965). Moulting in a parasitic nematode *Phocanema decipiens*—I. Cytological events. *Canadian Journal of Zoology* 43, 997–1003.

Davey, K. G. (1969). Moulting in a parasitic nematode *Phocanema decipiens*—V. Timing of feeding during the moulting cycle. *Journal of the Fisheries Research Board of Canada* 26, 935–939.

Davey, K. G. and Sommerville, R. J. (1974). Moulting in a parasitic nematode *Phocanema decipiens*—VII. The mode of action of the ecdysial hormone. *International Journal for Parasitology* 4, 241–259.

Denham, D. A. (1967). Application of the *in vitro* culture of nematodes, especially *Trichinella spiralis*. *Symposium of the British Society for Parasitology* 5, 49–60.

Denham, D. A. (1969). Secretion of metabolic antigens by *Nippostrongylus brasiliensis in vitro*. *Journal of Parasitology* 55, 676–677.

Denham, D. A. (1970). *Ostertagia circumcincta*: Methyridine and Tetramisole against third stage larvae. *Experimental Parasitology* 28, 493–498.

Dennis, R. D. (1976). Insect morphogenetic hormones and developmental mechanisms in the nematode *Nematospiroides dubius*. *Comparative Biochemistry and Physiology* 53A, 53–56.

Dennis, D. T., Despommier, D. D. and Davis, N. (1970). Infectivity of the new born larva of *Trichinella spiralis* in the rat. *Journal of Parasitology* **56**, 974–977.

Despommier, D. D. and Jackson, G. J. (1972). Actinomycin-D and puromycin-HCl in axenic cultures of the nematode *Neoaplectana glaseri*. *Journal of Parasitology* **58**, 774–777.

Despommier, D., Aron, L. and Turgeon, L. (1975). *Trichinella spiralis:* Growth of the intracellular (muscle) larva. *Experimental Parasitology* **37**, 108–116.

Dhar, D. N., Basu, P. C. and Pattanayak (1967). *In vitro* cultivation of *Dirofilaria repens* microfilariae up to the sausage stage. *Indian Journal of Medical Research* **55**, 915–919.

Dick, J. W. and Leland, S. E., Jr. (1973). The influence of pH on the *in vitro* development of *Cooperia punctata* (Ranson, 1907). *Journal of Parasitology* **59**, 770–775.

Dick, J. W., Leland, S. E., Jr. and Hansen, M. F. (1973). Hatching and *in vitro* cultivation of the nematode *Ascaridia galli* to the third-stage larva. *Transactions of the American Microscopical Society* **92**, 225–230.

Douvres, F. W. (1970). Cultivation of *Oesophagostomum radiatum* developmental stages in cell culture systems. *Journal of Parasitology* **56**, number 4, section 2, 83–84.

Douvres, F. W. and Malakatis, G. M. (1977). *In vitro* cultivation of *Ostertagia ostertagi*, the medium stomach worm of cattle—I. Development from infective larvae to egg laying adults. *Journal of Parasitology* **63**, 520–527.

Douvres, F. W. and Tromba, F. G. (1970). Influence of pH, serum, and cell cultures on development of *Ascaris suum* to fourth stage *in vitro*. *Journal of Parasitology* **56**, 238–248.

Douvres, F. W., Tromba, F. G. and Doran, D. J. (1966). The influence of NCTC 109, serum and swine kidney cell cultures on the morphogenesis of *Stephanurus dentatus* to fourth stage *in vitro*. *Journal of Parasitology* **52**, 875–889.

Dryushenko, E. A. and Berdyeva, G. T. (1974). A comparative study of amino acid consumption by *Ascaris suum* and *Ascaridia galli*. *Parazitologiya* **8**, 208–211 (Russian).

Fatt, I. (1976). "Polarographic Oxygen Sensors". CRC Press, Cleveland, Ohio.

Fedorko, A. and Stanuszek, S. (1971). *Pristionchus uniformis* sp.n. (Nematode, Rhabditida, Diplogasteridae), a facultative parasite of *Leptinotarsa decemlineata* Say and *Melolontha melolontha* L. in Poland. Morphology and biology. *Acta Parasitologie Polanska* **19**, 95–112.

Finney, J. R. (1976a). The *in vitro* culture of mermithid parasites of blackflies and mosquitoes. *Journal of Nematology* **8**, 284.

Finney, J. R. (1976b). Personal communication.

Fisher, J. M. (1975). Chemical stimuli for feeding by *Aphelenchus avenae*. *Nematologia* **21**, 358–364.

Fisher, J. M. and Triantaphyllou, A. C. (1976). Observations on development of the gonad and on reproduction in *Aphelenchus avenae*. *Journal of Nematology* **8**, 248–254.

Garrigues, R. M., Hockmeyer, W. T. and Balinas, J. C. (1975). Development of larval forms of *Brugia pahangi in vitro*. *American Society of Parasitology, Annual Meeting*, Abstract, 96.

Gevrey, J. and Gevrey, T. (1974). Culture *in vitro* d'*Haemonchus contortus* jusqu'au stade imaginal. *Annales de Recherches Veterinaires* **5**, 451–464.

Gonzalez de la Torre, P. and Sotolongo Guerra, F. (1973). Nuevo metodo de aislamiento y cultivo de larvas de *Ancylostoma caninum*. *Revista Cubana Medicina Tropical* **25**, 89–93.

Gordon, R. and Webster, J. M. (1972). Nutritional requirements for protein synthesis during parasitic development of the entomophilic nematode *Mermis nigrescens*. *Parasitology* **64**, 161–172.

Gorkhall, C. P. and Basir, M. A. (1968). Effect of temperature on the course of development of the free-living generation of *Strongyloides fülleborni* von Linstow, 1805. *Indian Journal of Helminthology* **20**, 25–29.

Grabda, J. (1976). Studies on the life cycle and morphogenesis of *Anisakis simplex* Rudolphi, 1809 (Nematoda: Anisakidae) cultured *in vitro*. (Badania nad cyklen rozwojowym i morfogeneza *Anisakis simplex* /Rudolphi, 1809/ /Nematoda:Anisakidae/ w hodowli *in vitro*). *Acta Ichthyologica et Piscatoria* **6**, fasc. 1, 119–141.

Guerrero, J. and Silverman, P. H. (1971). *Ascaris suum*: Immune reaction in mice. II. Metabolic and somatic antigens from *in vitro* cultured larvae. *Experimental Parasitology* **29**, 110–115.

Guerrero, J., Green, C., Silverman, P..H. and Mercadante, M. L. (1974). Simplification of the media used in cultivation of *Ascaris suum* larvae. *Third International Congress of Parasitology* **1**, 462–463.

Gulden, W. J. I. van der and Aspert-van Erp, A. J. M. van (1976). *Syphacia muris*: Response to environmental stimuli when hatching *in vitro*. *Experimental Parasitology* **39**, 45–50.

Gwadz, R. W. and Spielman, A. (1974). Development of the filarial nematode *Brugia pahangi*, in *Aedes aegypti* mosquitoes: Nondependence upon host hormones. *Journal of Parasitology* **60**, 134–137.

Hansen, E. L. (1974). Effect of serum on tissue culture from *Biomphalaria glabrata* (Mollusca). *In Vitro* **10**, 348.

Hansen, E. L. (1976). Application of tissue culture of a pulmonate snail to culture of larval *Schistosoma mansoni*. In "Invertebrate Tissue Culture: Applications in Medicine, Biology and Agriculture", pp. 87–93 (E. Kurstak and K. Maramorosch eds), Academic Press, New York and London.

Hansen, E. L. and Berntzen, A. K. (1969). Development of *Caenorhabditis briggsae* and *Hymenolepis nana* in interchanged media. *Journal of Parasitology* **55**, 1012–1017.

Hansen, E. L. and Buecher, E. J. (1970). Effect of insect hormones on nematodes in axenic culture. *Experientia* **27**, 859–860.

Hansen, E. L. and Cryan, W. A. (1966). Continuous culture of free-living nematodes. *Nematologica* **12**, 138–142.

Hansen, E. L., Yarwood, E. A., Jackson, G. T. and Poinar, G. O. Jr. (1968). Axenic culture of *Neoaplectana carpocapsae* in liquid media. *Journal of Parasitology* **54**, 1236–1237.

Hansen, E. L., Buecher, E. J. and Cryan, W. S. (1969). *Strongyloides fülleborni*: Environmental factors and free-living generations. *Experimental Parasitology* **26**, 336–343.

Hansen, E. L., Buecher, E. J. and Evans, A. A. F. (1970). Axenic cultivation of *Aphelenchus avenae*. *Nematologica* **16**, 328–329.

Hansen, E. L., Buecher, E. J. and Yarwood, E. A. (1973). Alteration of sex of *Aphelenchus avenae* in culture. *Nematologica* **19**, 113–116.

Hansen, E. L., Buecher, E. J. and Yarwood, E. A. (1975). *Strongyloides fülleborni*: Development in axenic culture. *Experimental Parasitology* **38**, 161–166.

Herlich, H., Douvres, F. W. and Romanowski, R. D. (1973). Vaccination against *Oesophagostomum radiatum* by injecting killed worm extracts and *in vitro*-grown larvae into cattle. *Journal of Parasitology* **59**, 987–993.

Hink, W. F. (1976). A compilation of invertebrate cell lines and culture conditions. In "Invertebrate Tissue Culture" pp. 319–369 (K. Maramorosch, ed.). Academic Press, New York and London.

Hirumi, H., Chen, T-A. and Maramorosch, K. (1968). Intra-uteral development of the root knot nematode in organ culture. *Second International Colloquium on Invertebrate Tissue Culture*, Pavia, 1968, 147–152.

Hirumi, H., Hung, C-L. and Maramorosch, K. (1969). Nematode cell culture. *Phytopathology* **59**, 1557.

Hitcho, P. J. and Thorson, R. E. (1971). Possible moulting and maturation controls in *Trichinella spiralis*. *Journal of Parasitology* **57**, 787–793.

Jackson, G. J. (1973a). The aging of *Neoaplectana glaseri*. *Proceedings of the Helminthological Society of Washington* **40**, 74–76.

Jackson, G. J. (1973b). *Neoaplectana glaseri*: Essential amino acids. *Experimental Parasitology* **34**, 111–114.

Jackson, G. J. and Platzer, E. G. (1974). Nutritional biotin and purine requirements, and the folate metabolism of *Neoaplectana glaseri*. *Journal of Parasitology* **60**, 453–457.

Jaffe, J. J. and Doremus, H. M. (1970). Metabolic patterns of *Dirofilaria immitis* microfilariae *in vitro*. *Journal of Parasitology* **56**, 254–260.

Jakstys, B. P. and Silverman, P. H. (1969). Effect of heterologous antibodies on *Haemonchus contortus* development *in vitro*. *Journal of Parasitology* **55**, 486–492.

Khalil, L. F. (1969). Larval nematodes in the herring (*Culpea harengus*) from British coastal waters and adjacent territories. *Journal of the Marine Biological Association of the United Kingdom* **49**, 641–659.

Kim, J. J. (1969). [The influence of various environmental conditions upon the eggs and larvae of hookworm.] (Korean, English summary.) *Korean Journal of Public Health* **6**, 245–254.

Klein, J. B. and Bradley, R. E. Sr. (1974). Introduction of morphological changes in microfilariae from *Dirofilaria immitis* by *in vitro* culture techniques. *Journal of Parasitology* **60**, 649.

Kozek, W. J. (1971). The moulting problem in *Trichinella spiralis*—I. A light microscope study. II. An electron microscope study. *Journal of Parasitology* **57**, 1015–1028; 1029–1038.

Kurashvili, B. and Kakulia, G. (1974). Cultivation of *Neoaplectana georgica* Kakulia and Veremtshuk, 1968. *Third International Congress of Parasitology* **1**, 465–466.

Lackie, A. M. (1975). The activation of infective stages of endoparasites of vertebrates. *Biological Reviews* (Cambridge) **50**, 285–323.

Leland, S. E. Jr. (1963). Studies on the *in vitro* growth of parasitic nematodes. I. Complete or partial parasitic development of some gastrointestinal nematodes of sheep and cattle. *Journal of Parasitology* **49**, 600–611.

Leland, S. E. Jr. (1967). *In vitro* cultivation of *Cooperia punctata* from egg to egg. *Journal of Parasitology* **53**, 1057–1060.

Leland, S. E. Jr. (1968). *In vitro* egg production of *Cooperia oncophora*. *Journal of Parasitology* **54**, 136.

Leland, S. E. Jr. (1969). Cultivation of the parasitic stages of *Hyostrongylus rubidus in vitro*, including the production of sperm and development of eggs through 5 cleavages. *Transactions of the American Microscopical Society* **88**, 246–252.

Leland, S. E. (1970a). *Oesophagostomum quadrispinulatum*: *in vitro* development of parasitic stages. *Transactions of the American Microscopical Society* **89**, 539–550.

Leland, S. E. (1970b). *In vitro* cultivation of nematode parasites important to veterinary medicine. *In* "Advances in Veterinary Science", Vol. 14, pp. 29–59. Academic Press, New York and London.

Leland, S. E. Jr. (1975). Resistance to *Cooperia punctata* in calves by culture products. *The American Society of Parasitologists, Annual Meeting*, Abstract 225.

Leland, S. E. Jr., Ridley, R. K., Slonka, G. F. and Zimmerman, G. L. (1975). Detection of activity for various anthelminthics against *in vitro*-produced *Cooperia punctata*. *American Journal of Veterinary Research* **36**, 449–456.

Levine, H. S. and Silverman, P. H. (1969). Cultivation of *Ascaris suum* larvae in supplemented and unsupplemented chemically defined media. *Journal of Parasitology* **55**, 17–21.

Love, R. J., Ogilvie, B. M. and McLaren, D. J. (1975). *Nippostrongylus brasiliensis*: further properties of antibody damaged worms and induction of comparable damage by maintaining worms *in vitro*. *Parasitology* **71**, 275–283.

Lower, W. R. and Buecher, E. J. Jr. (1970). Axenic culturing of nematodes: an easily prepared medium containing yeast extract. *Nematologica* 16, 563–566.

Mapes, C. J. (1969). The development of *Haemonchus contortus in vitro*. I. The effect of pH and pCO$_2$ on the rate of development to the fourth-stage larva. *Parasitology* 59, 215–231.

Mapes, C. J. (1970). The development of *Haemonchus contortus in vitro*. II. The effect of disulfide-reducing and sulfhydryl-blocking reagents on the rate of development to the fourth-stage larvae. *Parasitology* 60, 123–135.

Mapes, C. J. (1972). Bile and bile salts and exsheathment of the intestinal nematodes *Trichostrongylus colubriformis* and *Nematodirus battus*. *International Journal for Parasitology* 2, 433–438.

McClelland, G. and Ronald, K. (1974a). *In vitro* development of the nematode *Contracaecum osculatum* Rudolphi 1802 (Nematoda: Anisakinae). *Canadian Journal of Zoology* 52, 847–855.

McClelland, G. and Ronald, K. (1974b). *In vitro* development of *Terranova decipiens* (Nematoda) Krable 1878. *Canadian Journal of Zoology* 52, 471–479.

Meerovitch, E. (1970). Cultivation of *Trichinella spiralis in vitro*. *In* "Trichinosis in Man and Animals" (S. E. Gould, ed) pp. 102–108. Charles C. Thomas, Springfield, Ill.

Meza-Ruiz, G. and Alger, N. E. (1968). First parasitic ecdysis of *Haemonchus contortus in vitro* without stimulation by carbon dioxide. *Experimental Parasitology* 22, 219–222.

Moreau, J. P. and Lagraulet, J. (1972). Survie *in vitro* des larves de troisième stade d'*Angiostrongylus cantonensis*. *Annales de Parasitologie Paris* 47, 525–529.

Mráček, Z. (1977). *Steinernema kraussei*, a parasite of the body cavity of the sawfly, *Cephaleia abietis*, in Czechoslovakia. *Journal of Invertebrate Pathology* 30, 87–94.

Muria, R. (1975). Effects of temperature on hatching and development of *Nematospiroides dubius* in aerated water. *Journal of Helminthology* 49, 293–296.

Myers, B. J. (1975). The nematodes that cause Anisakiasis. *Journal of Milk, Food Technology* 38, 774–782.

Myers, R. F. (1967). Axenic cultivation of *Aphelenchoides sacchari* Hooper. *Proceedings of the Helminthological Society of Washington* 34, 251–255.

Myers, R. F. (1968). Nutrient media for plant parasitic nematodes: I. Axenic cultivation of *Aphelenchoides* sp. *Experimental Parasitology* 23, 96–103.

Myers, R. F. (1972). Assay of nematicidal and nematistatic chemicals using axenic cultures of *Aphelenchoides rutgersi*. *Nematologica* 18, 447–457.

Myers, R. F. and Balasubramanian, M. (1973). Nutrient media for plant parasitic nematodes. V. Nutrition of *Aphelenchoides rutgersi*. *Experimental Parasitology* 34, 123–133.

Myers, R. F., Buecher, E. J. and Hansen, E. L. (1971). Oligidic medium for axenic culture of *Aphelenchoides* sp. *Journal of Nematology* 3, 197–198.

Neilson, J. T. McL. (1969). Gel filtration and disc electrophoresis of a somatic extract and excretions and secretions of *Haemonchus contortus* larvae. *Experimental Parasitology* 25, 131–141.

Neilson, J. T. McL. (1972). Isolation of metabolic antigens of *Oesophagostomum columbianum* following *in vitro* cultivation in a simplified medium. *Journal of Parasitology* 58, 555–562.

Neilson, J. T. McL. (1975). Failure to vaccinate lambs against *Haemonchus contortus* with functional metabolic antigens identified by electrophoresis. *International Journal of Parasitology* 5, 427–430.

Nicholas, W. L. (1975). "The Biology of Free-living Nematodes". Clarendon Press, Oxford.

Obeck, D. K. (1973). Blood microfilariae: New and existing techniques for isolation. *Journal of Parasitology* 59, 220–221.

Olson, L. J. and Jones, F. R. (1974). Preparation of sterile *Toxocara canis* larvae. *Journal of Parasitology* 60, 941.

Ortiz-Valqui, R. E. and Lumbreras-Cruz, H. (1970). Larval hatching of eggs of *Trichuris trichura in vitro. Congresso Latin-american Parasitologica* (2nd), Mexico, p. 17.

Ozerol, N. H. and Silverman, P. H. (1972). Enzymatic studies on the exsheathment of *Haemonchus contortus* infective larvae. The role of leucine aminopeptidase. *Comparative Biochemistry and Physiology* **42B**, 109–121.

Pandey, V. S. (1972). Effect of temperature on development of the free-living stages of *Ostertagia ostertagi. Journal of Parasitology* **58**, 1037–1041.

Parker, S. and Croll, N. A. (1976). *Dictyocaulus viviparus*: The role of pepsin in the exsheathment of infective larvae. *Experimental Parasitology* **40**, 80–85.

Petersen, J. J. (1973). Role of mermithid nematodes in biological control of mosquitos. *Experimental Parasitology* **33**, 239–247.

Phillipson, R. F. (1973). Extrinsic factors affecting the reproduction of *Nippostrongylus brasiliensis. Parasitology* **66**, 405–413.

Platzer, E. G. (1978). Culture media for nematodes. *In* "CRC Handbook of Nutrition and Food" (M. Rechcigl ed). CRC Press, Cleveland, Ohio (in press).

Poinar, G. O. Jr. (1975). "Entomogenous Nematodes. A Manual and Host List of Insect-Nematode Associations" (317 pp.) pp. 22–28. E. J. Brill, Leiden, Netherlands.

Poinar, G. O. and Hansen, E. L. (1972). Clarification of the status of the DD-136 strain of *Neoaplectana carpocapsae* Weiser. *Nematologica* **18**, 288–290.

Poinar, G. O. and Hess, R. (1976). Uptake of ferritin particles through the body wall of a mermithid nematode. *International Research Communication System: Medical Science* **4**, 296.

Poinar, G. O. Jr. and Hess, R. (1977). *Romanomermis culicivorax*: Morphological evidence of transcuticular uptake. *Experimental Parasitology* **42**, 27–33.

Rajulu, G. S., Kulasekarapandian, S. and Krishnan, N. (1972). Nature of the hormone from a nematode *Phocanema depressum* Baylis. *Current Science* **41**, 67–68.

Ridley, R. K. and Leland, S. E. Jr. (1973). Efficacy of commercial bovine serum as a component of media used to grow *Cooperia punctata in vitro. Journal of Parasitology* **59**, 277–281.

Roche, M., Fecht, B. and Giménez, A. (1971). *Ancylostoma caninum*: Pharyngeal activity *in vitro. Experimental Parasitology* **29**, 417–422.

Rogers, W. P. and Brooks, F. (1976). Zinc as a co-factor for an enzyme involved in exsheathment of *Haemonchus contortus. International Journal for Parasitology* **6**, 315–319.

Rogers, W. P. and Head, R. (1972). The effect of the stimulus for infection on hormones in *Haemonchus contortus. Comparative and General Pharmacology* **3**, 6–10.

Rogers, W. P. and Sommerville, R. J. (1968). The infectious process, and its relation to the development of early parasitic stages of nematodes. *In* "Advances in Parasitology" (B. Dawes, ed), Vol. 6, pp. 327–348. Academic Press, London and New York.

Rose, J. H. (1973). The development *in vitro* of some gastro-intestinal nematodes of sheep, cattle and pigs. *Research in Veterinary Science* **14**, 326–333.

Rose, J. H. (1976). Preliminary results using metabolites and *in vitro* grown larvae of *Ostertagia circumcincta* to immunise lambs against oral challenge. *Research in Veterinary Science* **21**, 76–78.

Rothwell, T. L. W. and Love, R. J. (1974). Vaccination against the nematode *Trichostrongylus colubriformis*. I. Vaccination of guinea-pigs with worm homogenates and soluble products released during *in vitro* maintenance. *International Journal for Parasitology* **4**, 293–299.

Rutherford, T. A. and Webster, J. M. (1974). Transcuticular uptake of glucose by the entomophilic nematode, *Mermis nigrescens. Journal of Parasitology* **60**, 804–808.

Sanders, R. D., Stokstad, E. L. R. and Malatesta, C. (1973). Axenic growth of *Reesimermis nielseni* (Nematoda: Mermithidae) in insect tissue culture media. *Nematologica* **19**, 567–568.

Savel, H., Kim, C. W. and Hamilton, L. D. (1969). Synthesis of radioactive *Trinchiella spiralis* larval antigen *in vitro*. *Experimental Parasitology* **24**, 171–175.

de Savigny, D. H. (1975). *In vitro* maintenance of *Toxacara canis* larvae and a simple method for the production of *Toxacara* ES antigens for use in serodiagnostic tests for visceral larva migrans. *Journal of Parasitology* **61**, 781–782.

Sayre, F. W., Hansen, E. L. and Yarwood, E. A. (1963). Biochemical aspects of the nutrition of *Caenorhabditis briggsae*. *Experimental Parasitology* **13**, 98–107.

Schulz, H. P. (1967). Versuche zur Kultivierung der parasitischen Larvenstadien von *Haemonchus contortus in vitro*. *Berliner und Münchener Tierärztliche Wochenschrift* **80**, 89–96.

Schulz, H. P. (1974). *Anisakis* larvae from the herring: *in vitro* development and morphology. *Third International Congress of Parasitology* **3**, 1627.

Schulz, H. P. and Dalchow, W. (1967). Versuche zur Kultivierung dir parasitischen Larvenstadien von *Chabertia ovina* (Fabricus, 1788) *in vitro*. *Berliner und Münchener Tierärztliche Wochenschrift* **80**, 410–415.

Schulz, H. P. and Dalchow, W. (1969). Kultivierung der parasitischen Larvenstadien von *Oesophagostomum quadrispinulatum* (Marcone, 1901) *in vitro*. *Berliner und Münchener Tierärztliche Wochenschrift* **82**, 143–147.

Scott, H. L. and Whittaker, F. H. (1970). *Pelodera strongyloides* Schneider 1866: a potential research tool. *Journal of Nematology* **2**, 193–203.

Shanta, C. S. and Meerovitch, E. (1970). Specific inhibition of morphogenesis in *Trichinella spiralis* by insect juvenile hormone mimics. *Canadian Journal of Zoology* **48**, 616–620.

Shepperson, J. R. and Jordan, W. C. (1974). Observations on *in vitro* survival and development of Meloidogyne. *Proceedings of the Helminthological Society of Washington* **41**, 254.

Shishova, K. O. A., Mazhuga, N. H. and Sokhina, L. I. (1973). *In vitro* uptake of proteins of different structure and biological value by *Ascaridia galli*. *Trudy Gel'mintologicheskoi Laboratorii (Ekologiya i taksonomiya gel'mintov)* **23**, 211–217.

Silverman, P. H. and Hansen, E. L. (1971). *In vitro* cultivation procedures for parasitic helminths: Recent advances. *In* "Advances in Parasitology" (B. Dawes, ed), Vol. 9, pp. 227–258. Academic Press, London and New York.

Slocombe, J. O. D. (1974). Some analyses of exsheathing fluid from infective *Haemonchus contortus* larvae from Ontario. *International Journal for Parasitology* **4**, 397–402.

Slocombe, J. O. D. and Whitlock, J. H. (1969). Rapid ecdysis of infective *Haemonchus contortus cayugensis* larvae. *Journal of Parasitology* **55**, 1102–1103.

Slonka, G. F. and Leland, S. E. (1970). *In vitro* cultivation of *Ancylostoma tubaeforme* from egg to fourth-stage larva. *American Journal of Veterinary Research* **31**, 1901–1904.

Slonka, G. F., Ridley, R. K. and Leland, S. E., Jr. (1973). The use of *in vitro*-grown *Cooperia punctata* (Nematoda: Trichostrongyloidea) to study incorporation of carbon from D-glucose-U-^{14}C into major chemical fractions. *Journal of Parasitology* **59**, 282–288.

Solomon, G. B. and Soulsby, E. J. L. (1973). *Capillaria hepatica* (Bancroft, 1893): *in vitro* hatching. *Proceedings of the Helminthological Society of Washington* **40**, 159–160.

Sommerville, R. I. (1976). Influence of potassium ion and osmotic pressure on development of *Haemonchus contortus in vitro*. *Journal of Parasitology* **62**, 242–246.

Sommerville, R. I. and Davey, K. G. (1976). Stimuli for cuticle formation and ecdysis *in vitro* of the infective larva of *Anisakis* sp. (Nematoda: Ascaridoidea). *International Journal for Parasitology* **6**, 433–439.

Sommerville, R. I. and Weinstein, P. P. (1967). The *in vitro* cultivation of *Nippostrongylus brasiliensis* from the late fourth stage. *Journal of Parasitology* **53**, 116–125.

Soroczan, W. and Krauze, M. (1973). (The effect of pH and mineral salt concentration of the medium under laboratory conditions on the survival of the roundworm *Rhabditis strongyloides* Schneider, 1886.) *Annales Universitatis Mariae Curie-Sklodowska, D.D. Medicina Veterinaria* **28**, 207–214.

Stoll, N. R. (1973). Rudolf William Glasser and *Neoaplectana*. *Experimental Parasitology* **33**, 189–196.

Sylk, S. R., Stromberg, B. E. and Soulsby, E. J. L. (1974). Development of *Ascaris suum* larvae from the third to fourth stage *in vitro*. *International Journal for Parasitology* **4**, 261–265.

Taira, N. (1975). [Effects of KCl on the growth of the preinfective stage of *Dictyocaulus viviparus* larvae.] *Bulletin of the National Institute of Animal Health Tokyo* **71**, 44–46 (in Japanese).

Tarakanov, V. I. (1971). *In vivo* and *in vitro* development of *Trichinella spiralis.* *Trudy Vsesoyuznogo Instituta Gel'mintologii im K. I. Skryabina* **18**, 255–264 (in Russian).

Tarakanov, V. I. (1973). Axenic culture of *Metastrongylus elongatus* infective larva to 5th stage. *In* "Problemy Obschei I Prikladnoi Gel'mintologii" (V. G. Gagarin ed), pp. 145–150. Izdatel'stvo "Nauka", Moscow, USSR.

Tarakanov, V. I. and Andreeva, G. N. (1974). Axenic culture of three nematode species of the *Neoaplectana* genus. *Third International Congress of Parasitology* **1**, 466–467.

Tarakanov, V. I. and Kaarma, A. I. (1972). The growth and development of *Oesphagostomum dentatum* larvae under axenic conditions. *Trudy Vsesoyuznogo Instituta Gel'mintologii im K. I. Skryabina* **19**, 203–212.

Taylor, A. E. R. and Baker, J. R. (1968). "The Cultivation of Parasites *in vitro*". Blackwell Scientific Publications, Oxford.

Thirugnanam, M. (1976). *Aphelenchoides rutgersi*: Growth requirements for heme and purine nucleotides in crude chick embryo extract. *Experimental Parasitology* **40**, 149–157.

Timofeev, B. A. (1969). [Multiplication of parasitic microfilaria in arthropod tissue culture.] *Meditsinskaya Parazitologiya I Parazitaryne Bolenzi* (USSR) **38**, 489–491 (in Russian).

Townsley, P. M., Wight, H. G., Scott, M. A. and Hughes, M. L. (1963). The *in vitro* maturation of the parasitic nematode *Terranova decipiens* from cod muscle. *Journal of the Fisheries Research Board of Canada* **20**, 743–747.

Tromba, F. G. and Douvres, F. W. (1969). Survival of juvenile and adult *Stephanurus dentatus in vitro*. *Journal of Parasitology* **55**, 1050–1054.

Van Banning, P. (1971). Some notes on a successful rearing of the herring-worm, *Anisakis marina* L. (Nematoda: Heterocheilidae). *Journal du Conseil Permenant International Pour L'Exploration de la Mer* (Denmark) **34**, 84–88.

Wang, G. T. (1971). *Haemonchus contortus*: Food of preinfective larvae. *Experimental Parasitology* **29**, 201–207.

Weinstein, P. P. (1970). A review of the culture of filarial worms. *In* "H. D. Srivastava Commemoration Volume" (K. S. Singh and B. K. Tandan, eds), pp. 493–505. Izatnagar, Uttar Pradesh, India.

Weinstein, P. P. (1976). Personal communication.

Weinstein, P. P. and Jones, M. F. (1959). Development *in vitro* of some parasitic nematodes of vertebrates. *Annals of the New York Academy of Sciences* **77**, 137–162.

Weinstein, P. P., Newton, W. L., Sawyer, T. K. and Sommerville, R. J. (1969). *Nematospiroides dubius*: Development and passage in the germ free mouse and a comparative study of the free living stages in germ free feces and conventional cultures. *Transactions of the American Microscopical Society* **88**, 95–117.

Weiser, J. (1966). "Nemoci Hmyzu". Academia, Prague.
Weiser, J. (1976). *Steinernema krausei* as an insect pathogen. *In* "Proceedings of the first International Colloquium on Invertebrate Pathology", pp. 245–249. Queens University, Kingston, Canada.
Wilson, R. J. M. (1967). Homocytotropic antibody response to the nematode *Nippostrongylus brasiliensis* in the rat—Studies on the worm antigen. *Journal of Parasitology* 53, 752–762.
Wong, H. A. and Fernando, M. A. (1970). *Ancyclostoma caninum*: Uptake of ^{14}C-glucose *in vitro*. *Experimental Parasitology* 28, 253–257.
Yarwood, E. A. and Hansen, E. L. (1968). Axenic culture of *Pelodera strongyloides* Schneider. *Journal of Parasitology* 54, 133–136.
Yasuraoka, K. and Weinstein, P. P. (1969). Effects of temperature on the development of eggs of *Nematospiroides dubius* under axenic conditions relative to *in vitro* cultivation. *Journal of Parasitology* 55, 44–50.
Zimmerman, G. L. and Leland, S. E., Jr. (1971). Completion of the life cycle of *Cooperia punctata in vitro*. *Journal of Parasitology* 57, 832–835.
Zimmerman, G. L. and Leland, S. E., Jr. (1974). Isolation and partial characterization of exo- and endoantigens of *Cooperia punctata* cultured *in vitro*. *Journal of Parasitology* 60, 794–803.
Zuckerman, B. M. (1971). Gnotobiology. *In* "Plant Parasitic Nematodes" (B. M. Zuckerman, W. F. Mai and R. A. Rohde, eds), Vol. 2, pp. 159–184. Academic Press, New York and London.

Chapter 11

Acanthocephala

ANN M. LACKIE

Zoology Department, The University, Glasgow, Scotland

The Acanthocephala pose interesting problems with respect to *in vitro* culture since, apart from the egg or "shelled acanthor" as it is more properly called (Crompton, 1970), all stages of the life-cycle are parasitic. In addition, these parasites possess no gut and rely on uptake of nutrients through the body wall.

Depending on the stage of the life-cycle required for study, it may be necessary to hatch or activate dormant stages. Edmonds (1966) produced a method for hatching "eggs" of *Moniliformis dubius* to release acanthors, and cystacanths of *M. dubius* and *Polymorphus minutus* can be activated by the methods of Graff and Kitzman (1965) and Lackie (1974) respectively.

The importance of the correct osmotic pressure in activation and culture media was pointed out by Van Cleave and Ross (1944) for *Neoechinorhyncus emydis*, Branch (1970) for *M. dubius* and Lackie (1974) for *P. minutus*. Detailed reviews on physicochemical conditions in the alimentary tracts of birds (Crompton and Nesheim, 1976) and mammals (Mettrick and Podesta, 1974) and the haemocoeles of *Gammarus pulex* and *Periplaneta americana* (Crompton, 1970) exist, and should be useful to workers attempting to set up new methods of long-term culture. Nevertheless, few advances in culture techniques for adult Acanthocephala have occurred since the review by Taylor and Baker (1968). There are, however, many examples of short-term maintenance of parasites (e.g., up to 4 h) in simple balanced salt solutions (BSS), often for metabolic investigations. The choice of BSS appears to be fairly arbitrary and examples include: various modifications of Krebs–Ringer solution for *M. dubius* (Laurie, 1959; Beames and Fisher, 1964; Graff, 1964; Bryant and Nicholas, 1966); sodium chloride solution, 0.9% for *N. emydis* (Van Cleave and Ross, 1944); Hanks's saline and 8% carbon dioxide in nitrogen for *Macracanthorhyncus hirudinaceus* (Branch, 1970) and amphibian Ringer solution for *Acanthocephalus ranae* (Hammond, 1968). Ward (1952) kept *M. hirudinaceus* in Ringer–Tyrode solution with penicillin and streptomycin at 37°C for 43 h. In general, such simple media are suitable for short-term cultivation, but during experiments on glucose uptake with adult *P. minutus*, Crompton and Ward (1967) and Crompton and Lockwood (1968) found that the worms would stay alive and moving for up to 19 d if they were incubated in Hanks's solution containing 1 mg ml^{-1} penicillin and streptomycin and at least 2 mg ml^{-1} glucose, at pH 7·7 and 41–42°C. Worms were shown to be still viable by transplantation into the small intestine of ducks at the end of the experiments on glucose uptake.

Methods for long-term cultivation of adult Acanthocephala in complex media have hardly changed since 1967. Nicholas and Grigg's method (1965) still remains the most important for cultivation of adult *M. dubius* and is described again here. Adult *M. dubius* developed sufficiently during the 8-day cultivation period so that the female worms produced "eggs" *in vitro*; however, immature "eggs" were also released, suggesting that the uterine bell, which is responsible for sorting out and withholding immature "eggs" (Whitfield, 1969), was functioning incorrectly.

It would seem that worms with poikilothermic definitive hosts can be cultivated *in vitro* for longer periods than can worms from homeotherms. *P. minutus* (Crompton and Ward, 1967) and *M. dubius* (Nicholas and Grigg, 1965) became moribund after 19 d and 8 d *in vitro* respectively, but *N. emydis* from turtles (Dunagan, 1962) and *Pomphorhyncus bulbocolli* from freshwater fish (Jensen, 1952) remained alive for 96 d and 72 d respectively. The reader is referred to the discussion in Taylor and Baker (1968) and to the original papers for methods of cultivating adult *N. emydis* from turtles using M199 or Eagle's HeLa medium with turtle serum (Dunagan, 1962), and adult *Octospinifer macilentus* from fish (Harms, 1965). Jensen's (1952) method for the cultivation *in vitro* of *P. bulbocolli* from a late larval stage to a prepatent adult is described below. He tried several different media and found that in the most successful ones, the acanthellae, obtained from the crustacean *Hyalella azteci*, grew from 3·5 mm up to 10·0 mm in length. Ovarian balls developed in female worms after 7 weeks, but although the worms were very active and the males were regularly observed to evert their copulatory bursae, copulation was never observed and fertilized eggs were not produced.

One possibility which does not seem to have been explored is the use of diphasic media: it is conceivable that copulation is made difficult by the absence of a substrate to which the worms could attach, and this could easily be tested by the provision of agar slopes within the culture flasks. A point which Jensen (1952) omits to mention is the activation of the cystacanth. In most Acanthocephala the last larval stage is semi-dormant with an invaginated proboscis, and physicochemical conditions in the gut of the final host stimulate proboscis eversion and development of the larva to the adult form. Presumably such an event must have occurred during Jensen's cultivation period without involvement of the kinds of stimuli usually associated with activation of worms in homeothermic hosts (A. M. Lackie, 1975).

Long-term cultivation of young larvae of *M. dubius* was attempted by J. M. Lackie (1975) using a modification of the medium developed by Landureau and Jollès (1969) for culture of embryonic cells of the cockroach *Periplaneta americana* (see Section 3°). All cultures were incubated at 24°C or 28°C, a temperature range more suitable for the survival of both cockroaches and parasite larvae; higher temperatures, such as 37°C used by Horvath (1971) in his short-term studies on cystacanths *in vitro*, are inconsistent with the temperatures to which the larvae are exposed in their intermediate hosts. Lackie and Rotheram's (1972) attempt at cultivating hatched acanthors was not very successful, for although the larvae remained alive and showed some movement over 19 d, no visible growth occurred. A halo of membranous streamers developed around the larvae, similar to the microvillar coat they produce *in vivo*. Greater success was obtained with Stage I acanthellae dissected from the cockroach haemocoele (J. M. Lackie, 1975), which developed to Stage V acanthellae within 25 d at 28°C. However, further development to the cystacanth stage did not occur, there being an apparent block to development at the stage of proboscis retraction.

In conclusion, it would seem that few successful methods have yet been found for long-term cultivation of Acanthocephala *in vitro*; most of the methods mentioned above do little more than provide a technique whereby Acanthocephala may be maintained in a healthy state *in vitro* for short periods only. Also the relatively slow changes in size and development of worms may mask the natural variations found within a sample of worms of the same age—more information on the growth of worms *in vivo* may yet be required before useful knowledge can be gained from attempts to cultivate Acanthocephala *in vitro*.

1° Adult *Moniliformis dubius*
(Nicholas and Grigg, 1965)

STERILIZATION AND MEDIUM

(i) *Sterilization.* Worms, obtained from the small intestine of rats 6–12 weeks post-infection, are incubated at 37–39°C for 24 h in sterilizing solution [Hanks's (Chapter 1: IV, B, 1, g) saline with 5 mg penicillin, 5 mg streptomycin, 2 mg nystatin and 7 mg crude tetracycline litre^{-1} gassed with

5% carbon dioxide in nitrogen] and then rinsed in sterile Ringer's (presumably Ringer–Locke, see Chapter 1: IV, B, 1, d) solution.

(ii) *Medium*. Eagle's (Taylor and Baker, 1968, p. 355).

TECHNIQUE

Worms are transferred, 1 worm per flask, to 160 ml square flasks containing 11 ml of Eagle's medium. Flasks are then gassed with 5% CO_2 in N_2, sealed with screw caps and incubated at 37–39°C in the horizontal position. Worms are transferred, using a plantinum wire hook, to new flasks containing fresh medium daily, and re-gassed as above.

The worms survived, producing viable "eggs", throughout the 8 d experimental period but as some eggs were immature, the uterine bell may not have been functioning properly.

2° *Pomphorhyncus bulbocolli*
(Jensen, 1952)

STERILIZATION AND MEDIUM

(i) *Sterilization*. *Hyalella azteci* are placed in 1:1000 aqueous "Zephiran chloride" (dilute bleach solution, supplier not given) for 5 min, then are washed in sterile water and dissected under sterile saline. The acanthellae are allowed to settle and are washed 4 times in sterile saline (Chapter 1: IV, B, a).

(ii) *Media*. "No. 7": 3 volumes of peptone broth, 1 volume of horse serum, plus 1% glucose.

"No. 8": 3 volumes of peptone broth, 1 volume of horse serum, 1 volume of yeast extract, plus 1% glucose.

Both media also contain 100 iu crystalline penicillin-G and 50 µg streptomycin sulphate ml^{-1}.

TECHNIQUE

1–3 larvae are incubated in 4 ml of either medium contained in 8 ml test tubes with rubber stoppers, at 20–22°C; the medium is changed daily.

Maximum growth of acanthellae was 3·5–10·0 mm length, ovarian balls developed in female worms after 7 weeks and males regularly everted their copulatory bursae (medium 7 or 8). However, copulation was not seen and fertilized eggs were not produced.

3° *Moniliformis dubius larvae*
(Lackie and Rotheram, 1972; J. M. Lackie, 1975)

STERILIZATION

(i) *Acanthors*: adult worms are removed from the rat, surface-sterilized in 70% ethanol and macerated in sterile 0·9% saline. Eggs and ovarian balls are surface sterilized in 0·5% "Chloros" (Imperial Chemical Industries) for 5 min.

(ii) *Acanthellae*. Cockroaches are surface-sterilized in 1% "Chloros" for 1–2 min and dissected under sterile 0·9% saline, without rupturing the gut. The larvae are washed in sterile culture medium, allowed to settle, then transferred to fresh sterile culture medium and rewashed; this is repeated 10 times.

MEDIUM

(J. M. Lackie, 1973)

	mg litre^{-1}		mg litre^{-1}
L-Aspartic acid	250	$CaCl_2$	490
L-Glutamic acid	1500	KCl	1050
α-Alanine	120	$MgSO_4.7H_2O$	1260
L-Arginine HCl	800	$MnSO_4.H_2O$	65
L-Cysteine HCl	260	NaCl	8500
L-Glutamine	300	$NaHCO_3$	360
Glycine	750	H_3PO_3	900
L-Histidine	300		
L-Isoleucine	120	Glucose	3000
L-Leucine	250		
L-Lysine HCl	160	Folic acid	0·01
L-Methionine	500	*d*-biotin	0·01
L-Phenylalanine	200	Choline HCl	0·40
L-Proline	750	Inositol	0·05
L-Serine	80	Nicotinamide	0·03
L-Threonine	200	Pantothenate (Ca salt)	0·10
L-Tryptophan	200		
L-Tyrosine	180	Pyridoxine-HCl	0·03
L-Valine	150	Riboflavin	0·05
		Thiamine-HCl	0·01
Penicillin	50	Vitamin B_{12}	0·02
Streptomycin	70		

10% calf serum (inactivated, 56°C; 30 min)

TECHNIQUE

Larvae are placed in 3 ml of medium per Falcon flask (25 cm^2) and incubated at 24°C (acanthors) and 28°C (acanthellae).

Acanthors survived 19 d, stage I acanthellae developed to stage V in 25 d at 28°C.

ACKNOWLEDGEMENT

I am extremely grateful to Dr. D. W. T. Crompton for his advice and comments on this chapter.

REFERENCES

Beames, C. G. and Fisher, F. M. (1964). A study on the neutral lipids and phospholipids of the Acanthocephala *Macracanthorhyncus hirudinaceus* and *Moniliformis dubius*. *Comparative Biochemistry and Physiology* 13, 401–412.

Branch, S. I. (1970). *Moniliformis dubius* and *Macracanthorhyncus hirudinaceus*: Na, K, Ca and Mg content and Na and K active transport. *Experimental Parasitology* 27, 33–43.

Bryant, C. and Nicholas, W. L. (1966). Studies on the oxidative metabolism of *Moniliformis dubius* (Acanthocephala). *Comparative Biochemistry and Physiology* 17, 825–840.

Crompton, D. W. T. (1970). "An Ecological Approach to Acanthocephalan Physiology". Cambridge University Press.

Crompton, D. W. T. and Lockwood, A. P. M. (1968). Studies on the absorption and metabolism of D-(μ-^{14}C)-glucose by *Polymorphus minutus* (Acanthocephala) *in vitro*. *Journal of Experimental Biology* 48, 411–425.

Crompton, D. W. T. and Nesheim, M. C. (1976). Host-parasite relationships in the alimentary tract of domestic birds. *Advances in Parasitology* 14, 96–194.

Crompton, D. W. T. and Ward, P. V. (1967). Lactic and succinic acids as excretory products of *Polymorphus minutus in vitro*. *Journal of Experimental Biology* 46, 423–430.

Dunagan, T. T. (1962). Studies on *in vitro* survival of Acanthocephala. *Proceedings of the Helminthological Society of Washington* 29, 131–5.

Edmonds, S. J. (1966). Hatching of eggs of *Moniliformis dubius*. *Experimental Parasitology* 19, 216–226.

Graff, D. J. (1964). Metabolism of C^{14}-glucose by *Moniliformis dubius* (Acanthocephala). *Journal of Parasitology* 50, 230–234.

Graff, D. J. and Kitzman, W. B. (1965). Factors influencing the activation of acanthocephalan cystacanths. *Journal of Parasitology* 51, 424–429.

Hammond, R. A. (1968). Some observations on the role of the body wall of *Acanthocephalus ranae* in lipid uptake. *Journal of Experimental Biology* 48, 217–225.

Harms, C. E. (1965). The life cycle and larval development of *Octospinifer macilentus* (Acanthocephala: Neoechinorhyncidae). *Journal of Parasitology* 51, 286–293.

Horvath, K. (1971). Glycogen metabolism in larval *Moniliformis*. *Journal of Parasitology* 57, 132–136.

Jensen, T. (1952). The life-cycle of the fish Acanthocephalan *Pomphorhyncus bulbocolli* (Linkins) van Cleave, 1919, with some observations on larval development *in vitro*. Ph.D. dissertation, University of Minnesota.

Lackie, A. M. (1974). The activation of cystacanths of *Polymorphus minutus* (Acanthocephala) *in vitro*. *Parasitology* 68, 135–146.

Lackie, A. M. (1975). The activation of infective stages of endoparasites of vertebrates. *Biological Reviews* 50, 285–323.

Lackie, J. M. (1973). *Moniliformis* and its intermediate hosts. Ph.D. dissertation, University of Cambridge.

Lackie, J. M. (1975). The host-specificity of *Moniliformis dubius* (Acanthocephala), a parasite of cockroaches. *International Journal of Parasitology* 5, 301–307.

Lackie, J. M. and Rotheram, S. (1972). Observations on the envelope surrounding *Moniliformis dubius* (Acanthocephala) in the intermediate host, *Periplaneta americana*. *Parasitology* 65, 303–308.

Landureau, J. C. and Jollès, P. (1969). Etude sur des exigences d'une lignée de cellules d'insectes (souche EPa). I. acides amines. *Experimental Cell Research* 54, 391–398.

Laurie, J. S. (1959). Aerobic metabolism of *Moniliformis dubius* (Acanthocephala). *Experimental Parasitology* 8, 188–197.

Mettrick, D. F. and Podesta, R. B. (1974). Ecological and physiological aspects of helminth-host interactions in the mammalian gastrointestinal tract. *Advances in Parasitology* 12, 183–279.

Nicholas, W. L. and Grigg, H. (1965). The *in vitro* culture of *Moniliformis dubius* (Acanthocephala). *Experimental Parasitology* 16, 332–340.

Taylor, A. E. R. and Baker, J. R. (1968). "The Cultivation of Parasites *in Vitro*". Blackwell Scientific Publications, Oxford and Edinburgh.

Van Cleave, H. J. and Ross, E. L. (1944). Physiological responses of *Neoechinorhyncus emydis* (Acanthocephala) to various solutions. *Journal of Parasitology* 30, 369–372.

Ward, H. L. (1952). Glycogen consumption in Acanthocephala under aerobic and anaerobic conditions. *Journal of Parasitology* 38, 493–494.

Whitfield, P. J. (1969). Studies on the reproduction of Acanthocephala. Ph.D. dissertation, University of Cambridge.

Appendix
NAMES AND ADDRESSES OF SUPPLIERS
MENTIONED IN THE TEXT
(All addresses are in England unless shown otherwise)

ABBOTT LABORATORIES LTD, Queenborough, Kent.

ALBRIGHT AND WILSON LTD, Detergents and Chemicals Group, P.O. Box 3, Hagley Road West, Oldbury, Warley, West Midlands, B69 4LN.

ALCONOX LTD, 215 Park Avenue South, New York, New York, 10003, USA (British Agency: Piper Chemicals).

AQUARIUM SYSTEMS INC., East Lake, Ohio, USA.

ARMOUR PHARMACEUTICAL CO. LTD, Hampden Park, Eastbourne, E. Sussex.

C. ASH SON AND CO., Amalco House, 26–40 Broadwick Street, London W.1.

BAIRD AND TATLOCK LTD, P.O. Box 1, Romford, Essex, RM1 1 HA.

BALTIMORE BIOLOGICAL LABORATORIES (BBL), Division of Bioquest, P.O. Box 175, Cockeysville, Maryland, 21030, USA (British Agency: Becton, Dickinson UK Ltd).

ÉTS BEAUDOUIN, 1 et 3 Rue Rataird, Paris Ve, France (British Agency: Camlab Ltd).

BECTON, DICKINSON UK LTD, York House, Empire Way, Wembley, Middlesex, HA9 0PS.

BECTON, DICKINSON AND CO. (CANADA) LTD, 2464 South Sheridan Way, Clarkson, Ontario, Canada.

BIO-RAD LABORATORIES, Richmond, California, 94804, USA.

BRINKMAN, Cantiague Road, Westbury, New York, 11590, USA (British Agency: Camlab Ltd).

BRITISH DRUG HOUSES (BDH CHEMICALS LTD), Broom Road, Poole, Dorset, BH12 4NN.

BRITISH OXYGEN CO. (BOC), Deer Park Road, London S.W.19.

BUCKLEY MEMBRANES LTD, 24 Clifton Road, Chesham Bois, Amersham, Buckinghamshire.

CALBIOCHEM, 10933 North Torrey Pines Road, La Jolla, California, 92037, USA.

CALBIOCHEM LTD, Thorpe House, King Street, Hereford, HR4 9BQ.

CAMLAB LTD, Nuffield Road, Cambridge, CB4 1TH.

CHEMICAL CONCENTRATES LTD, 39 Webb's Road, London, SW11 6RX.

CORNING LTD, Laboratory Division, Stone, Staffordshire, ST15 0BG.

CURFEW APPLIANCES LTD, Ottershaw, Chertsey, Surrey.

DIFCO LABORATORIES, P.O. Box 1058-A, Detroit, Michigan, 48232, USA.

DIFCO LABORATORIES LTD, P.O. Box 14B, Central Avenue, East Molesey, Surrey, KT8 0SE.

DOW CORNING CORP., Midland, Michigan, 48640, USA (British Agency: Hopkin and Williams).

DURHAM CHEMICALS DISTRIBUTORS LTD, Birtley, County Durham.

ESCO (RUBBER) LTD, 14–16 Great Portland Street, London, W1N 5AB.

EUROLAB (UK) LTD, Loomer Road, Chesterton, Newcastle-under-Lyme, Staffordshire, ST5 7PQ.

FALCON PLASTICS, Los Angeles, California, USA (British Agency: Gibco Bio-Cult Ltd).

FISHER SCIENTIFIC CO., 711 Forbes Avenue, Pittsburgh, Pennsylvania, 15219, USA.

FISONS LTD (MSE SCIENTIFIC INSTRUMENTS), Manor Royal, Crawley, Sussex, RH10 2QQ.

FLOW LABORATORIES LTD, Victoria Park, Heatherhouse Road, Irvine, KA12 8NB, Scotland.

FLOW LABORATORIES INC., Rockville, Maryland, USA.

GALLENKAMP LTD, P.O. Box 290, Technico House, Christopher Street, London, EC2P 2ER.

GENERAL ELECTRIC CORP., Pleasanton, California, 94566, USA (British Agency for Nuclepore membranes: Buckley Membranes).

GIBCO BIO-CULT LTD (AND GIBCO BIO-CULT DIAGNOSTICS LTD), 3 Washington Road, Paisley, PA3 4EP, Scotland.

GIBCO, 3175 Staley Road, P.O. Box 68, Grand Island, New York, 14072, USA.

GIBCO DIAGNOSTICS, P.O. Box 4385, 2801 Industrial Drive, Madison, Wisconsin, 53711, USA.

GIBCO DIAGNOSTICS, 2392 Industrial Street, Burlington, Ontario, L7P 1A5, Canada.

GILLETTE SURGICAL, Great West Road, Isleworth, Middlesex.

GLAXO LTD, Greenford, Middlesex.

GRAND ISLAND BIOLOGICAL CO.—*see* Gibco.

HARRISONS AND CROSSFIELD (AMERICA) INC., P.O. Box 39, 1 Stone Place, Bronxville, New York, 10708, USA.

HOPKIN AND WILLIAMS, P.O. Box 1, Romford, Essex, RM1 1HA.

ARNOLD R. HORWELL LTD, 2 Grangeway, Kilburn High Road, London, NW6 2BP.

V. A. HOWE AND CO. LTD, 88 Peterborough Road, London, SW6 3EP.

HUMKO-SHEFFIELD CO., Norwich, New York, 13815, USA.

IMPERIAL CHEMICAL INDUSTRIES LTD (ICI), Pharmaceuticals Division, Alderley Park, Macclesfield, Cheshire, SK10 4TG.

INSTITUTO LLORENTE, S.A., Madrid, Spain.

JANSSEN PHARMACEUTICALS LTD, Janssen House, Chapel Street, Marlow, Buckinghamshire, SL7 1ET (Agents for "Hypnorm": Crown Chemical Co. Ltd., Lamberhurst, Kent, TN3 8DJ).

JENCONS (SCIENTIFIC) LTD, Mark Road, Hemel Hempstead, Hertfordshire, HP2 7DE.

LABORATORY THERMAL EQUIPMENT, Green Bridge Lane, Greenfield, Near Oldham, Lancashire, OL3 7EN.

LEE ENGINEERING CO., Milwaukee, Wisconsin, USA.

LINBRO CHEMICAL CO. INC., New Haven, Connecticut, USA.

MAY AND BAKER LTD, Dagenham, Essex.

MEDICAL PHARMACEUTICAL DEVELOPMENTS LTD, Ellen Street, Portslade, Brighton, East Sussex, BN4 1EQ.

MICKLE LABORATORY ENGINEERING CO., Mill Works, Gomshall, Guildford, Surrey, GU5 9LJ.

288

288 APPENDIX

MICROFLOW LTD, Fleet Mill, Minley Road, Fleet, Aldershot, Hampshire, GU13 8RD.
MILLIPORE CORP., Bedford, Massachusetts, 01730, USA.
MILLIPORE (UK) LTD, Millipore House, Abbey Road, London, NW10 7SP.
MINNESOTA MINING AND MANUFACTURING CO. (3M), 3M House, Wigmore Street, London, W1A 1ET.
MOCHIDA PHARMACEUTICAL MANUFACTURING CO., Tokyo, Japan.
MSE SCIENTIFIC INSTRUMENTS—see Fisons.

NORTH AMERICAN BIOLOGICALS, INC., 16500 North-West 15th Avenue, Miami, Florida, 33169, USA.
NUTRITIONAL BIOCHEMICAL CORP. (NBC)—see United States Biochemical Corp.

OLYMPUS OPTICAL CO., Tokyo, Japan (British Agency: Gallenkamp).
OXOID LTD, Wade Road, Basingstoke, Hampshire, RG24 0PW.

PAINES AND BYRNE LTD, Pabyrn Laboratories, 177–179 Bilton Road, Perivale, Greenford, Middlesex, UB6 7HG (US Agency: Harrisons and Crossfield).
PFANSTIERL CHEMICAL CO., Waukegan, Illinois, USA.
PIPER CHEMICALS, 270 Neville Road, London, E7 9QN.
C. J. PLUNKNETT AND CO., LTD, Charlton Village, London, S.E.7.
PORTEX LTD, Hythe, Kent (Orders for Portex autoclavable nylon film to be placed with Southern Syringe Services).
W. PREWETT LTD, Stone Flour Mills, Horsham, Sussex.

ROCHE PRODUCTS LTD, P.O. Box 8, Welwyn Garden City, Hertfordshire, AL7 3AY.
ROHM AND HAAS CO., Washington Square, Philadelphia, Pennsylvania, 19105, USA.
ROHM AND HAAS (UK) LTD, Lennig House, 2 Mason's Avenue, Croydon, Surrey.

SARTORIUS MEMBRAN FILTER GmbH, 34 Göttingen, Postfach 142, West Germany (British Agency: V. A. Howe).
SCIENTIFIC SUPPLIES CO. LTD, Scientific House, Vine Hill, London, EC1R 5EB.
SIGMA LONDON CHEMICAL CO. LTD, Norbiton Station Yard, Kingston-upon-Thames, Surrey, KT2 7BH.
SIGMA CHEMICALS, P.O. Box 14508, Saint Louis, Missouri, 63178, USA.
SOUTHERN SYRINGE SERVICES LTD, New Universal House, 303 Chase Road, London, N14 6JB.
E. R. SQUIBB AND SONS, INC., Princeton, New Jersey, 08540, USA.
STERILIN LTD, 43–45 Broad Street, Teddington, Middlesex, TW11 8QZ.

TISSUE CULTURE SERVICES LTD, 10 Henry Road, Slough, Buckinghamshire, SL1 2QL.

UNITED STATES BIOCHEMICAL CORP., P.O. Box 22400, Cleveland, Ohio, 44122, USA (British Agency: V. A. Howe).

WATSON-MARLOW LTD, Falmouth, Cornwall, TR11 4RU.
WELLCOME REAGENTS LTD, Wellcome Research Laboratories, Beckenham, Kent, BR3 3SB.
WILSON LABORATORIES, Chicago, Illinois, USA.

Index of Authors whose Cultivation Techniques have been Described in Detail

Subject Index

Page numbers in bold type refer to detailed descriptions of cultivation techniques